MORAL DESIGN and TECHNOLOGY

MORAL DESIGN and TECHNOLOGY

edited by:

Bart Wernaart

Wageningen Academic
Publishers

EAN: 9789086863709
e-EAN: 9789086869220
ISBN: 978-90-8686-370-9
eISBN: 978-90-8686-922-0
DOI: 10.3920/978-90-8686-922-0

First published, 2022

P.O. Box 220,
NL-6700 AE Wageningen,
The Netherlands.
www.WageningenAcademic.com
copyright@WageningenAcademic.com

Table of contents

Preface

Bart F. W. Wernaart

School of Business and Communication, Fontys Hogescholen, P.O. Box 347, 5600 AH Eindhoven, the Netherlands; b.wernaart@fontys.nl

The idea to compose a book on Moral Design and Technology coincided with the idea to set up a research group with a similar theme. In June 2021, I became the chair of the research group Moral Design Strategy at Fontys University of Applied Sciences, based in Eindhoven, the Netherlands. On 20 January 2022 I will deliver my inaugural speech, which will coincide with the official launch of this book.

Setting up a research group, and simultaneously editing a corresponding book cannot be done alone. In fact, I owe a debt of gratitude to many. First, naturally, the authors who contributed to this book. It has been an enormous pleasure to work with them in a friendly, collegial and constructive spirit, creating a comprehensive work on something that makes our hearts beat faster, on a subject we urgently wish to address, and that we strongly believe can help overcome the moral confusion we sometimes feel when we consider the potential of future technologies.

Next, I would like to thank my colleagues at Fontys University of Applied Sciences for their huge support in our research endeavours. In particular, the researchers in our team, Moral Design Strategy (at the time of writing: Iris van Hest, Jo-An Kamp, Brishna Nader, Anne-Marie Sweep and Roberta Vaznytė), for stepping up and joining me in this – already very exciting – research journey. Willemijn Pouwels, Michelle Breemans and Mendy Franken: this would not have been manageable without your ongoing efforts. Also, a special thank-you to Cees-Jan Pen and Mignon van Halderen, who strongly supported the instalment of our research group within the Economic Domain; Danielle Arets, who has been a partner-in-crime from the start; and Mark de Graaf and Gerard Schouten who could not have been more supportive in establishing a strong moral fundament in ICT-related research. Furthermore, a special thanks to our management, in particular Anja Klomps and Ronald Philipsen, who were extremely encouraging in realising my research ambitions, and Richard Boon and Mariette Mambwe, who did not hesitate to give me the necessary space.

Our students are the 'fuel' of design processes and moral research. With their enthusiasm, creativeness, and sometimes simply crazy ideas, researching morality becomes as vivid as it can be. A lot can be learned from reading journal articles and conducting experiments, but I dare say that more can be learned from our students.

A special thanks to Wageningen Academic Publishers, and in particular Mike Jacobs and Renate Smallegange, who embraced the idea of composing a broad work on moral design and technology from the start. It really has been a pleasure working with you.

On a more personal note, I would like to spend a few words on my family. Our children Vik and Bo are, in essence, why Moral Design is important to me: we cannot afford to build a society that is grounded in an unethical technological infrastructure. Their enthusiasm for life, their love, honesty, humour and cleverness make me feel confident that the future can be bright as long as we can keep nourishing these characteristics and design the future in a way that amplifies rather than restricts such important virtues. Then, my wife, Sylvia: she sets a bright example of what academia can look like, and is a partner in so many ways that I have lost count. I am extremely grateful to have her by my side. Furthermore, a special word of thanks to my brothers and family-in-law. They are a huge support and I am lucky to have them around.

Without a past, there is no future. My mother and father are responsible for a fair share of my moral capacity, character and moral imagination. They offered – and still offer – a strong, loving and safe foundation in my life, which is an indispensable given in so many ways. It therefore makes total sense to dedicate my editorial and writing work in this publication to them.

1. An introduction to moral design and technology

Bart F.W. Wernaart

School of Business and Communication, Fontys University, P.O. Box 347, 5600 AH Eindhoven, the Netherlands; b.wernaart@fontys.nl

Abstract

In this chapter we introduce the central issue in this book: moral design and technology. We will first explore what technology is, how it has always been a main driver for societal change, and why such changes usually elicit new moral challenges. We then focus on an unconstructive distinction between techno-optimists and techno-pessimists, and why there is an urgent need for moral design approaches to overcome this contrast. Finally, we introduce the structure and chapters of this book, and dedicate some thought to inclusive authorship.

Key concepts

- Technology is 'the use of science in industry, engineering, etc., to invent useful things or to solve problems'
- A moral problem can be defined as a situation that requires a value judgement in which more, mutually excluding, value judgements can be right
- Societies are transforming into smart societies in which algorithms, artificial intelligence and robotisation are main themes
- Techno-optimists emphasise the potential and possibilities of technological innovation; techno-pessimists warn us about the immoral side of new technologies
- In a moral design approach we incubate morality as design principles when constructing new technologies

1.1 Technology: the driver of revolutions (and moral problems)

technology

The Merriam-Webster dictionary defines technology as:

'the use of science in industry, engineering, etc., to invent useful things or to solve problems';

or:

'…a machine, piece of equipment, method, etc., that is created by technology'.[1]

In some languages, a distinction can be made between the first and the second concept (e.g. in German: 'Technologie' and 'Technik'). In English, both are translated as 'technology'. The first concept can be traced back to the Greek words 'techne' and 'logos'. Freely translated when combined, this means 'the discourse on how to create things'. Other definitions are used as well. In particular, there could be some debate about the difference between technology and a tool, whether or not the organisational structure to shape technology is part of technology design (Friedman and Hendry, 2019: 10-20), and whether or not technology in itself is an applied science (MacKenzie and Wajcman, 1999). What most definitions have in common though is the emphasis on applying scientific knowledge for innovation that solves human problems.

societal change

Morality and technology have always been interrelated concepts. This is not surprising, since new technology can profoundly change societies. For instance, the various revolutions in agriculture were always driven by new technologies (see, for instance, Klerkx and Rose, 2020; Rasmussen, 2011). And with the development of the steam engine societies started to industrialise since – roughly – 1760. More recently, the shift from analogue technologies to digital technologies (the digital revolution) in the post-World War II era opened up many possibilities in connectedness and the sharing of information. What these revolutions have in common is that technological innovation drives societal change, which leads to new social challenges and opportunities. For instance, the industrial revolution led to far-reaching urbanisation, more unequal distribution of wealth (the so-called 'social question') and serious pollution (Wernaart, 2021a). It also led to remarkable achievements and opportunities in the field of e.g. science, architecture, fashion and the arts (Trinder, 2013).

[1] https://www.merriam-webster.com/dictionary/technology.

moral problems

With each societal change, new moral problems present themselves. A moral problem involves a situation that requires a value judgement in which more, mutually excluding, value judgements can be right. The industrial revolution led to the so-called 'social question': wealth was in the hands of those who possessed the means of production, while a large part of the population lived in poverty because their jobs were replaced by machines, and they had no future prospects for adapting their skills to match the new technological reality. This led to moral problems regarding how to distribute wealth in a society (Vercherand, 2014).

smart societies

When we focus on the last decade, the pace and impact of technological innovation has been unprecedented. Societies are transforming into smart societies (Mazali, 2018), in which algorithms, artificial intelligence and robotisation are main themes. This is profoundly changing the way we participate and interact in society, and at such a rapid pace that moral problems struggle to get noticed. These are sometimes dramatic, science fiction-like life-or-death scenarios in which moral problems are immediately recognisable. Think about the self-driving vehicle that needs to decide what to do in the event of an imminent crash (Awad *et al.* 2018), or the programming of a war-bot that needs to decide when and whom to shoot (Elands *et al.*, 2019). However, there are also challenges that are not immediate 'life-or-death' type moral problems, but nevertheless represent ethical challenges that affect important human values, especially in the longer term (Wernaart, 2021a). Just to give some random examples: when should a surveillance system that is used in preventive policing sacrifice the privacy of citizens to prevent criminality? To what extent should an insurer use the personal data of farmers in smart technology to prevent factory farm fires? What should be the impact of individual moral expectations when a social media platform designs an algorithm? And to what degree is programmed nudging for good acceptable when a business uses automated communication to reach their customers?

1.2 Technology and philosophy

In philosophy, the moral aspects of tools and artifacts were discussed by Aristotle in ancient Greece, and have been a continuous source of moral discourse ever since (Gómez, 2010). Aristotle started from the assumption that technological tools are a means to an end, and that technology (in itself value-neutral) is valuable because it helps realise human ends. In a way, this approach can be seen as the first version of the value-neutrality thesis that is now widely rejected. The philosophy of technology as a specific approach in philosophy was used for the first time by Ernst Kapp, in his famous work 'Grundlinien einer Philosophie der Technik' (1877). He focused on the interaction between humans and technology, and *vice versa*. After the Second

World War, Martin Heidegger can be seen as the founding father of modern philosophy of technology. In his groundbreaking work, 'Die Frage nach der Technik', Heidegger proposes a determinist perspective, suggesting that technologies are autonomous entities that have the potential to change or even command mankind – even against their own will (Ziemann, 2019). So, in the history of the philosophy of technology, scholars moved from technology as a value-neutral tool that serves as a means to an end, to an entity that can change men against their will, and all shades of grey in between. The enthusiasm with which technology is regarded also changed accordingly.

techno-optimists

In general, we could say that there are techno-optimists, who emphasise the potential and possibilities of technological innovation, and techno-pessimists, who warn us about the immoral side of new technologies. Techno-optimism is typically connected to the Enlightenment. During that era, technology was considered a valuable asset for realising important values for mankind. Also, after the Second World War, we saw a period if techno-optimism. As the smoke cleared from the war, the idea that science and technology could enable humans to do things we could barely imagine was very much alive in, for instance, USA politics. The report 'Science, the endless frontier', written by Vannevar Bush, Director of the Office of Scientific Research and Development, is exemplary, as was the 'space race' in the sixties and early seventies. After that time, however, when the downsides of rapid technological progress became more visible and more widely noticed, techno-optimism became more of a corporate, Silicon Valley-oriented mindset (O'Mara, 2019), while techno-pessimism became the leading voice, in particular in academia.

techno-pessimists

This does not mean that there are no optimists amongst scholars today. Erik Brynjolfsson (2014) and Andrew McAfee (2017), for example, emphasise how technology can help create opportunities for mankind, and share prosperity; Marco Bevolo and Filiberto Amati (2020) show us how AI can help design the future; and Mariusz Maciejewski (2017) underlines how big data can be a game changer for better public administration. However, the pessimistic voices take the upper hand. For instance, Cathy O'Neil (2016) warns us about the potential of Big Data to increase inequality and threaten democracy; Carl Benedict Frey (2019) warns us about the role of technology in the unequal distribution of power and welfare; José van Dijk, Thomas Poell and Martijn de Waal emphasise the destructive effects of the power position of big tech while creating a platform society (2018); Sheila Jasanoff (2016) emphasises that innovation changes the way we can have (and lose) democratic control; and Shoshana Zuboff (2019) warns us about an age of surveillance capitalism.

While both extremes (techno-optimism and techno-pessimism) appear to be difficult to reconcile (Dorrestijn, 2017; Russo, 2018), in reality, the moral side of technology is not that clear-cut and requires a nuanced approach.

1.3 Towards moral design

We therefore need to bridge the gap between techno-optimists and -pessimists, and make sure we consider the moral implications of new technology. It helps if we are better able to narrow down what precisely are the moral issues involved in the design of technology. Technology is not either right or wrong. It depends. And to fully understand what it depends on, we need to focus on how we design the technology we talk about in the first place (Friedman and Hendry, 2019; Wernaart, 2021a).

For this, we need to step away from the mindset that we can organise ethics primarily in a top-down manner using laws, ethical codes or ethical theories. Such instruments are useful of course, but will in the end fall short of dealing sustainably with moral issues that come with new technology. After all, what they have in common is that they focus on setting standards to review against, which most of the time comes down to assessing 'ethics at the backdoor' of innovation. Ethics is considered at the end of a design process, when new technology-in-action appears to affect values that were unforeseen or are considered undesirable or unethical. To illustrate: there is only so much the law can do. Legislators can never match the speed of innovation in creating rules and regulations that are adopted to guarantee moral technology. The EU's attempts to regulate AI safety are praiseworthy, but the process is slow; the political reality requires compromise, which leads to ambiguous and sometimes vague norms that need to be firmly revised once they have come into force to match the current technological reality (Floridi, 2021). On the other hand, what law can do is set minimum standards (e.g. human rights) that technology should comply with, regardless of the particular characteristics of the technology, and make sure that the consequences of breaking these rules are predictable (the legal certainty principle). What ethical codes can do is set general standards on ethics and install procedures by which to discuss ethics in technology, or guide us in normative directions that we believe are important. Efforts to formalise moral norms in technology are superfluous (e.g. Boddington, 2017; Floridi *et al.*, 2018; Patenaude *et al.*, 2015; Radanliev *et al.*, 2021). However, these top-down initiatives are – just like law – always lagging behind the technological reality, and mostly offer a yardstick for reviewing moral impact only, or regulate general moral practices within an organisation. In addition, ethical theories, such as utilitarianism, deontology, virtue ethics, relation ethics, post-modernist ethics and discourse ethics (e.g. Wernaart, 2021b: 69-96) can inspire but probably not directly solve moral challenges that

law

ethical codes

ethical theories

come with new technology. Not least due to practical limitations (Aliman and Kester, 2019; Wernaart, 2021a), and beyond that, not at the pace we would require to match the speed of innovation (Friedman and Hendry, 2019: 7; Vallor, 2016: 10).

design approach

Therefore, we *also* need ethics at the design table, at the start of the process ('ethics at the front door'), see also Figure 1.1. This is a bottom-up, design approach that incubates ethics when constructing new technologies. When we use this approach, we will learn that the ethics of technology is a multi-faceted phenomenon. The ethics of the designer influences how the design process evolves. The artifact that is designed will have its own moral ladenness that exists regardless of the ethics of the designer, and the moral impact of the technology on society may have different characteristics still from the first two. This means that we need to make sure there is a certain value alignment throughout these stages in order to have stable moral technology. This also means we need to address the question of who should decide on this morality throughout this process. Individuals may feel that technology is a 'take-it-or-leave-it' product and they are not able to influence the moral consequences for them when they are affected by such technology. This sentiment is amplified by the fact that technology is developed and used in the context of a globalised world with a more prominent role for the private sector. This can result in moral confusion or even panic: individuals who alternately take on the role as citizen and consumer feel unable to influence the design of technology that has a strong impact on their core values. Sustaining this moral confusion is in nobody's long-term interest.

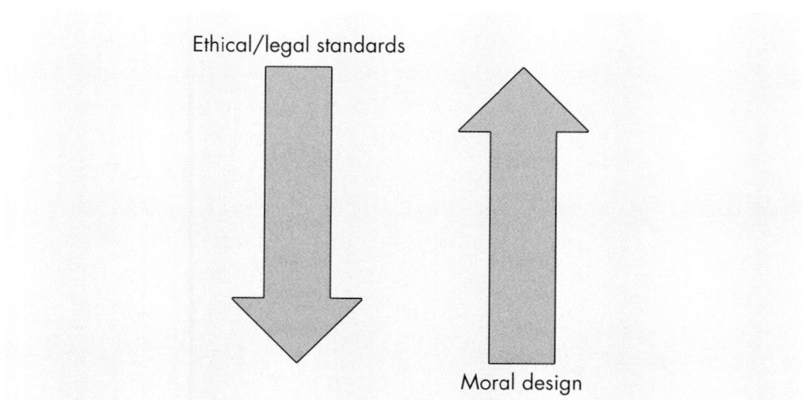

Figure 1.1. Top-down and bottom-up approaches through ethical/legal standards and moral design.

1.4 The structure and chapters of this book

In this book, we offer practical solutions in order to overcome this moral confusion, and look for ways to integrate individual moral expectations in the design of new technology. This book is composed of four parts.

moral design

In the first part we address moral design as a key solution for dealing with moral confusion by focusing on how to integrate moral decisions and morality in the design process of new technology. In this part, we discuss who should decide on the moral ladenness of technology (Bart Wernaart, in Chapter 2), how we can consider moral impact in a design process (Jo-An Kamp and Rens van der Vorst, in Chapter 3), and how we can encode morality in a machine-readable form using Augmented Utilitarianism (Nadisha-Marie Aliman and Leon Kester, in Chapter 4).

existing discourse

In the second part we assess how moral design relates to existing discourse in adjacent disciplines. First, we will discuss how key values and virtues change as a result of technological innovation and are at the core of new narratives in ethical debate (Bart Wernaart, in Chapter 5). Furthermore, we discuss how the related fields of business ethics (Bart Wernaart, in Chapter 6), law (Colette Cuijpers, in Chapter 7) and policy (Maarten van Veen and Piek Visser-Knijff, in Chapter 8) relate to ethics in technology.

case studies

In the third part of this book various case studies are highlighted that focus on particular moral issues at the crossroads of moral design and technological innovation in the public and private sector. In each case, an applied moral challenge is used as a starting point for moral debate, and possible solutions in moral design are offered in order to deal with these challenges. In Chapter 9, the responsible design of audience metrics in journalism is discussed (Danielle Arets, Jessy de Cooker, Jeroen van de Nieuwenhof, and Hasib Shirbaz). In Chapter 10, we focus on deep-fake science attacks and how to design 'cyborgnetic' defences in order to become resilient to such attacks (Nadisha-Marie Aliman and Leon Kester). In Chapter 11, we explore how a technological infrastructure can help to optimise moral behaviour (David Bamps, Veronique De Roeck and Lize Verbeke). In Chapter 12, we envision how we can design a griefbot for good (Rens van der Vorst and Jo-An Kamp). In Chapter 13, we reflect on persuasive design strategies to stimulate responsible online behaviour (Danielle Arets, Ruben Bastiaanse and Anne Giesen). In Chapter 14, we debate how to deal with benevolent lies by design in dementia care (Rens Brankaert, Ans Tummers-Heemels and Wijnand IJsselsteijn). Finally, in Chapter 15, the design of processes for the ethical usage of data in long-term care facilities is explored (Sil Aarts).

the future

In the final part of this book, we look ahead and discuss what the future might look like if we use moral design as a central approach in creating new technology. Mark de Graaf reflects on the position and interest of citizens in light of the fast developments in AI and machine learning (Chapter 16). Then, Gerard Schouten (Chapter 17) warns us that AI is not ready for use yet, and emphasises the need for a more coherent approach to organising AI safety. Furthermore, Frédérick Bruneault and Andréane Sabourin Laflamme emphasise the importance of information ethics, and how this can be used as a theoretical foundation for ethical assessment and moral design (Chapter 18). Next, David Bamps, Veronique De Roeck and Lize Verbeke show us a moral compass for the future in a network society where technology is increasingly important for making (moral) decisions based on complex information exchange (Chapter 19). Bart Wernaart concludes this book by wrapping up the discourse on Moral Design, and adds what we can learn from Science Fiction (Chapter 20).

1.5 Inclusive authorship

In this book we tried to bridge many worlds. This bridging is done on various levels. First, there is the disciplinary level: the morality of technology is discussed in many academic fields, including philosophy, technology, business and public governance. However, from the perspective of these sectors, researchers professionals do not speak on the same level; e.g. while looking at similar problems, the jargon in business ethics is very managerial, organisation-focused, and is not comparable with the programming language used in the field of technology or among engineers. To overcome the 'jargon-gap', we invited authors from those various domains who are involved with the morality of technology.

Second, we tried to bridge the academic world and the world of applied sciences. The morality of technology is not just something that needs to be tackled on an abstract level; it is also not something that can be solved solely by practising ethics. Both of these things need to be done. Therefore, this book falls somewhere between science and applied sciences, which means that the reflection on moral issues is done on high abstract levels, but also in the context of case studies, best practices and local examples.

Third, as also widely discussed throughout the chapters, the moral ladenness of technology (partly) depends on those who create it. Amongst others, the female voice does not always resonate in the moral design of technology. To this end, we tried to find a proper gender balance. Not only in the authorship of the contributing authors, but also in our references to literature and how we write about individuals. We tried either to use gender neutral language, or make sure to use he/she references in equal amounts. We also realise that this does not fully cover the diverse gender types that exist, and would welcome suggestions to improve that. Writing about the morality of technology in a gender-sensitive way is but a first step in realising inclusive authorship. In this book, the geographic scope of authors is predominantly European and North American, with a professional focus on the Netherlands. We are aware that this could be seen as a shortcoming. However, throughout the chapters the reader may recognise a global mind-set and a particular attention for racial biases in technology.

References

Aliman, N.M. and Kester, L., 2019. Requisite variety in ethical utility functions for AI value alignment. In Workshop on Artificial Intelligence Safety 2019 co-located with the 28th International Joint Conference on Artificial Intelligence AISafety@IJCAI 2019. Available at: http://ceur-ws.org/Vol-2419/paper_12.pdf.

Awad, E., Dsouza, S., Kim, R., Schulz, J., Henrich, J., Shariff, A., Bonnefon, J. and Rahwan, I., 2018. The moral machine experiment. Nature, 563: 59-64. https://doi.org/10.1038/s41586-018-0637-6

Benedict Frey, C., 2019. The technology trap. Princeton University Press, Princeton, NJ, USA.

Bevolo, M., and Amati, F., 2020. The potential role of AI in anticipating futures from a design process perspective: from the reflexive description of 'design' to a discussion of influences by the inclusion of AI in the futures research process. World Futures Review, 12: 198-218. https://doi.org/10.1177/1946756719897402

Boddington, P., 2017. Towards a code of ethics for artificial intelligence. Springer, Cham, Switzerland.

Brynjolfsson, E. and McAfee, A., 2014. The second machine age: work, progress, and prosperity in a time of brilliant technologies. W.W. Norton & company, New York, NY, USA and London, UK.

Dorrestijn, S., 2017. The care of our hybrid selves: ethics in times of technical mediation. Foundations of Science, 22: 311-321. https://doi.org/10.1007/s10699-015-9440-0

Elands, P.J.M., Huizing, A.G., Kester, L.J.H.M., Peeters, M.M.M. and Oggero, S., 2019. Governing ethical and effective behaviour of intelligent systems – a novel framework for meaningful human control in a military context. Militaire spectator, 188(6): 302-313.

Floridi L., 2021. The European legislation on AI: a brief analysis of its philosophical approach. Philosophy & technology, 34: 215-222. https://doi.org/10.1007/s13347-021-00460-9

Floridi, L., Cowls, J., Beltrametti, M., Chatila, R., Chazerand, P., Dignum, V., Luetge, C., Madelin, R., Pagallo, U., Rossi, F., Schafer, B., Valcke, P. and Vayena, E., 2018. AI4People – an ethical framework for a good AI society: opportunities, risks, principles, and recommendations. Minds & Machines, 28: 689-707. https://doi.org/10.1007/s11023-018-9482-5

Friedman, B. and Hendry, D.G., 2019. Value sensitive design, shaping technology with moral imagination. The MIT press, Cambridge, MA, USA and London, UK.

Gómez, R.J., 2010. What is this thing called philosophy of technology? In: Lorenzano, P., Rheinberger, H., Ortiz, E. and Delfino Galles, C. (eds.) History and philosophy of science and technology, volume IV, Eolss Publishers, Oxford, UK, pp. 47-81.

Jasanoff, S., 2016. The ethics of innovation. W.W. Norton & company, New York, NY, USA and London, UK.

Kapp, E., 1877. Grundlinien einer Philosophie der Technik. Westermann, Braunsweig, Germany.

Klerkx, L. and Rose, D., 2020. Dealing with the game-changing technologies of Agriculture 4.0: how do we manage diversity and responsibility in food system transition pathways? Global Food Security, 24: 100347. https://doi.org/10.1016/j.gfs.2019.100347

Maciejewski M., 2017. To do more, better, faster and more cheaply: using big data in public administration. International Review of Administrative Sciences, 83: 120-135. https://doi.org.10.1177/0020852316640058

MacKenzie, D. and Wajcman, J. (eds.), 1999. The social shaping of technology. Open University Press, Buckingham, UK.

Mazali, T., 2018. From industry 4.0 to society 4.0, there and back. AI & Society 33: 405-411. https://doi.org/10.1007/s00146-017-0792-6

McAfee, A. and Brynjolfsson, E., 2017. Machine platform crows, harnessing our digital future. W.W. Norton & company, New York, NY, USA and London, UK.

O'Neil, C., 2016. Weapons of math destruction. Penguin Random House, London, UK.

Patenaude, J., Legault, G.-A., Beauvais, J., Bernier, L., Béland, J.-P., Boissy, P., Chenel, V., Daniel, C.-É., Genest, J., Poirier, M.-S. and Tapin, D., 2015. Framework for the analysis of nanotechnologies' impacts and ethical acceptability: basis of an interdisciplinary approach to assessing novel technologies. Science and Engineering Ethics, 21: 293-315. https://doi-org.fontys.idm.oclc.org/10.1007/s11948-014-9543-y

Radanliev, P., De Roure, D., Ani, U. and Carvalho, G., 2021. The ethics of shared Covid-19 risks: an epistemological framework for ethical health technology assessment of risk in vaccine supply chain infrastructures. Health and Technology 11: 1083-1091. https://doi.org/10.1007/s12553-021-00565-3

Rasmussen, W.D., 2011. The impact of technological change on American agriculture, 1862-1962. The journal of economic history, 22: 578-591. https://doi.org/10.1017/S0022050700066778

Russo, F., 2018. Digital technologies, ethical questions, and the need of an informational framework. Philosophy &.Technology, 31: 655-667. https://doi.org/10.1007/s13347-018-0326-2

Trinder, B., 2013. Great cities, old and new. In: Britain's industrial revolution: the making of a manufacturing people, 1700-1870. Carnegie Publishing, Lancaster, UK.

Vallor, S., 2016. Technology and the virtues, a philosophical guide to a future worth wanting. Oxford University Press, New York, NY, USA.

Van Dijk, J., Poell, T. and De Waal, M., 2018. The platform society. Oxford University Press, Oxford, UK.

Vercherand, J., 2014. The 'social question' since the 19th century. In: Labour. Palgrave Macmillan, London, UK.

Wernaart, B., 2021a. Developing a roadmap for the moral programming of smart technology. Technology in Society, 46: 101466. https://doi.org/10.1016/j.techsoc.2020.101466

Wernaart, B., 2021b. Ethics and business: a global introduction (1st ed.). Routledge, Oxford, UK. https://doi.org/10.4324/9781003193951

Ziemann A., 2019. Martin Heidegger: Die Frage nach der Technik (1954). In: Ziemann A. (ed.) Grundlagentexte der Medienkultur. Springer, Wiesbaden, Germany. https://doi.org/10.1007/978-3-658-15787-6_7

Zuboff, S., 2019. The age of surveillance capitalism, the fight for a human future at the new frontier of power. Profile Books, London, UK.

Media

O'Mara, M., 2019, 28 September. The church of techno-optimism, Neither liberal nor conservative, the true ideology of Silicon Valley is an unwavering belief in the power of technology. The New York Times. Available at: https://www.nytimes.com/2019/09/28/opinion/sunday/silicon-valley-techno-optimism.html.

Part 1
MORAL DESIGN PROCESS

2. Moral authority

Bart F.W. Wernaart

School of Business and Communication, Fontys University, P.O. Box 347, 5600 AH Eindhoven, the Netherlands; b.wernaart@fontys.nl

Abstract

In this chapter, we will introduce the concept of moral authority in the context of technology. Technology can be value-laden, and humans increasingly make use of technology in ethical decision-making. This raises the question of moral authority: who should make the decisions on the moral ladenness of technology? In this chapter, we will first reject the existence of moral truth and incomparability of moral solutions, which leads us to the assumption that moral problems exist and can be solved. Then, we will introduce theories on moral decision-making processes, and how individual, situational and issue-related factors play an important role in these processes. In these theories lies the basis for the assumption that it matters who has the moral authority to decide on the moral charge of technology. Since engineers represent a rather homogeneous group, there is a high risk of producing biased technology that reflects the morality of this group. Therefore, the person who programmes the value-laden technology does not necessarily have the moral authority to decide on this programming. Moral authority can be found by further analysing the moral impact of technology and the concentration of that impact. It is proposed to make a distinction between low-stake and high-stake settings on the one hand, and specific concentration of effect on individual users and broad societal effects on the other. These elements can lie at the core of allocating moral authority.

Key concepts

- A moral problem can be defined as a situation that requires a value judgement in which more, mutually excluding, value judgements can be right
- When we assume moral problems exist, this means we reject the notion that moral truth exists and that moral solutions are comparable
- If moral problems are – partially or fully – dealt with by technology, the matter of moral authority becomes even more relevant: who gets to decide on the moral programming of technology?

- Ethical decision-making can be done in four stages: (1) recognising the moral issue; (2) making a moral judgement; (3) establishing moral intent; and (4) moral acting
- The stages of ethical decision-making are influenced by individual, situational and issue-related factors
- The engineer/designer plays a key role in the value alignment between intended, embodied and realised when new technology is designed, by learning how to recognise moral problems in the design of technology, and how to internalise external values to technology
- Moral authority depends on the moral stakes and the concentration of the effects
- In case of low-stake settings, the engineer can be entrusted with the moral authority in the moral design process
- In case of high-stake settings, the concentration of the effects is decisive:
  - when the moral consequences relate to individual users, they have the moral authority to decide on the moral ladenness of technology
  - when the moral consequences are societal, society in itself has the moral authority to decide on the moral ladenness of technology

2.1 Moral programming, moral impact and moral authority

The development of the self-driving vehicle as described in Case 2.1 raises various moral questions. While not all innovation involves immediate life-or death situations (Wernaart, 2021), those that do are excellent cases that help us to reflect on the relevant moral issues that we need to deal with before we can safely make use of such innovation. One major issue is the question of what morality should be programmed into the car to adequately deal with imminent crashes. Various ethical approaches could be translated into moral programming (Gerdes and Thornton, 2016). In essence, the car should be able to calculate safety risks and decide to act in such a way that it contributes to the programmed safety goals. As described in Chapter 4, this would require AI safety programming that involves meta-ethical calculations, such as Augmented Utilitarianism (Aliman and Kester, 2019) or other forms of comparative methods (Wernaart, 2021). This leads to at least two main related questions. The first is the question of moral authority, which will be dealt with in this chapter. Who should decide on the moral programming of – for instance – a self-driving vehicle? The programmer or company on whose behalf the programmer works? The user of the car? Society, through law- and policymaking? As we will see below, part of the answer to this question depends on the expected moral impact) of the technology. This is discussed in further detail in Chapter 3.

Case 2.1. A modern trolley dilemma.

Multiple companies have been working on the design of self-driving vehicles. The idea is that cars can drive autonomously, without the need for a human driver. For a long time, the self-driving vehicle has been used as a modern, upgraded version of the famous 'Trolley Problem'. Originally, the Trolley Problem was casually introduced by Phillipa Foot (1976) in a series of ethical dilemmas to discuss the moral challenges that come with the matter of abortion:

> To make the parallel as close as possible it may rather be supposed that he is the driver of a runaway tram which he can only steer from one narrow track on to another; five men are working on one track and one man on the other; anyone on the track he enters is bound to be killed.

Later, the Trolley Problem was restated countless times, adding moral layer after layer, and is mostly discussed to emphasise the difference between moral intent, moral action and moral consequences. A challenging restatement of the Trolley Problem is the situation in which you find yourselves on a bridge next to a fat man, watching a runaway train coming in your direction. Would you throw the fat man off the bridge, in front of the train, to save the lives of the people who are in the train? Or in the restatement of Judith Jarvis Thomson (1985):

> Consider a case – which I shall call Fat Man – in which you are standing on a footbridge over the trolley track. You can see a trolley hurtling down the track, out of control. You turn around to see where the trolley is headed, and there are five workmen on the track where it exits from under the footbridge. What to do? Being an expert on trolleys, you know of one certain way to stop an out-of-control trolley: Drop a really heavy weight in its path. But where to find one? It just so happens that standing next to you on the footbridge is a fat man, a really fat man. He is leaning over the railing, watching the trolley; all you have to do is to give him a little shove, and over the railing he will go, onto the track in the path of the trolley. Would it be permissible for you to do this?

The Trolley Problem was criticised for offering a laboratory type of case study that would not resonate with any real-life scenario (Nyholm and Smids, 2016). However, with the developing technologies around self-driving cars, the theorising that relates to the Trolley Problem might suddenly become more than a wild thought-provoking moral fantasy that is predominantly used for didactic purposes. A self-driving vehicle should after all be programmed in such a way that it is able to process relevant data-input so that it can act in the event of an imminent crash. This means that when confronted with a Trolley Problem kind of situation it should be able to act (Lin, 2016). An autonomous vehicle should be able to choose between crashing into a cyclist and a car, between two cyclists, one of whom is wearing a helmet, between a pedestrian and a dog, and between an adult with and an adult without a child in a stroller. In practice, matters are more complicated, and more data can be considered as relevant. For instance, does the age of those who are at risk matter? Should a car be able to sacrifice its owners? Does it matter whether those who are at risk comply with traffic rules? And, above all, who gets to decide on all these matters?

2.2 Moral problems

When we assume that we need to discuss the matter of moral authority, we automatically assume that moral problems exist and can be solved. A moral problem can be defined as a situation that requires a value judgement in which more than one, mutually excluding, value judgements can be right (loosely based on definitions used by Toffler (1986) Maclagan, (2003) and Kvalnes (2019)). The consequence is that when choosing moral option A over moral option B, you thereby violate the morality that supports option B. Examples are offered in the various Trolley Dilemmas discussed in Case 2.1. There are plenty of moral arguments for throwing the fat man off the bridge, but also enough moral arguments for not doing such a thing. From a utilitarian perspective we could say that we should always take a decision that leads to the greatest happiness for the greatest number. This would mean bad luck for the fat man, but a morally justified course of action in killing him. From a deontological perspective, we could say that killing a fellow-man is per definition wrong – 'thou shalt not kill' is a non-negotiable principle – and leaving the fat man alone would be the moral thing to do (resulting in the unfortunate death of the five working men on the trolley). The moral options are at least twofold, and mutually exclusive. You cannot – after all – kill the fat man and not kill the fat man at the same time. This means that when I choose to kill the fat man, I comply with the moral considerations that are at the core of utilitarianism, but at the same time violate deontological principles.

moral truth

Please note that the existence of moral problems in itself is not a given fact but a moral viewpoint. We could for instance argue that moral problems do not exist in the first place, because moral truth can be discovered (Dubbink, 2018). Such an approach is typically found in absolutist normative theories. These theories claim to lead exclusively to moral truth and are always applicable, excluding all other normative theories. Examples are duty ethics as proposed by Immanuel Kant (his famous categorical imperative), or forms of utilitarianism. If one accepts such theories, it also means that theoretically a moral problem cannot exist because each moral problem can per definition be solved using that all-inclusive theory. In other words, when I consider myself a utilitarianist, the Trolley Problem is not a problem: I simply kill the fat man because by definition the moral thing is to strive for the greatest happiness for the greatest number. The greatest happiness for the greatest number principle is then the moral truth that is always applicable. If we assume there is such thing as moral truth, there is theoretically barely a need to discuss the matter of moral authority, because – if applied correctly – morality is a uniform concept that ultimately leads to one moral conclusion, regardless of who takes the responsibility for reaching that conclusion.

However, moral truth is a problematic concept that is widely rejected (for instance, Johnson, 1993) From cognitive sciences and the resulting moral decision-making models (Wernaart, 2021), we can learn that it is probably more accurate to assume morality depends on the actor and is not a universal concept independent of the actor. This is a moral relativist approach. It matters who takes the call in case of a moral problem, because morality depends on many personal, situational and issue-related factors, as will be further discussed in Section 2.3.1.

comparability

When we assume moral problems exist, we need to discuss whether they are solvable. In other word: can we make a meaningful choice between two or more morally justifiable but mutually exclusive moral solutions? What should we do in case of such moral uncertainty? For this we would need to be able to compare moral solutions. The comparability of moral solutions (and their supporting moral theories) is also called 'the Problem of Inter-theoretic Value Comparison' (Sepielli, 2008). One approach could be that we assume moral solutions are incomparable (Messerly and Reuter, 2017), and we cannot establish meta-ethical frameworks that can be used for such a comparison. After all, how can we establish meaningful criteria with which to measure the morality of a utilitarian approach versus a deontological approach (Raz, 1986)? How can we compare the value of killing the fat man versus not killing the fat man? If indeed moral solutions are incomparable, a moral problem may exist but be unsolvable. Interesting as this discussion is, it is perhaps not necessary to establish criteria for comparing moral solution A with moral solution B. Instead, a hybrid approach that can include multiple ethical viewpoints might solve the Problem of Inter-theoretic Value Comparison. For instance, Augmented Utilitarianism, as proposed in Chapter 4, solves this problem by translating safety goals into functional programming. Another example is the Action-Deed-Consequence (ADC) model (Dubljević, 2020), that can lead to calculative methods in which normative theories such as virtue ethics, deontology and consequentialist ethics are integrated. Yet another solution is offered through the Expected Moral Value Approach (Lockard, 2000), in which relevant moral solutions are reviewed against the degree to which various moral theories are fulfilled based on the values that are supposed to be fulfilled by that moral theory. In other words: to what extent does moral solution A fulfil the values of, say, utilitarianism, and to what degree does it comply with the categorical imperative. A similar analysis can be done in light of moral solution B. After comparing the results, the solution that has the highest degree of fulfilment of moral theories should be chosen.

What these approaches have in common is that they offer a framework in which various ethical approaches can be used in a comparative setting to evaluate what is most valuable in light of, for instance, specified safety functions. In Case 2.2, the ADC is used to solve the 'kill the fat man' restatement of the Trolley Problem. As we can see, the calculation leads to a clear output: we should kill the fat man. If we would translate this to the self-driving vehicle, it would mean that it is theoretically possible to program the car so that it can decide on moral issues in such a way that the programming in itself cannot lead to conflicting results. At the same time, there is reason for caution: these meta-ethical calculative methods may seem neutral, but they are in fact not. It is still a choice of preference to award '+1' and not '+2' or '-1' to any of the elements in the ADC model. Similar concerns could be raised regarding the other models (Augmented Utilitarianism and the Expected Value Approach, see also Wernaart, 2021). The moral understanding of those who set the calculative frameworks heavily influences the way these models can be used in moral decision-making. In the case of the self-driving vehicle, this is most probably the programmer or engineer. All the more reason to discuss the matter of moral authority: it matters who gets to decide on the moral programming of technology.

Case 2.2. We should kill the fat man.

When we consider using the ADC model in the context of the Trolley Problem, we need to analyse the available moral solutions at three levels: the moral intention of the actor (which corresponds to virtue ethics), the morality of the moral action (which corresponds to deontology) and the consequences of the moral action (which corresponds to utilitarianism).

Using the 'kill the fat man' restatement of the Trolley Problem, it could look like this:

Action A: kill the fat man. We could focus on the intention of the actor first (the actor does not intend to kill, but save), the deed itself (killing is generally wrong), and the consequences (more people are saved when we kill the fat man).

If we were to translate this into a calculative model, we could award positive and negative credits to each stage: A: + 1 D: – 1 C: + 1 = 1

Action B: do not kill the fat man. The intention of the actor is neither to kill nor to save; the deed would be the absence of killing, and the consequence is four casualties instead of one.

If we were to translate this into a calculative model, we could award positive and negative credits to each stage: A: 0 D: + 1 C: – 1 = 0

In this calculation, the conclusion – based on three combined moral theories – is that we should kill the fat man.

2.3 Moral decision-making and the engineer

In the previous section we discussed the urgency of exploring the concept of moral authority. Before we looking at methods for establishing the appropriate moral authority when developing new technologies, we first discuss how ethical decision-making works, since this closely relates to how we can conclude what moral authority is suitable in the design of what technology. In this section we first explore the rich literature that exists on ethical decision-making (Section 2.3.1). After that, we will briefly discuss how we can put these models in the context of design processes (Section 2.3.2).

2.3.1. Ethical decision-making models

Theories on ethical decision-making (EDM) are typically developed in the business ethics literature. In general, two issues are discussed. First, the stages of ethical decision-making, and second, the factors that influence ethical decision-making (see Figure 2.1 for a visualisation).

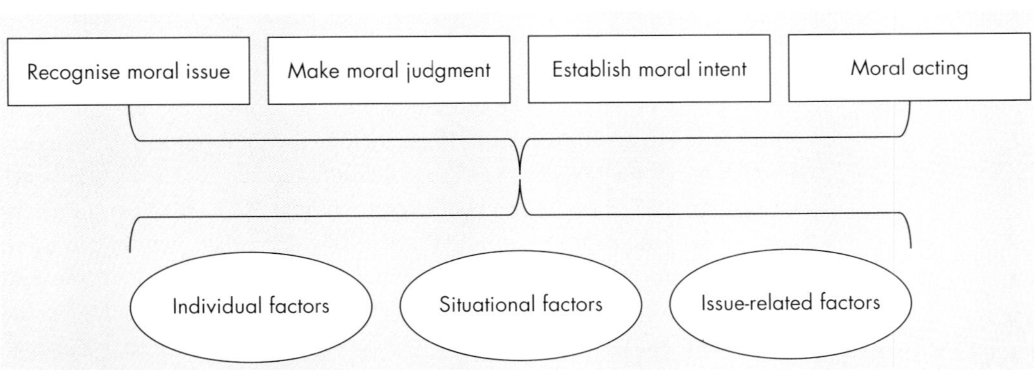

Figure 2.1. Ethical decision-making (EDM).

Ethical decision-making stages

A widely embraced starting point in EDM theory is offered by James R. Rest (1986). He introduced an ethical decision-making process composed of four main steps: (1) recognising the moral issue; (2) making a moral judgement; (3) establishing moral intent; and (4) moral acting. This four-step approach – with some variations (see: Jones, 1991) – is deeply embedded in business ethics literature (Wernaart, 2021), and introduces crucial stages that are required for moral action. For example, an engineer should have the moral skills to recognise that programming a self-driving vehicle could involve moral issues. If these skills are absent, the discourse on the moral programming in itself will not follow. If the engineer recognises a moral issue, she will make a value judgement on that issue. For instance, the programming should primarily result in the least casualties per situation. Then there should be the moral intent to actually engage in a moral action. There can be a difference between moral intent and moral action. The moral intent may never be realised due to a superior in the organisation who instructs the programmer to adjust the programming in such a way that the self-driving vehicle is not led by the number of casualties but instead by the level of legal liability, and who threatens to fire the engineer if she does not include that in the programming. Moral intent and moral action are thus two distinct matters. Another example could be that there is an unintentional discrepancy between the moral intention and the moral action. This could happen if there are flaws in the programming or there are unexpected side-effects of the programming. The moral action – the poor or incomplete programming – then does not match with the moral intention.

Influential factors

Then there are various factors that influence how individuals can go through these four stages in ethical decision-making. In the rich literature around this theme (Craft, 2013; Schwartz, 2016) a wide variety of influential factors is categorised. To be able to see the wood for the trees, a workable sub-categorising is proposed by Thomas M. Jones (1991): individual factors, situational factors and issue-related factors. The individual factors relate to the person that is confronted with an ethical issue. These include demographics, psychological features and ethical experience. These factors together are the cornerstones of the moral capacity or moral character of an individual. When we consider our engineer who is programming the self-driving vehicle, gender can be an important demographical feature. Gender plays a role in how people deal with ethical issues, as famously described by Carol Gilligan (1982). Due to the overrepresentation of males in the engineering profession, there is a serious risk of gender biased technology (Leavy, 2018).

individual factors

situational factors Situational factors relate to the organisational and personal context in which the moral actor operates (Trevino *et al.*, 2006). Organisational characteristics, such as the reward system, code of conduct or business culture, influence the way employees deal with moral issues. The example of the superior who threatens to fire an engineer is illustrative for a situational factor that may influence the way an engineer addresses a moral issue. Someone's personal context is also important, such as living conditions or financial situation. If the engineer is very much depending on the job, it has an effect on how seriously she deals with the threat from her superior.

issue-related factors Furthermore, there are issue-related factors. These factors relate to the particularities of the moral issues that presents themselves. The moral intensity of a situation affects the way individuals deal with moral problems. (Jones, 1991). In his issue-contingent model, Thomas M. Jones further describes the various elements that altogether define moral intensity. These include: (1) the magnitude of the consequences; (2) the level of social consensus; (3) the probability of the effect; (4) the temporal immediacy; (5) the proximity; and (6) the concentration of the effect. For instance, in the case of the Trolley Problem applied to the self-driving vehicle: (1) the magnitude of the consequences is large, since the car may have to decide in life-or-death situations; (2) there is limited social consensus on how to deal with the moral issues relating to the self-driving vehicle; (3) there is a large probability that the moral programming will have to be put to use; (4) the effects of the programming – when used in practice – are immediate; (5) the proximity towards the programmer is limited (the programmer will not personally know the individuals involved in the event of an imminent crash); and (6) the concentration of the effect is wide: self-driving vehicles will probably change the way society participates in traffic, including the moral aspects to it. All these factors influence the way the engineer will go through the four stages of ethical decision-making, and – in the case of the self-driving vehicle – decide on its moral programming.

2.3.2 Ethical decision-making and the design process

As we have seen in the previous section, individual, situational and issue-related factors influence the way people go through a moral decision-making process. The particular context of the engineering profession requires special attention. As we will discuss further in Section 5.2, internal values typically relate to the technology and engineering profession itself. Not surprisingly, engineers may have a natural tendency to focus on internal values, such as effectiveness, reliability, robustness, maintainability, compatibility, quality and rationality. These values are not per se the same as external values that are deemed important in our society, such as safety, health, well-being, sustainability, justice, democracy and inclusiveness. In the case of the self-driving vehicle, the engineering focus will probably be on effectiveness of data processing, the reliability of the programming so it is operational, or the robustness of the design. These could include moral considerations, but are mostly not the grand themes that are discussed in the Trolley Problem:. the fact that self-driving vehicles may have an impact on how people generally participate in traffic, which may affect people's sense of safety on public roads or their wellbeing; or the programming of the safety measures may harm the principle of equality (justice) by favouring some traffic participants (e.g. the owners of the self-driving cars) over others. In other words, external values are not naturally internalised by engineers (Van de Poel, 2015). This is an important factor to take into consideration when deciding on matters that relate to moral authority, especially in cases in which external values are – let's assume unintentionally – negatively affected by new technology. This can be caused by a failed value alignment, where there is a discrepancy between the intended values pursued by the engineer, the values embodied in technology, and the values that are –in the end – realised. For instance, let's assume it is the intention of an engineer to contribute to traffic safety, and as such he programmes the self-driving vehicle in such a way that it is led by the utilitarian 'greatest happiness for the greatest number' principle. This would mean that in the case of an imminent crash between either a cyclist with a helmet or a cyclist without one, the car would crash into the cyclist with the helmet, maximising the overall chances of survival of all traffic participants involved. However, this may result in rewarding unsafe participation in traffic, since those who do not wear helmets are favoured over those who do, and are consequently better protected. In other words, the intention of the engineer (safety) does not match with the consequential moral output (unsafe traffic behaviour) caused by the values embodied in the technology.

Considering the factors that influence moral decision-making, it is important to note that engineers typically represent a rather homogeneous group (Auernhammer, 2020). There is a risk that their moral programming is laden with stereotypical moral viewpoints. For instance, testing AI in an isolated engineering environment may result in feeding the AI with the unintentional biases of the homogenous group (Garcia, 2016). A striking example is the gender and race biases that were found in face recognition algorithms: a comparative study across 189 algorithms revealed that the technologies involved were considerably less accurate in the case of women of colour (Grother *et al.*, 2019). Furthermore, the engineer is in practice part a member of an organisation, such as a private business entity. This means that the situational factors, concerning e.g. how the business is organised, how the engineer is rewarded for doing the job, and how ethics plays a role in that organisation, are important influential factors that will determine how the engineer will decide on moral aspects of the technology he or she designs.

2.4 Moral authority and technology

From the previous sections we can see that it matters who decides on the morality of technology. While in practice engineers or designers will carry out the moral programming of technology, it does not automatically mean they have the moral authority to decide on this morality as well (Dubljević, 2020). Considering that technology may have a serious impact on others, it is reasonable to assume that those who are likely to bear the consequences of new technology should play a role in deciding on its moral programming. A useful distinction can be made between low-stake and high-stake ethical settings (Millar, 2017). In the case of a low-stake setting, the impact of the technology is fairly negligible for those who are affected by it. In such cases, it is most practical to assume that the engineers that design the new technology have the moral authority to decide on its moral ladenness.

low-stake setting

high-stake setting

In the case of a high-stake setting however, the impact of technology is more serious, and important values of those who are involved can be – positively and negatively – affected. This means that those involved are – or should be – no longer indifferent to the moral signature of technology. It is therefore reasonable to assume that it should not be the engineer who has the moral authority when it comes to moral programming. After all, this could lead to an undesirable paternalistic relationship on the part of the engineer, who can impose moral viewpoints through technology design on its users (Millar, 2014). Also, it may greatly influence the autonomy of those who are affected by technology when they do not have a say in its moral ladenness (Verbeek, 2006). Jason Millar argues that the moral authority to decide on the moral programming of technology depends on who is affected. The effects of moral decisions can be highly concentrated on small groups or individuals, or have a broader impact on society as such (Jones, 1991). In case of the first, it is most reasonable to assume that the individuals involved have the moral authority to decide on moral programming. This will mostly be the end user of this technology. This could be done through e.g. a moral interface that enables the user to adjust the settings of the technology so that it can be used as they see fit. Think of the various degrees in which privacy settings can be adjusted when consumers use a smartphone or tablet. In case of more societal moral consequences, it seems more reasonable to assume that society as such has the moral authority to decide on the moral ladenness of technology. This could be done at the relevant governmental level, using appropriate means to take the moral responsibility that comes with moral authority. This could be by means of the instrument of law or policy (see also Chapters 6 and 7), but is not necessarily limited to such means. Increasingly, government institutions muster a hybrid approach towards realising public values, e.g. through public-private partnerships, privatisation policies and – in some shape or form – market participation (Pen and Wernaart, 2022). The surveillance system discussed in Case 2.3 demonstrates at least two challenges that accompany high-stake moral issues with societal consequences. The first is that there could be a discrepancy between using (seemingly) valid public powers to regulate moral issues in technology and the public acceptance thereof. The fact that something is legal does not necessarily mean that it is perceived to be ethical. In this case, the response of the responsible minister focussed on the legal aspects of using the algorithmic systems, and did not respond to the moral issues that come with biased algorithmic indicators. A second challenge is how to involve the public in the moral decision-making. A government institution can wield power because it is competent to do so, and the public can be involved through democratic processes, e.g. by electing individuals or political parties per office term. This general 'fiat' does not usually involve the particular moral issues that come with specific technologies. The Sensing project (Case 2.3) led to widespread societal discomfort and protest, not only due to the alleged biases in the system, but also due to the top-down nature of the decision-making. A moral design approach would suggest the opposite: a bottom-up approach that involves relevant stakeholders at the design table when the algorithmic systems are created. The positive feedback loop that was emphasised by the minister is not a design feedback loop: there is a difference between accepting false algorithmic accusations as 'collateral damage' while operating a preventive policing system with the promise to delete (and learn from) data of those falsely accused on the one hand, and validating the alignment between intended and realised values of new technology on the other. The first is a way to mitigate between collateral damage and operating preventive police systems. The latter is part of a moral design approach.

Case 2.3. I sense trouble – racist biases in local surveillance systems.

A Dutch Municipality – Roermond – used a surveillance system in the fight against mobile banditry (such as pick-pocketing and shoplifting). An advanced surveillance system – called the Sensing Project – was used to monitor traffic into and out of the inner-city. The relevant authorities (such as the police) were now able to systematically collect this data through cameras and other sensors, which would include information about the vehicles that were used, and their movement patterns. This data was fed into an algorithmic model that could calculate to what extent individual vehicles represented what kind of risk in terms of mobile banditry. This so-called 'preventive policing' was presented as a neutral piece of technology. In reality however, some of the algorithmic indicators were used to establish whether a person was of Eastern European origin. This data was incorporated in the equation to decide on the level of risk an individual vehicle user would represent. In practice, the heritage of vehicle users was used as an important predictive indicator for potential crime. In 2020, Amnesty International published a critical report on these practices, and held that the way the algorithmic systems were used was a violation of human rights, due to its discriminatory nature (Amnesty International, 2020).

The Dutch Minister of Justice and Security disagreed with this assessment, and held that the technology – while targeting a certain group was used in a proportional way, not in conflict with any laws. Furthermore, he emphasised that the technology contained a positive feedback loop: if an individual vehicle appeared to be wrongly put in a high-risk category, the data would be filtered out of the system (Grapperhaus, 2020).

2.5 Conclusions

Moral authority in technological innovation is an important concept. As we have emphasised in this chapter, rejecting the notion of moral truth and incomparability of moral solutions has consequences regarding how we see moral authority. Due to the highly personal nature of morality, it matters who gets to decide on the moral ladenness of technology. Individual, situational and issue-related factors all influence the way individuals take moral decisions, also in moral design processes.

A remaining question is how to draw a line between low- and high-stake settings. A leading factor here is the moral impact technology can have on those who are involved, and the concentration of that effect. Moral authority and moral impact are therefore interrelated issues (as visualised in Figure 2.2). In Chapter 3 we will further explore how we can measure moral impact, and use that in a moral design process.

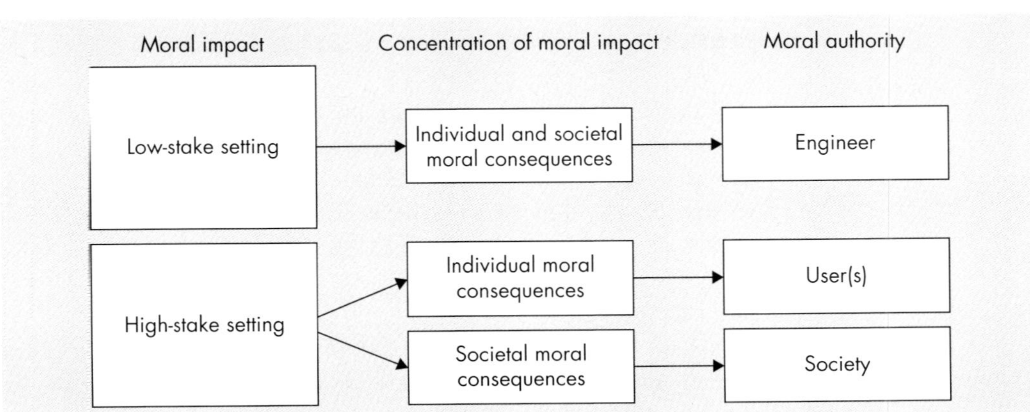

Figure 2.2. Moral impact and moral authority.

The engineer plays a key role in recognising the moral issue in morally laden technology. If she does not recognise the moral issue in the first place, it has consequences for the entire design process of new technology. Therefore, to avoid biased moral programming, it is important that engineers are increasingly trained in internalising external values of technology (see also Chapter 5) and knowing at what point to raise a flag when moral decision-making is involved that is beyond the moral authority of the individual engineer or the company she represents. For instance, to help the engineer to structurally review her moral authority, Bart Wernaart proposed a moral intensity dashboard (Wernaart, 2021) that builds further on the issue-related moral intensity framework offered by Thomas M. Jones (1991). This way, engineers can reflect on important issue-related features, such as the moral impact of new technology and the concentration of these effects.

When the engineer does not have the moral authority to decide on the moral charge of technology, it means that either individual users or society itself must be involved in the (moral) design process. It is helpful not to confuse moral authority with formalised consent. It is unlikely that individual moral authority can be deduced from e.g. accepting general terms or the buying behaviour of consumers. For societal input in the moral design process, it is important to recognise that the legal or democratic validity of government decisions is not the same as exercising moral authority. In both cases (individual moral authority and societal moral authority) it is important to move beyond a 'tick the box to approve' concept. Instead, moral design means to actively involve those whom it concerns at the design table to help engineers to internalise the relevant external values throughout the entire design process. This includes feedback loops during the use of the new technology to systematically reflect on the value alignment between intended, embodied and realised values of the technology (see also Chapter 3 and 4).

References

Aliman, N.M. and Kester, L., 2019. Transformative Al governance and Al-empowered ethical enhancement through preemptive simulations. Delphi – Interdisciplinary Review of Emerging Technologies, 2: 23-29. https://doi.org/10.21552/delphi/2019/1/6

Amnesty International, 2020. We sense trouble: automated discrimination and mass surveillance in predictive policing in the Netherlands. Available at: https://www.amnesty.nl/content/uploads/2020/09/Report-Predictive-Policing-RM-7.0-FINAL-TEXT_CK-2.pdf?x53356

Auernhammer, J. (2020). Human-centered AI: the role of human-centered design research in the development of AI. DRS2020, August 2020, Brisbane, Australia. https://doi.org/10.21606/drs.2020.282

Craft, J.L., 2013. A review of the empirical ethical decision-making literature: 2004-2011. Journal of Business Ethics 117: 221-259. https://doi.org/10.1007/s10551-012-1518-9

Dubbink, W., 2018. A typology of ethical problems. Ethical perspectives, 25: 683-714.

Dubljević, V., 2020. Toward implementing the ADC model of moral judgment in autonomous vehicles. Science and Engineering Ethics, 26: 2461-2472. https://doi.org/10.1007/s11948-020-00242-0

Foot, P., 1976. The problem of abortion and the doctrine of the double effect. The Oxford Review, 5: 5-15.

Garcia, M., 2016. Racist in the machine: the disturbing implications of algorithmic bias. World Policy Journal, 33: 111-117

Gerders, J.C. and Thornton, S.M., 2016. Implementable ethics for autonomous vehicles. In: Maurer M., Gerdes J., Lenz B. and Winner H. (eds.) Autonomous driving. Springer, Berlin/Heidelberg, Germany. https://doi.org/10.1007/978-3-662-48847-8_5

Gilligan, C., 1982. In a different voice. Harvard University Press, Cambridge, MA, USA.

Grother, P., Ngan, M. and Hanaoka, K., 2019. Face recognition vendor test (FRVT) Part 3: demographic effects. National Institute of Standards and Technology, 8280. https://doi.org/10.6028/NIST.IR.8280

Johnson, M. (1993). Moral imagination, Implications of cognitive science for ethics. The University of Chicago Press, Chicago, IL, USA.

Jones, T.M., 1991. Ethical decision making by individuals in organizations: an issue-contingent model. Academy of Management Review, 16: 366-395. https://doi.org/10.2307/258867.

Kvalnes Ø., 2019. Moral dilemmas. In: Moral reasoning at work. Palgrave Pivot, Cham, Switzerland, pp. 11-19. https://doi.org/10.1007/978-3-030-15191-1_2

Leavy, S., 2018. Gender bias in artificial intelligence: the need for diversity and gender theory in machine learning. Proceedings of the 1st International Workshop on Gender Equality in Software Engineering (GE '18). ACM, New York, 14-16.

Lin P., 2016. Why ethics matters for autonomous cars. In: Maurer M., Gerdes J., Lenz B. and Winner H. (eds.) Autonomous driving. Springer, Berlin/Heidelberg, Germany. https://doi.org/10.1007/978-3-662-48847-8_4

Lockard, T., 2000. Moral uncertainty and its consequences. Oxford University Press, Oxford, UK.

Maclagan, P., 2003. Varieties of moral issue and dilemma: a framework for the analysis of case material in business ethics education. Journal of Business Ethics 48: 21-32. https://doi.org/10.1023/B:BUSI.0000004364.63317.73

Messerli, M. and Reuter, K., 2017. Hard cases of comparison. Philosophical Studies, 174: 2227-2250. https://doi.org/10.1007/s11098-016-0796-y

Millar, J., 2014. Technology as moral proxy: Autonomy and paternalism by design. 2014 IEEE International Symposium on Ethics in Science, Technology and Engineering, Chicago, IL, USA pp. 1-7. https://doi.org/10.1109/ETHICS.2014.6893388.

Millar, J., 2017. Ethics settings for autonomous vehicles. In: Lin, P., Jenkins, R. and Abney, K. (eds.) Robot ethics 2.0. Oxford University Press, Oxford, UK, pp. 20-34.

Nyholm, S. and Smids, J., 2016. The ethics of accident-algorithms for self-driving cars: an applied trolley problem? Ethical Theory and Moral Practice, 19: 1275-1289. https://doi.org/10.1007/s10677-016-9745-2

Pen, C-J. and Wernaart, B., 2022. Decentralized government and global challenges: how the Dutch datacenter debate demonstrates the need for powerful and hybrid governance. Bestuurskunde, ###: ###-###.

Raz, J., 1986. Value incommensurability: some preliminaries. Proceedings of the Aristotelian Society, 86: 117-134. https://doi.org/10.1093/aristotelian/86.1.117

Rest, J.R., 1986. Moral development: advances in research and theory. Praegerm, New York, NY, USA.

Schwartz, M.S., 2016. Ethical decision-making theory: an integrated approach. Journal of Business Ethics 139: 755-776. https://doi.org/10.1007/s10551-015-2886-8

Sepielli, A., 2008. Moral uncertainty and the principle of equity among moral theories. Philosophy and Phenomenological Research, 86: 580-589. https://doi.org/10.1111/j.1933-1592.2011.00554.x

Thomson, J., 1985. The trolley problem. The Yale Law Journal, 94: 1395-1415. https://doi.org/10.2307/796133

Toffler, B.L., 1986. Tough choices: managers talk ethics. Journal of Business Ethics, 8: 252-288.

Trevino, L.K., Weaver, G.R. and Reynolds, S.J., 2006. Behavioural ethics in organizations. Journal of Management 32: 951-990. https://doi.org/10.1177/0149206306294258

Van de Poel, I., 2015. Values in engineering and technology. In: Gonzalez, W.J. (ed.) New perspectives on technology, values, and ethics. Boston Studies in the Philosophy and History of Science, Volume 315, Springer, Cham, Switzerland, pp. 29-46.

Verbeek, P., 2006. Materializing morality, design ethics and technological mediation. Science, Technology & Human Values, 31: 361-379. https://doi.org/10.1177/0162243905285847

Wernaart, B., 2021. Developing a roadmap for the moral programming of smart technology. Technology in Society, 46: 101466. https://doi.org/10.1016/j.techsoc.2020.101466

Legal texts, policy documents and case law

Grapperhaus, F., 2020 11 December. Reactie op rapport Amnesty International over massasurveillance. Available at: https://www.rijksoverheid.nl/documenten/kamerstukken/2020/12/11/tk-reactie-op-rapport-amnesty-international-over-massasurveillance.

3. Moral impact

Why we need to think harder about the impact of technology (and how we put this into practice)

Jo-An M. Kamp[1]* and Rens van der Vorst[2]

[1]School of Information and Communication Technology, Fontys University of Applied Sciences, P.O. Box 347, 5600 AH Eindhoven, the Netherlands; [2]Fontys University of Applied Sciences, Corporate IT department, P.O. Box 347, 5600 AH Eindhoven, the Netherlands; j.kamp@fontys.nl

Abstract

In a world where technology is becoming increasingly dominant, it is crucial to reflect on its societal impact. Whether you are a student or professional, technology will play a significant role in not only technological fields, but in any field. Therefore you will probably make decisions about the design, use, or implementation of technology. It would be a pity if the decisions you make now were to make you one of the tech regrets of the future. That is why it is crucial for everyone to have a good understanding of the moral impact: the way moral design influences the values of the stakeholders of new technology. At Fontys University (Eindhoven, the Netherlands), we conducted two years of research into the practical side of reflecting on technology and its societal impact. We did that through the practice of ethics, embedded in the product. The result is the Technology Impact Cycle Tool (TICT): a free, open, online tool that helps you ask questions about the impact of a particular technology and even helps you to improve (the use of) your product along the way. The Technology Impact Cycle Tool consists of quick, full and improvement scans and crash courses in ten different categories that together provide a broad picture of the potential positive and negative impact that a technology can have. The categories are based on social and external values like impact on society, human values, stakeholders (and platforms), privacy, data, bad actors (and alternate reality), inclusivity, transparency (and AI), sustainability and future scenarios. *We strongly believe that the use, implementation and development of technology is the practice of ethics, but with different means: embedded in the product.* The tool is created by a multidisciplinary team of people from various lectureships, students, teachers and professionals from the field; it was launched in January 2021 and has already been used by thousands of engineers, students and professionals. The tool is open, online, free to use and available for everyone at www.tict.io. The Technology Impact Cycle Tool is an example of a moral design toolkit that helps you think critically about the impact of technology on society. It is also an example of practising ethics so that it functions as a driving force for innovation.

Key concepts

- To prevent unconscious and unintentional biases, multidisciplinary and inclusive teams are preferred during the development or implementation stages of developing new technologies
- Testing your product in multiple stages with a broad, inclusive range of people is also highly recommended to prevent biases and to make a better product
- Technological products can be constructed through a moral design process: a circular, recurring process that helps to internalise external (terminal) values
- Practising ethics can be seen as a driving force for innovation
- We argue that it is better not to judge but to guide technological development
- The moral design process is part of context-driven applied ethics
- The TICT is an example of a moral design toolkit

3.1 Introduction

As discussed elsewhere in this book, technology can have a serious impact on social and environmental values. The impact on such values can be constructive, but also destructive (Chapter 1). It is becoming increasingly apparent that ethics is something that needs to be added to the equation when we design new technology, or should even be at the core of the design process if we want to accomplish meaningful and sustainable innovation. In a healthy moral design process we need to address who has the moral authority to decide on the ethical charge of technology (Chapter 2) and what the nature of ethically laden technology can be (Chapter 4). However, to be able to do this, we also need to reflect on the impact of technology on society. In this chapter, we do this from a techno-ethical perspective. After all, part of the answer to the question of moral authority or the nature of moral programming depends on the moral impact of a specific technology in a specific context. To this end, we have developed a tool that helps in assessing the moral impact of new technology: the Technology Impact Cycle Tool (TICT). In this chapter, we will share our experience in building this tool, and how we used this in our education at Fontys University, the Netherlands, as discussed in Case 3.1.

Case 3.1. Building the Technology Impact Cycle Tool at Fontys University of Applied Sciences.

Since 2016 we have become really serious about teaching digital ethics at Fontys University of Applied Science. During lessons (see Case 3.2 for an example) we noticed that our students were building nice products without considering their potential impact on society, so we encouraged our students to think more broadly about the impact of the things they were creating. At first, our didactics focussed on programmers, engineers, designers and other tech-students. However, we soon realised that these weren't the only students involved in projects in which they were looking for technological solutions to contribute to a better society. Students of journalism, health care, business economics, agriculture, real estate and pedagogy were also active in this area. At the same time, we noticed that there was a real need for concrete tools to help them think better and in a more structured way about the impact of technology.

These two reasons (i.e. everyone must be able to think about the impact of technology and the need for concrete tools) prompted us at the beginning of 2019 to start a research project to develop a concrete tool for use by students and engineers from all fields of study.

The Technology Impact Cycle Tool was officially launched in January 2021 and is available to everyone at www.tict.io.

In this chapter, we will first explain why it is so important to be able to carefully consider the impact of technology (Section 3.2). Then we discuss the impact of technology on humans from a philosophical viewpoint (Section 3.3). Finally, we discuss the creation, features, use and experiences with the tool so far (Section 3.4).

3.2 Why we need to think harder about the impact of technology

As teachers in IT we know all too well what many of our students want. They are part of a makers' community. Most of them just want to build cool stuff with whatever (new) technology is available. Our students are *grosso modo* millennials or even post-millennials: a synthetic generation for whom the concept of digital 'transformation' is alien (Van Doorn *et al.*, 2019). So most of our students are digital natives: they use IT as a tool to make things for society. But does that mean they use their (IT) skills to build the right things? Or more specifically: do they automatically build things in a morally correct way? We don't think so, as demonstrated in Case 3.2: IT enthusiasts have a natural tendency to focus on the so-called internal values (Van de Poel, 2015) that typically relate to the technology and engineering profession itself (see for a further discussion Section 5.2.3). This results in a focus on values such as the effectiveness and quality of technology, and not per se on external – societal – values, such as justice, equality or environmental values. Or, as Van de Poel stated, '*those who create new technology should internalise the external (terminal) values in order to design meaningful innovation*'. Another issue is that the values that are realised through technology are not necessarily the same as the intended values of the designer. This is reflected in Case 3.2 in which the tool designed by the student affected values he did not think of when he constructed his new technology. So, if IT enthusiasts do not naturally reflect on the external values of technology then technology may fulfil different values from those intended by its creators. Both represent important arguments for thoroughly reflecting on the impact of technology to avoid becoming the tech regrets of the future.

Case 3.2 Student case: Facebook Messenger's Hidden Gems.

In one of our courses in 2019, one of our students came up with a tool that could extract data from Facebook messenger. It enabled him to see which of his friends was online or offline at a specific moment. He therefore used and combined some off-the-shelf software in a creative manner and then transferred his data into an Excel sheet, plotted against time.

After doing so, he told us he could now easily find out when and how long his friends spent time on social media. Since he had made use of publicly available data, he thought it was legitimate to process this data in the way he did, arguing this was similar to looking at the status information of his friends every once in a while. We explained to him that although it was indeed public information, his friends would probably not know about this and would probably not be happy if they saw his statistics of their status updates over time.

Seen from the viewpoint of his friends, it is not easy to exclude yourself from participating in a project like this, especially if you are not aware of the fact that you are actually part of it. First of all, as has been said, you have to know that you are exposing personal information and to what extent. For example, even if you do not look at Facebook for a while, Facebook's Messenger app will tell others the length of time that you have not been active. So even if you've only opened the App but are not using it, you might be sharing more about yourself than you know or want to.

The tool designed by our student could have unintended side effects. For instance, his friends might look for ways to hide their online status. Currently, there are a lot of online tutorials (see, for instance, Rahman, 2017 or check your own settings on

Facebook) on how to turn off your online status. In practice, only a select group with knowledge on this issue will search on this topic. And when you do decide to switch off this function, you can no longer see the status of your friends either, which might limit your own social media experience (Figure 3.1).

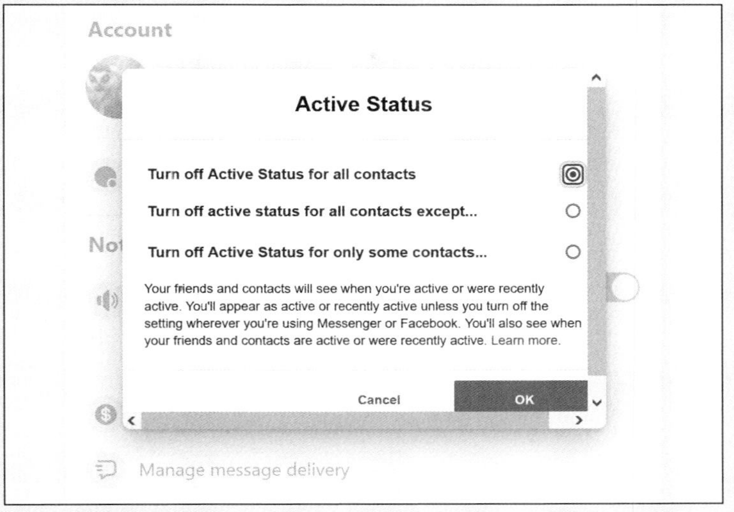

Figure 3.1. Facebook Messenger Online Status (picture taken from Jo-An's account).

At the end of the project we told our student that he might have come up with a Trojan horse, because after a second look at his data he might conclude that his father has been online every night between 2 and 3 am without his mother knowing about it. And then even he, as an innocent visitor of his own generated data, could not go back to not knowing about what might be concluded from that. What a dilemma this could have created in his family! In other words, the tool created by our student was legal and could do what it was designed for. However, if our student had designed from an explicitly moral perspective, he would probably have come up with a different application.

If our student had thought not only about the internal – technical – values (Is it possible to make a tool like this? Can I make simple data come to life if I plot them in time?) but also more about external – societal – values, such as justice, equality or inclusivity, he might have come up with a better tool: for instance, one that informed his friends and shared all the data, so that his friends could see what he could see (and could draw their own conclusions). And, not least, our student could have saved himself from being a tech regret in the future. Although his intentions might have been good, products can turn out differently, as we have seen in this case, so it is crucial to think of all the things that could happen (intended or unintended) to prevent unintended surprises at a later stage.

One of the most striking examples of tech regrets came from Tim Berners-Lee, the inventor of the World Wide Web. While working at CERN in 1990, Tim wrote the three technologies that remain the foundation of the web today (HTML, URL and HTTP). As the web began to grow, Tim realised that its true potential would only be unleashed if anyone, anywhere, could use it without paying a fee or having to ask for permission (History of the Web, n.d.). The flipside of this fundamental and *de facto* noble choice is that anyone could post whatever they liked without consulting an editor or expert, resulting in phenomena we now know as anonymous postings and online trolling and giving conspiracy theorists endless opportunities to spread fake news, without the need to edit or fact-check their posts. The fundamental starting points of the medium also gave companies almost endless possibilities to earn money from advertising and monitoring people's behaviour online. In 2009 Lee therefore started the World Wide Web Foundation, to protect human rights across the digital landscape. In his later work at MIT, Berners-Lee has also been working on a new platform – called Solid – (Berners-Lee, n.d.) that lets people store their data securely in decentralised data stores called Pods, so that they themselves can control which people and applications can access it, to reclaim the Web from corporations and return it to its democratic roots (Brooker, 2018).

This is not the only example out there. When intended, embodied and realised values don't match, things can turn out different than expected (see also Chapter 5 for the theory behind this): Zuckerberg probably didn't found Facebook with the intention of manipulating elections (Confessore, 2018); Jack Dorsey and the other Twitter founders didn't intend to give Donald Trump a digital bullhorn (Zhang *et al.*, 2018); and the founder of 8chan, one of the internet's darkest corners, expressed regret after the Christchurch mosque shooting, taking responsibility for the possible controversial role the platform might have had in this case (McMillan, 2019).

3.3 The impact of technology on humans from a philosophical viewpoint

As more and more engineers leaving Silicon Valley reach the headlines because of their tech regrets, the impact of technology on society can no longer be ignored. The influence can be positive (utopian) as well as negative (dystopian) (see also Chapter 1). But, to really get a grip on this, let's see what we can learn from some philosophers who investigated this earlier:

Karl Jaspers

In the age of rising industrialisation, Karl Jaspers (1883-1969) was one of the first philosophers who had to deal with upcoming technologies. To support this industrial system, people had to perform with the help of technology which, according to Jaspers, alienated them from their natural human environment. Jaspers' statement at that time was that technology is neutral; but in current times, this statement might not be tenable (Verbeek, 2000: 47).

Martin Heidegger

German philosopher Martin Heidegger (1889-1976) opposes the image of technology as a neutral means. Heidegger calls that image 'instrumental' (Sesink, 2012). He agrees with the analysis of technology as instrumental, but says that we need to look deeper: he investigated technology more as a social and cultural phenomenon. Heidegger's theory can be defined as determinism (see also: Dafoe (2015): he sees technology as an unstoppable force of nature (see also Section 5.2.2).

Both perspectives have the disadvantage of getting stuck in extreme utopian or dystopian scenarios, which will lead to black and white discussions where it will be difficult to see things from the other's perspective. Like Jaspers' theory, Heidegger's theory can be helpful, but it might not be helpful enough when answering the more complex questions of our era.

Peter-Paul Verbeek

Current philosophers of technology like Peter-Paul Verbeek (1970-present) and Don Ihde (1934-present) stress the importance of looking at the intertwined relationship between humans and artifacts. As things are shaping the user and vice versa, we should focus more on the intertwined relationship between humans and technology. Ihde states that both people and technology are constantly changing in interaction with each other. Technologies are multi-stable: the same artefact can have different meanings or identities in different contexts (Verbeek and de Jong, 2017).

Leaning on the work of Ihde, Verbeek describes multiple ways in which technology, people and the world mediate (Verbeek 2000: 141-144). He calls this the theory of technological mediation: its central idea is that technologies, when they are used, help us to shape the relationships between human beings and the world. Rather than approaching technologies as material objects opposed to human subjects, or as mere extensions of human beings, he sees them as mediators of human-world relations. This reminds us of a famous quote: *'First we shape our tools, and thereafter our tools shape us'* (this quote is often credited to Marshall McLuhan, but might be evolved from a remark made by the UK Prime Minister Winston Churchill during a speech in the House of Commons in October 1943).

Verbeek's mediation theory can help us think critically about the entangled relationship we have with technology. His theory is important, because if we keep on discussing things from a utopian or dystopian perspective it might not get us any further. His theory sketches a field in which people and technology both appear as actors – intimately connected – sometimes even without distinction, but both with their own dynamics. From this viewpoint, technology can no longer be seen as neutral or passive: it does something to the world just as we do something to technology. There is continuous interaction and it is not entirely predictable in which direction these interactions will take us.

3.3.1 A common misunderstanding: technology is only for engineers

The term 'philosophy of technology' seems to suggest that it is mainly applicable in the technological field. However, as technology is influencing our lives in many ways, an increasing number of people are involved in these kinds of mediations (sometimes knowingly, but most of the time unknowingly: note the people around our student in the student case example above).

stakeholders

The stakeholders of new technology are – of course – not limited to its creators: the engineers. Non-technical people are also involved because they can affect or are affected by the creation of new technology: for instance, people who make use of technology, business leaders who adopt strategies for the use of technology, politicians who adopt laws and policies that set the playing field in which technological companies can operate, and journalists, who critically report on the state of affairs in our technological world.

But there is a difference. '*One of the main differences between engineers and non-technical persons is that engineering is not just about better understanding the world but also about changing it. Many engineers believe that such change improves, or at least should improve, the world. In this sense engineering is an inherently morally motivated activity*' (Van de Poel and Royakkers, 2011).

3.3.2 Translating theory into action

Back to our research. What does all of the above mean in a practical sense? In our opinion it is crucial to let engineers be aware of the possible (un)intended social (side) effects of their design choices. Therefore, we need to structurally assess the impact of technology on social and environmental values, such as human rights, privacy, transparency, inclusivity and sustainability.

Case 3.3. How to design a practical toolkit for students and engineers?

When we developed our own toolkit we first looked at the toolkits that already existed. We saw that a lot of them were either *too specific* (for instance, mainly targeted at data or privacy, like De Ethische Data Assistent ((DEDA), n.d.). or Privacy Design Strategies (Hoekman, 2020)); *too broad* (e.g. the Sustainable Development

Goals (United Nations, n.d.) which are in our opinion too broad to tackle in the smaller cases that we reflect upon); *too theoretical* (if seen from a solely philosophical perspective, the practical question of what decisions to make based on that remained unanswered); or *too negative* (e.g. the Ethical OS (Institute for the Future and Omidyar Network, 2018) is very practical, but mainly focusses on risks (seen as threats to external values, with the intention of not regretting the things you will build in the future. This can in fact be very useful; but if you only focus on risks, the pitfall can be that you only see a negative impact and you might tend to think that the world will be doomed due to technology). None of these perspectives were a perfect fit, so we decided to build on the good things that we noticed and develop our own toolkit (incorporating all the others!)

- From the MOOC from the University of Twente (2017) and the works of Van de Poel and Royakkers (2011) and Rijnboutt and Heerink (2020), we learned about multiple theoretical viewpoints that showed us a philosophical way to look at ethical dilemmas.
- The ethical OS (Institute for the Future and Omidyar Network, 2018) from Silicon Valley taught us how to turn these kinds of philosophical theories into a practical toolkit. And also how to use future-proofing to pick a strategy to embed learnings and best practices from now onwards.
- And from Spiekermann (2015) we learned about human values, and that building future scenarios and storytelling could help us to imagine and share stories of what could happen in the future.
- We added the categories of impact on society, data, hateful and criminal actors and futuring to our previously defined categories of human values, privacy, transparency, inclusivity and sustainability to further broaden our view.

When we combined all our insights we noted that external values might be as important as technical (internal) values, and thus our toolkit should not only focus on developing a broad perspective from the positive side (the desirable impact on society) but also from the negative (impact) side. We then added future thinking to broaden our perspectives and to be able to balance between the positive and negative sides, like yin and yang, as can be seen in Figure 3.2.

From this point we finally came to the concept of a tool as a cycle with the following categories: impact on society, human values, stakeholders (and platforms), privacy, data, bad actors (and alternate reality), inclusivity, transparency (and AI), sustainability and future scenarios.

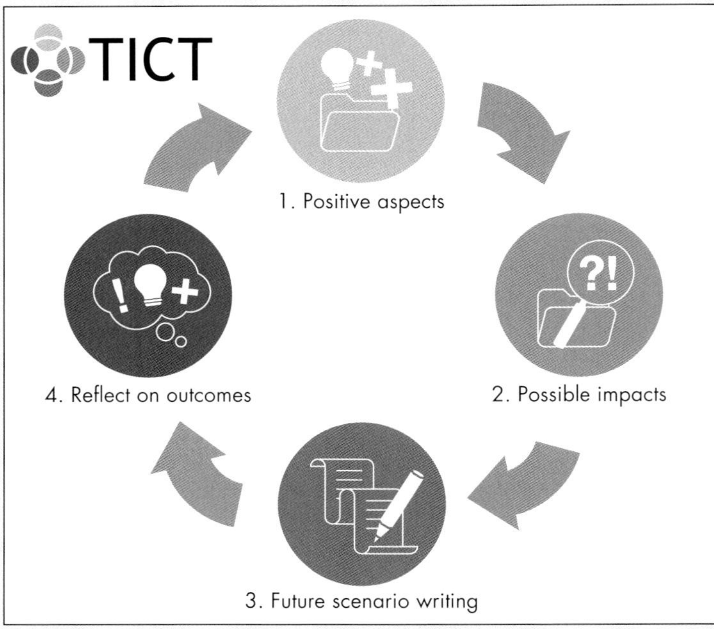

Figure 3.2. Balancing positive and negative impact with the help of storytelling.

In Chapter 12 we will show you how you can use the toolkit on the basis of a concrete case (in this example, a griefbot), where we first try to reshape the question and then use the questions in the categories to come up with our final conclusion about the desirability and the practical concerns that can arise around this concept.

Technology has not only made decision-making more difficult but also more complex. The difficulty in predicting interactions requires designers who will take responsibility at all stages of the design process and who can arrange their designs in such a way that policymakers, citizens, media, politicians (especially at the municipal level) or even other engineers or designers can easily understand the choices made and intervene when necessary.

There are so many processes that we do not have a grasp of, so we might have to accept that we will never be able to control everything. Nevertheless, we have to keep trying. And we must spring into action.

But on what basis should we continue? What should be our guiding lights? Here are some suggestions:

multidisciplinary perspective

1. Form teams from a multidisciplinary perspective and test from multiple perspectives too

Freely interpreting Verbeek, we have learned that we shape our tools and therefore our tools shape us: they are intertwined. It is therefore of the utmost importance that we shape our tools in a way that benefits us as humans and as a society. We know that biases can arise in technology when products are mainly designed by highly educated white technicians (European Commission, 2020). A multidisciplinary approach might therefore be a good idea if you want to develop tools for a broad spectrum of possible users. We know that demographic factors have a major impact on how ethics can be applied, so it makes sense to work with a multidisciplinary and inclusive team right from the beginning. Multidisciplinary teams with different backgrounds can benefit us all, because the people you make your products for might be diverse too, and it is a good idea if they are represented by your team members.

Not only different backgrounds, but also gender can play a role here: individual factors relating to the person that is confronted with an ethical issue, including demographics, psychological features and ethical experience, can make a difference. These factors together are the cornerstones of the moral capacity or moral character of an individual. Gender plays a role in how people deal with ethical issues, as famously described by Carol Gilligan (1982). Due to the overrepresentation of males in the engineering profession, there is also a serious risk of gender-biased technology (Leavy, 1982).

Just as team members can only see things from their own perspective, the same goes for your test subjects. Testing your product in multiple stages with a broad, inclusive range of people is also highly recommended if you want to avoid biases in test subjects and make a better product that suits the needs not only of the people that look and think like you, but of all of us. This way we can also ensure that people are not neglected or overlooked, which can ultimately lead to unforeseen ethical questions.

moral design process

2. Technological products can be constructed on the basis of a moral design process

Technological products can be constructed on the basis of a moral design process: a circular, recurring process that helps internalise external (terminal) values. It doesn't matter if you are used to working with agile/scrum, design thinking or with a PDCA cycle (to name just a few); the main thing is that you can shape and reshape your product at each stage of the process to make it better and avoid becoming one of the tech regrets of the future. Consciousness and discussion are important at each stage: recognising human values and asking whether human values are central or are at stake are good things to remind yourself of continuously. So, we have to apply ethics 'from the inside out', and not 'from the outside in' or only at the end, as is often the case in ethical discussions. The TICT is an example of a moral design toolkit that can guide you through all stages of this process.

ethics drive innovation

3. Ethics can be seen as a driving force for innovation

In our meetings with representatives in the professional field we noticed that there was initially a slight hesitation about practising ethics: companies thought that practising ethics would result in ethical viewpoints that could lead to people not wanting to do something because of an ethical problem that might arise. We think that ethics should not be seen in that way. The question should not be what to do or what not to do, but how to make or do things in a better way. So, the question we have to ask here is: 'How are we going to design or redesign human-technology interactions in favour of our well-being?' Reframing ethics in that sense can give companies a commercial advantage over competitors who don't practise ethics. After two years of research we already see that perspectives have changed: companies are now asking us to tell them more about our research so that they can benefit from putting ethics into practice. This is the point at which techno-ethics and business ethics (see Chapter 6) meet.

postpone judgement

4. Try to postpone judgement during the guiding process of technological development

Peter-Paul Verbeek argues that it is better to guide rather than judge technological development. We should add here that our role as a university is not to decide what students or engineers should think or what the morally correct answer should be to an ethical dilemma, but to let them make up their own minds. The questions we ask should therefore be non-judgemental: there might not be such a thing as an absolute right or wrong, but there are a lot of options and choices to make. In the TICT this way of thinking shows in that sense that some categories may have a normative aspect to them, but you can fill it in as good or evil as you like: we won't judge.

Of course, it is possible to draw a final conclusion at the end of the process, but our suggestion would be to do that only when you have run through all the categories and are able to make up your mind based on multiple viewpoints.

context-driven applied ethics

5. The moral design process is part of context-driven applied ethics

There are no 'one size fits all' solutions here, as every case has a different context. From the perspective of Mediation Theory, ethics should not be limited to whether or not a technology is acceptable (a yes/no question), but should be concerned with how a new technology could assume its place in society and how that affects our human values, considering the impact of this specific technology within its applied context. The underlying philosophy that we choose here is a techno-philosopher's view: a relativistic (case-based) form of applied ethics.

So, to avoid possible future tech regrets like Berners-Lee, Zuckerberg, Dorsey a.o., we should ask ourselves a lot of questions and embark on a discussion at each stage of the design process (for instance, in the beginning, middle and end phase, in each iteration stage or in the emphasise, design, idea, prototype and testing phase in the case of design thinking), called 'ethics from within'. In so doing we can design, invent or use technology with a positive impact on society. Note here that the interpretation of what a 'positive impact' is depends on our own moral compass and the chosen moral authority as discussed in Chapter 2.

3.4 The Technology Impact Cycle Tool

In this section we present the creation (design process), features and evidence-based use of the technology impact cycle toolkit.

creation process

Our multidisciplinary team of students, teachers and professionals involved in developing and implementing the educational innovation had two important starting points:

1. Our realisation that in education it is important to facilitate not only engineers, but all kind of students from all study programmes, lecturers and the surrounding professionals from the field to systematically reflect on the impact of technology;
2. The wish to construct a concrete, free, open, online toolkit that can be used in any given technological context by those who want to.

design-based research

The ambition of our initiative is to provide students, engineers, teachers, professionals and other interested parties with the means to think about the impact of technology in a better, fun and more structured way.

We have strived for this through various research processes, always starting from research-by-trying. From the first moment, prototypes of teaching materials and tools have been built, which have been adapted step by step based on input from the various stakeholders.

Determining the impact of technology is an inclusive issue. We are proud to say that our project team consists of people with diverse backgrounds and expertise, who have all been able to put together a piece of the puzzle (although we realise that of course a team can always be more inclusive). The experience from all sides has been converted into a first version of the toolkit, which has been continuously improved, incrementally and based on feedback from all stakeholders.

Literature research was also carried out and existing tools and frameworks were looked at. Where existing tools and frameworks often look at one aspect of the impact of technology (e.g. ethics or privacy) or one type of technology (e.g. data or AI), the strength of our toolkit is that it helps people look at the total impact of any technology in an accessible way. And we included these other toolkits, to make it a library of cross-references as well.

A graduate of Applied Psychology conducted research into the effectiveness of the tool and how it is perceived by students and teachers. The tool has been adapted based on that information and our student has graduated with honours with his research.

Finally, the tool was launched in January 2021. We continually collect feedback and will adapt the tool accordingly, so the toolkit is never really 'finished'.

The technology impact cycle tool consists of five major components, the most important of which are shown in Figure 3.3.

features of the toolkit

Smart scans (quick, full or improvement scan)

The heart of the tool consists of questions that are asked to determine the impact of technology. Because we think it is important that engineers learn to look at the total impact of technology, the questions are divided into ten categories. We have linked a specialist from our team (the ambassador) to each category to keep the material up-to-date, to keep improving and for being the contact point for anyone with questions and comments. The challenge was to formulate the questions in such a way that they are relevant to as many users as possible. This has been optimised by constantly putting them to the

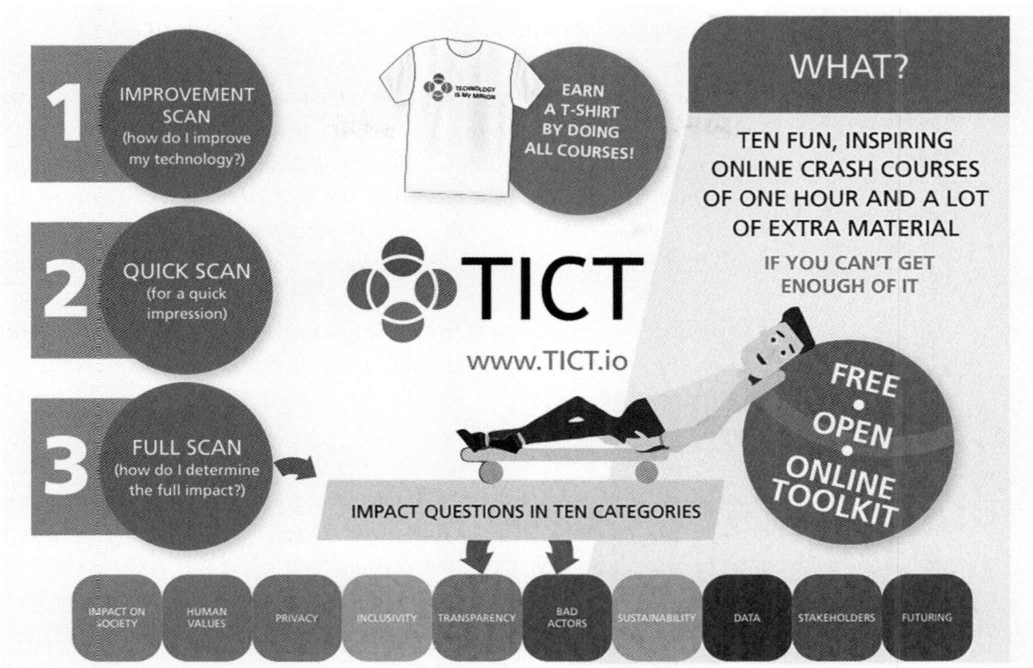

Figure 3.3. Short introduction TICT.

test, collecting and adjusting feedback and, together with students, engineers, professionals, teachers and lecturers, by further refining the questions (a process that is never finished).

So that the tool is accessible, it also offers the option of answering only one question per category (Quick Scan) or answering the questions that concern the improvement of a particular technology (Improvement Scan). Reports can be printed or downloaded as a PDF after completion. Progress can be saved, shared and collaborated on.

Open, free, online crash courses

The second important component is the open, online courses that are offered in each of the ten categories. To be able to think carefully about the impact of technology, you need to be inspired and have the means to think critically. The masterclasses are offered in the form of ten open courses of exactly one hour. These courses have been developed and tested in teaching situations, with professionals and together with a team of teachers.

Best practices (see www.tict.io)

The third component is best practices. We have built the tool so that it can be used in any kind of education and in any way. From briefly, fragmentary, to complete and extensive. We do think there are a number of ways in which the tool works best. To give you some examples of which viewpoint you can use the tool from, we came up with best practices, divided into specific roles that users can move themselves into (such as The Designer, The Teacher and The Critic).

All best practices came from our own practice and range from a teacher who organises a one-hour workshop for awareness to a designer who improves his product by using the tool for six months (and in multiple cycles), to a journalist who wants to write a critical article.

To give you an example: when we tested our toolkit with journalists, we noticed that they were a bit insecure at first when asking critical questions about technology, knowing that their own knowledge at this point might not be perfect. We noticed that, with the help of the questions in the toolkit, they regained confidence because they could ask questions that were already pre-formulated. In this way they broadened their view and at the same time reduced the chances of overlooking important things (Arets and Kamp, 2019).

Additional (teaching) material (see below each category in the full scans online)

The fourth component is extra (teaching) material. The TICT offers inspiration material, exercises and links to other frameworks and publications. The tool also offers a number of completed cycles (such as that of a mourning robot which we will talk about in Chapter 11) for inspiration and as an example. This also makes it a real gathering place for anyone who wants to help think about the future of technology. Or for anyone who likes to think along!

Community building (see the contact page at www.tict.io or meet us at LinkedIn)

As the fifth important component, the tool facilitates a community. All users can make their own answers (in the TICT we call them cycles) available to the community, so that users can inspire each other and you can respond to each other's analysis. You can also meet and connect to the ambassadors of each category via LinkedIn.

Use of the toolkit

Below we give an idea of the impact of the TICT on the basis of a number of concrete practical examples. First, the tool has been used in the Netherlands in the discussion about the effectiveness of the corona contact tracing app. The versatility of the tool provided a complete picture, which contributed greatly to the quality of the discussion.

Second, the tool is already widely used by thousands of students within Fontys University of Applied Science in diverse ways:

- Students of the EdTech minor worked for six months on a prototype with the aim of improving education with technology. These students were inspired by the open material in the toolkit to discuss this in workshops with their teachers and then use the tool to improve their EdTech solution.
- Education technology companies went through an EdTech validation programme for several months, and used the toolkit to test and improve their own solutions.
- Movement designers from the Sports College used the toolkit for the same applications as above, but in their own field.
- Smart Mobile students quickly tested their intended apps with a Quick Scan, so that they could make the choices with the best impact from the start.
- Students of the minors Data Driven Business Lab and Artificial Intelligence followed a workshop about the toolkit and learned how to reflect on their own cases with it.
- Employees from companies with a partner covenant followed workshops about TICT and were challenged to reflect on their own cases with it. After the workshop they wanted to spread the word and wanted to ask these kinds of questions multiple times during the design- and development cycle at their company.
- Students of the minor Embrace Technology used the Improvement Scan to see what they could improve about their technological solution.
- Marketing Management students used the courses offered to gain a general impression of the impact of technology. They took an exam using the tool.
- The new Master DTE (digital technology engineering: a two-year long technical programme that tackles the challenges posed by the world's digital transformation) will use the tool as a quality tool for the 'solutions to be delivered'.
- Journalism students used the toolkit to ask or prepare critical questions. For example, an analysis of the Dutch CoronaMelder App was drawn up using the toolkit.

In summary, it can be said that the TICT has an impact on all kinds of stakeholders in many different ways:

- Students who use technology with the aim of improving society, think more broadly and in a more structured way about the impact of their solutions after working with TICT.
- Students who use technology in their professional practice are given tools to think about the effectiveness of their technology.
- Students who have to be able to critically examine technology (journalists, nurses, managers) become better at it.
- The many teachers who want to add 'something with ethics' or 'something with impact' to their teaching are provided with the means to do so.
- The open, online and free courses combined with best practices provide space and guidelines for working hands-on with students.
- The professional field that collaborates with Fontys is inspired to examine its own products with the toolkit and can now come up with new and ethically well-thought-out solutions that will serve as a business benefit for now and in the future.
- Anyone who wants to know more about his or her relationship to technology will have access to access to publicly available and inspiring courses.

It is a conscious choice to offer the toolkit openly and for free. We think that giving everyone unlimited access allows the toolkit to have more impact, sometimes even in surprising ways that are as yet unknown to us.

We certainly hope that initiatives like this will contribute to a world where moral impact is not just a concept we debate, but is something we act upon: creating moral impact by continuously shaping a future we want to live in!

References

Arets, D. and Kamp, J., 2019. Rapportage slimme luis in de pels. HBO kennisbank. Available at: https://hbo-kennisbank.nl/details/sharekit_fontys:oai:surfsharekit.nl:9944bcc9-8bf0-4a62-a823-d90d8400b36f.

Confessore, N., 2018. Cambridge analytica and facebook: the scandal and the fallout so far. The New York Times. Available at: https://www.nytimes.com/2018/04/04/us/politics/cambridge-analytica-scandal-fallout.html.

Dafoe, A., 2015. On technological determinism. Science, Technology, & Human Values, 40(6): 1047-1076. https://doi.org/10.1177/0162243915579283

De Ethische Data Assistent (DEDA). (n.d.). Utrecht data school. Available at: https://dataschool.nl/deda/.

European Commission, 2020. Gendered Innovations 2: how inclusive analysis contributes to research and innovation. Available at: https://ec.europa.eu/info/sites/default/files/research_and_innovation/strategy_on_research_and_innovation/documents/ki0320108enn_final.pdf.

Gilligan, C., 1982. In a different voice: psychological theory and women's development. Harvard University Press, Cambridge, MA, USA.

Hoekman, J.-H., 2020. Privacy design strategies (The little blue book). Radboud Universiteit. Available at: https://www.cs.ru.nl/~jhh/publications/pds-booklet.pdf.

Institute for the Future and Omidyar Network, 2018. Ethical OS. Available at: https://ethicalos.org/.

Leavy, B., 1982. Wilkie Collins's Cinderella: the history of psychology and 'The woman in white'. Dickens Studies Annual, 10: 91-141. Available at: http://www.jstor.org/stable/44372398.

McMillan, R., 2019. After New Zealand shooting, founder of 8chan expresses regrets. The Wall Street Journal. Available at: https://www.wsj.com/articles/after-new-zealand-shooting-founder-of-8chan-expresses-regrets-11553130001.

Rijnboutt, J. and Heerink, M., 2020. Ethiek in ICT en techniek (1st ed.). Pearson Benelux B.V, Amsterdam, the Netherlands.

Sesink, W., 2012. Heideggers Frage nach der Technik ... bildungstheoretisch nachgedacht (Bildung und Technik 2) (German Edition) (1ste ed.). Available at: https://www.academia.edu/37564540/Heideggers_Frage_nach_der_Technik_bildungstheoretisch_nachgedacht.

Spiekermann, S., 2015. Ethical IT innovation: a value-based system design approach. Taylor & Francis, Palm Bay, FL, USA.

Van de Poel, I. and Royakkers, L., 2011. Ethics, technology, and engineering: an introduction. Wiley, West Sussex, UK.

Van de Poel, I., 2015. Values in engineering and technology. In: Gonzalez, W.J. (ed.) New perspectives on technology, values, and ethics. Boston Studies in the Philosophy and History of Science, Volume 315, Springer, Cham, Switzerland, pp. 29-46.

Van Doorn, M., Duivestein, S. and Pepping, T., 2019. The synthetic generation: growing up in an uncertain and changing world. Digital Happiness 03/04. Available at: https://www.sogeti.com/explore/reports/digital-happiness-reports/the-synthetic-generation/.

Verbeek, P., 2000. De daadkracht der dingen: over techniek, filosofie en vormgeving. Boom, Amsterdam, the Netherlands.

Verbeek, P., and De Jong, R., 2017. MOOC philosophy of technology and design. University of Twente. Available at: https://www.futurelearn.com/courses/philosophy-of-technology.

Zhang, Y., Wells, C., Wang, S. and Rohe, K., 2018. Attention and amplification in the hybrid media system: the composition and activity of Donald Trump's Twitter following during the 2016 presidential election. New Media & Society, 20(9): 3161-3182. https://doi.org/10.1177/1461444817744390

Media

Berners-Lee, T. (n.d.). Solid. Available at: https://solidproject.org/.
Brooker, K., 2018. 'I was devastated': Tim Berners-Lee, the man who created the World Wide Web, has some regrets. Available at: https://www.vanityfair.com/news/2018/07/the-man-who-created-the-world-wide-web-has-some-regrets.
History of the Web (n.d.). World Wide Web Foundation. Available at: https://webfoundation.org/about/vision/history-of-the-web/.
Rahman, A., 2017. Facebook Messenger Online Status. Available at: http://www.createregisteraccount.com/2017/05/facebook-messenger-online-status.html.

Codified standards and case law

United Nations (n.d.). THE 17 GOALS | Sustainable Development. Available at: https://sdgs.un.org/goals.

4. Moral programming

Crafting a flexible heuristic moral meta-model for meaningful AI control in pluralistic societies

Nadisha-Marie Aliman[1] and Leon Kester[2*]

[1]*Utrecht University, Faculty of Science, Department of Information and Computing Sciences, Princetonplein 5, 3584 CC Utrecht, the Netherlands;* [2]*TNO Netherlands, Intelligent Autonomous Systems, Postbus 96864, 2509 JG The Hague, the Netherlands; leon.kester@tno.nl*

Abstract

Artificial Intelligence (AI) permeates more and more application domains. Its progress regarding scale, speed, and scope magnifies potential societal benefits but also ethically and safety relevant risks. Hence, it becomes vital to seek a meaningful control of present-day AI systems (i.e. tools). For this purpose, one can aim at counterbalancing the increasing problem-solving ability of AI with boundary conditions core to human morality. However, a major problem is that morality exists in a context-sensitive steadily shifting explanatory sphere co-created by humans using natural language – which is inherently ambiguous at multiple levels and neither machine-understandable nor machine-readable. A related problem is what we call epistemic dizziness, a phenomenon linked to the inevitable circumstance that one could always be wrong. Yet, while universal doubt cannot be eliminated from morality, it need not be magnified if the potential/requirement for steady refinements is anticipated by design. Thereby, morality pertains to the set of norms and values enacted at the level of a society, other not nearer specified collectives of persons, or at the level of an individual. Norms are instrumental in attaining the fulfilment of values, the latter being an umbrella term for all that seems decisive for distinctions between right and wrong – a central object of study in ethics. In short, for a meaningful control of AI against the background of the changing context-sensitive and linguistically moulded nature of human morality, it is helpful to craft descriptive and thus sufficiently flexible AI-readable heuristic models of morality. In this way, the problem-solving ability of AI could be efficiently funnelled through these updatable models so as to ideally boost the benefits and mitigate the risks at the AI deployment stage with the conceivable side-effect of improving human moral conjectures. For this purpose, we introduced a novel transdisciplinary framework denoted augmented utilitarianism (AU) (Aliman and Kester, 2019b), which is formulated from a meta-ethical stance. AU attempts to support the human-centred task to harness human norms and values to explicitly and traceably steer AI before humans themselves get unwittingly and unintelligibly steered by the obscurity of AI's deployment. Importantly, AU is descriptive, non-normative, and explanatory (Aliman, 2020), and is not to be confused with normative utilitarianism. (While normative ethics pertains to 'what one ought to do', descriptive ethics relates to empirical studies on human ethical decision-making.) This chapter offers the reader a compact overview of how AU coalesces elements from AI, moral psychology, cognitive and affective science, mathematics, systems engineering, cybernetics, and epistemology to craft a generic scaffold able to heuristically encode given moral frameworks in a machine-readable form. We thematise novel insights and also caveats linked to advanced AI risks yielding incentives for future work.

Key concepts

- For meaningful AI control, one needs to channel AI's ability using human morality
- Human morality is context-sensitive, shifting, and ambiguously embedded in language
- It is difficult to make human morality machine-readable
- Descriptive ethics is the study of how human morality is enacted and not what one ought to do
- Augmented utilitarianism (AU) is not to be confused with utilitarianism. AU is a transdisciplinary framework for meaningful AI control of a meta-ethical, descriptive, explanatory, and non-normative nature
- AU offers a generic AI-readable scaffold for heuristic moral models facilitating AI control
- Thereby, AU coalesces elements from descriptive ethics, epistemology, and a wide range of scientific disciplines leading to novel insights that are relevant for future work on AI control

4.1 A theoretical solution for meaningful artificial intelligence control

Under augmented utilitarianism (AU), one can subdivide the core practical task of meaningfully controlling AI via machine-readable heuristic moral models[2] into three challenging subtasks. The first two subtasks require theoretical epistemic and scientific foundations. The third subtask pertains to realising a complex multi-stakeholder implementation. While this section focuses on the first two theoretical subtasks, the latter is presented in the next Section 4.2. Firstly, while it is habitual for humans to mentally extract morally relevant events from the ongoing experience to which moral judgements are applied, AI neither automatically performs such assessments nor the kind of moral extraction procedure required in the first place. One could call the latter a moral chunking problem, i.e. the problem of locating a meaningful abstraction level for heuristic moral models. As an illustrative analogy from a different context, one can consider the example of human language acquisition where infants face the linguistic chunking problem of dividing incoming information, such as, for instance, auditory information, into meaningful chunks (Isbilen and Christiansen, 2020). A similar problem is also encountered by adults in the process of second language acquisition. Without prior semantic knowledge it is often daunting to identify the boundaries of individual words or multi-word expressions from speech. Secondly, even if an AI were programmed with an appropriate moral abstraction level for a selected ethical framework and the parameters of relevance within that chosen framework were filled in, you are faced with an old epistemic: what if you were wrong and changed your mind or what if you were uncertain about your choices? Below, we look at some theoretical solutions to these two AU subtasks.

moral chunking problem

4.1.1 Moral chunking

As mentioned earlier, AU is formulated from a non-normative meta-ethical stance which signifies that instead of discussing what one ought to do (as is the case in utilitarianism), it is instead located at a higher level focusing on what morality itself is, how it is applied and why. However, instead of having philosophical aspirations, AU engages in these deliberations for quintessentially pragmatic safety-oriented socio-technological reasons. To this end, AU employs knowledge from different scientific disciplines to get a clearer view of the nature of human morality, and integrates an engineering-relevant systems

---

[2] Note that these models are heuristics since one is always only provisionally modelling utterly complex moving targets of morality. As a result, both moral targets and abstract models thereof are fallible and not necessarily optimal at any point in time. In short, it is acknowledged that one could always be wrong.

perspective to finally work out technically feasible solutions. It is thus in such a pragmatic safety-oriented context that AU studies this first subtask of moral chunking for meaningful AI control. To start with, AU can be conceptually understood as a sort of input-agnostic interface between humans equipped with a certain instance of an ethical framework and the AI they intend to align with this framework. To be able to fulfil this generic interface role, AU must search for efficient moral chunking with the capacity to ideally be applicable to most ethical frameworks that society would select to control AI. In the context of autonomous vehicles, recent work found that 'it is essential to increase the public's confidence that the values of a pluralistic society are accounted for' (Dubljević *et al.*, 2021). This may apply more generally to any AI deployed at societal scales. In this vein, AU targets what one could conceive of as a possible smallest heuristic moral superset (SHMS) capturing the plurality of candidate ethical frameworks available in practice for moral programming – with the pre-condition that AU must flag known formal inconsistencies to forestall predictable practical safety problems in AI deployment. The latter shows why moral chunking in AU cannot merely be descriptive but must be simultaneously explanatory. From a systems engineering perspective, to specifically strive for an SHMS is not only of interest by virtue of being simpler but it is also more energy-efficient since one does not integrate more parameters than necessary (which would e.g. afford more computing power) while still acting on the requirement to integrate moral pluralism phenomena whose existence has been often corroborated in moral psychology studies (Schein and Gray, 2018).

smallest heuristic moral superset

The three classical normative ethical frameworks that are often mentioned as candidates for moral programming are virtue ethics, deontic ethics, and utilitarianism. Note thereby that moral chunking in AU is agnostic to whether the candidate ethical framework is in fact of a normative, relativistic, comparative, purely pragmatic, or other nature. Here, we use the example of the three normative frameworks due to their salience in AI ethics and AI safety debates. While virtue ethics focuses on the virtues exhibited by the agent, deontology foregrounds the action itself and utilitarianism the consequences of the action to the receiver of the action, also called the patient. In moral programming, the agent carrying out the action is often the AI while the patient is a human. However, in human-machine teaming contexts, agents and patients can be either humans or AIs depending on the situation at hand, which means that sometimes a task can in theory also be purposefully limited to AIs (e.g. if the collaborative task were too dangerous to be performed by human agents). In the AI field, the widespread use of utility-maximising AI systems in the form of reinforcement learning agents, modelled in the fashion of utilitarian economics, established utility functions as default standard for moral programming in their characteristic of reflecting supposed rational decision-making. However, there can be a difference between: (1) what humans think they ought to do; and (2) what they want an AI to do in practice. Furthermore, there can be a difference between the former and what humans can actually do as well as between the latter and what an AI itself can do in practice. The practical consequences of such caveats (re)appear throughout our analysis. In the following paragraph, we first briefly shed light on key scientific insights that are of relevance for moral chunking under AU.

Recent moral psychology studies suggest that human morality is based on constructions, is quintessentially perceiver-dependent, context-sensitive, multi-faceted, and is linked to diverse often implicit mind perception phenomena and unconscious assessments. For instance, empirical evaluations in the context of the Agents, Deeds, and Consequences (Dubljević *et al.*, 2018) model (ADC-model) corroborated that moral judgements unify intuitions from all three common normative frameworks mentioned earlier, which according to the authors 'provides an explanation for the intuitive appeal of dominant moral theories' (Dubljević *et al.*, 2018). Following the theory of dyadic morality (TDM) (Schein and Gray, 2018), human moral judgements are harm-based constructions that relate to a perceiver-dependent cognitive template encoding how much damage an intentional agent causes to a vulnerable patient. In this way, TDM frames morality as exhibiting unity in the variety of perceived harm. In view of the moral judgment as categorisation (MJAC) framework (McHugh *et al.*, 2021), 'understanding the making of moral judgments requires accounting for the full complexity and variability of our moral judgments'. Namely, MJAC conjectures (in the spirit of constructionist frameworks in psychology (Oosterwijk *et al.*, 2012)) that moral judgements are perceiver-dependent cognitive-affective constructions based on domain-general elements (not limited to harm) – describing morality as one of the potentially infinite set of situated embodied conceptualisations, i.e. categories that human minds can bring forth and learn to bring forth. Generally, constructionist theories in psychology and affective neuroscience reveal that human cognition is inseparably affective (Hoemann and Feldman Barrett, 2019; Kleckner *et al.*, 2017), and so the assumption that humans are cold rational agents acting according to an affect-free mathematical strategy is a myth. As stated by Barrett (2017), the human brain is anatomically structured in such a way that no action is free of interception and affect.

Overall, one can conclude that in order to meaningfully address the moral chunking subtask, AU, in its pursuit of an SHMS integrating submitted candidates, must try to accommodate the potential breadth and variety of widespread human morality. Parameters must be specifiable (which means they do not need to be but can optionally be specified) not only for the agent (as e.g. in virtue ethics), not only for the action (as e.g. in deontology), not only for the consequences (as e.g. in utilitarianism) but also for the first-person perspective of the perceiver performing the moral judgement (as e.g. in secular care ethics, in some schools of Indian philosophy but also in Christian ethics), which adds further relational considerations. But since the human perceiver exists within unique socio-psycho-techno-physical strata, has a unique developmental history, and is located in a specific spatial and temporal context, parameters as heterogeneous as socio-cultural background, psychological features, peculiarities in mind perception, current physiological and health state, and environmental conditions can have a decisive effect on the judgement of moral events (see also e.g. Haidt (2001) and Schwartz (2016) for more related perspectives on human ethical decision-making). Interestingly, even the language in which a moral event is presented can have an impact on moral judgements via differences in affective associations, which can lead to a tendency for more distance in second-language contexts (McHugh *et al.*, 2021). Beyond that, similar crashes with self-driving cars were considered more severe compared to those involving human drivers irrespective of cause (Liu *et al.*, 2019).

To sum up, it becomes clear that for sense-making regarding human morality across multiple different ethical frameworks, AU-based moral programming must at the minimum be able to accommodate a variety of parameters with regard to the following generic elements that can provisionally serve as an SHMS basis for moral chunking: perceiver, agent, action, patient. These four elements build up a tentative SHMS moral chunking tetrad for AU (hereafter referred to as SHMS chunking tetrad). That being said, it is important to note that, in addition, a breadth of epistemically relevant biases can arise in the context of moral judgements. This includes especially mind perception distortions (Gray *et al.*, 2012; Wegner and Gray, 2017) and formal inconsistencies. For instance, in empirical studies, people based moral judgements of artificial agents, among others on perceived intentionality and anthropomorphism (Bigman *et al.*, 2019), perceived feelings (Yam *et al.*, in press), and perceived agency, whereby the latter can even affect the sense of agency of the human perceiver (Ciardo *et al.*, 2020).

Instead of imposing a choice, AU provides a supportive generic interface as scaffold for candidate ethical frameworks. However, AU encourages critical thinking and flags epistemic issues related to (formal) inconsistencies that can affect AI safety post-deployment. For instance, for AI in ethical high-stake contexts, it has been shown that utilitarian and consequentialist utility functions face safety-relevant impossibility theorems (Eckersley, 2018), which can be derived formally when applying utilitarianism to population ethics (Arrhenius, 2000; Greaves, 2017). These impossibility theorems are linked to a mind perception problem that has been called 'the perspectival fallacy of utility assignment'[3] (Aliman, 2020; Aliman and Kester, 2019a). Hence, AU flags classical utilitarian utility functions as problematic for high-stake AI. To solve this utility function problem in AI safety, different solutions have been proposed in recent years, among which the two options that we briefly enumerate next. As a first option, it is notable that AU itself facilitates the framing of utility functions at a higher abstraction level that is not subject to the utilitarian impossibility theorems. In fact, AU gains its name from this point of departure. Namely, in its quest for an SHMS, it was also considered whether ethical mathematically formulated functions such as AU candidates should always be problematic on theoretical grounds or whether a meaningful alternative was possible. Classical utilitarianism assigns utilities to states of the world, i.e. to time snapshots encoding merely the consequences of actions on the patient. Extending beyond that, AU facilitates the crafting of utility functions that are not affected by those impossibility theorems. This is possible under AU via dynamically updatable utility functions formulated at the level of time integrals covering the entire SHMS chunking tetrad, i.e. with information on perceiver, agent, action, and patient (Aliman, 2020). A second option to bypass utilitarian impossibility theorems is to probabilistically encode moral uncertainty into AI utility functions (which mathematically means to consider functions that are either partial orders or probability distributions over total orders instead of the classical total orders that utilitarianism uses) as advanced by Eckersley (2018). The latter leads us to the next subtask pertaining to the inescapable question: what if one is uncertain about one's ethical framework?

dynamically updatable at the level of time integrals

4.1.2 Epistemic dizziness

Under AU, it is not only acknowledged that one could be wrong in the context of moral decision-making but that one could always be wrong. This is in line with the epistemic assumptions of Karl Popper that all knowledge is fallible (Popper, 1995). Human beings experience embodied and socially modulated, cognitive-affective perspectival projections of the world. Also, under a deflationary account where truth is not equated with consensus, humans do not inhabit a 'post-truth' era (Aliman and Kester, 2020) since it is clear that even in the past sensory experience did not directly provide access to truth by virtue of being embedded in a web of prior knowledge from within an egocentric first-person perspective of the world shaped by cognitive-affective dynamics – making 'post-truth' a misnomer. Generally, well-tested unfalsified conjectures could be wrong, and previously empirically falsified conjectures could be true if the assumptions underlying the empirical observations were wrong (Frederick, 2020a). Affective realism, the phenomenon that one tends to

[3] For exemplary scenarios on how this fallacy manifests, see Aliman, 2020.

see what one believes via affect being interwoven in perception (Fridman *et al.*, 2019; Siegel *et al.*, 2018), adds up to what we call epistemic dizziness. Morality is of a conjectural nature; even unconscious components stem from a generative model of the world with certain priors encoding assumptions and associations from the past, which is key to affective realism. In short, morality too is affected by this inescapable epistemic dizziness. Fortunately, our general epistemic aim can still be to create ever better explanations (Frederick, 2020a), i.e. to tease out better from worse explanations (on the basis of theoretical criticism and if possible empirical falsification) while steadily creating new ones. The crucial question now becomes: what does it signify for moral programming?

Firstly, due to the fallibility of human knowledge, any framework for meaningful AI control must be updatable-by-design. (Since one cannot scientifically predict the future of knowledge creation, even entities that claim to know objective moral truth today cannot predict that they would not be able to change their own view in the future.) When applied to an AI that is deployed on societally relevant scales, such a generic update mechanism has been termed to instantiate a socio-technological feedback loop (Aliman, 2020). Secondly, the starting point for such an AI-enacted moral feedback mechanism (even if on smaller scales) need not and cannot logically be proven right. Justifications are logically invalid, as shown by Popper long ago (Bartley, 1976). For instance, when contemplating the statement that all swans are white, no amount of repeated observations of white swans can prove its veracity. Hence, it is sufficient to choose reasonable, well-tried starting points without truth-related claims. Depending on the scale of the AI deployment, one can select from a vast array of internally formally consistent options comprising e.g. normative theories from one's culture, the AU-enhanced tetradic utility function scaffold (SHMS) filled in with parameters from a self-selected ethical framework, legal rules for criminalised contexts, some parameters from the UN Sustainable Developmental Goals, other international norms, comparative theories, pragmatic alternatives, any tailored combinations of those or weighted sums, and so forth. Thirdly, AU can then be utilised to refine moral reasoning if desired. For instance, with the AU-enhanced utility function option, one can test AI-enacted moral knowledge specified at the beginning of an iteration of a socio-technological feedback loop or organisational feedback loop. Using simulation environments, one can exploit counterfactual reasoning (Aliman and Kester, 2019a) and craft adversarial AU-enhanced utility functions either to let different moral conceptions compete against each other for self-education or purely in order to explore novel candidates for one's own moral values.

The question at the end of the last section asked: what if one is uncertain about one's ethical framework? Two remaining caveats relevant to answering this question are addressed in this and the next paragraph. Firstly, irrespective of the ethical framework chosen, there is in general an inherent limitation to the injection of moral models in AI systems. In fact, depending on the framework, one assumes pre-deployment that the entirety of options, alternatives, parameters, and/or outcomes is known in advance. However, human moral decision-making is unpredictable, often disruptive, and can even be adversarial as one can deduce from the example with the adversarial AU-enhanced utility function. As stated by Deutsch (2011): 'It is a mistake to conceive of choice and decision-making as a process of selecting from existing options according to a fixed formula. That omits the most important element of decision-making, namely the creation of new options.' In short, in moral decision-making, it can be that people create additional yet unknown alternatives. It can be that they delete and replace initial moral assumptions if they adversarially experiment with novel differing values that they create or discover. From this point of view, 'rationality permits us to act in accord with our best-tested theories, since they may be true; but it also permits us to act against them, precisely because our best-tested theories may be false and may, indeed, be refuted when we act against them' (Frederick, 2020b). A deployed present-day AI is not able to perform a creative adversarial task requiring explanatory knowledge. It is for this reason that we depicted AU-based moral programming as facilitating the design of heuristic moral models, ephemeral approximative shadows of morality to be necessarily updated with time.

heuristic moral models

Secondly, the idea of crafting moral models by assigning probabilistic uncertainty to moral values themselves (which is different from weighted sums) in the hope of outsourcing its resolution to an AI system post-deployment brings about epistemic and safety-relevant caveats. From epistemic dizziness it is clear that objective probabilities for conjectures, i.e. absolute certainty, are never guaranteed for lack of experiencing truth directly. However, the case of subjective probabilities for normative ethical frameworks can lead to self-contradictory implications for AI, among others because it is epistemically permissible to act against one's prior assumptions (Frederick, 2020a) (e.g. in an attempt to test them). Therefore, this also includes: (1) the option to act against one's own subjective probabilities; and (2) to act on novel differing spontaneously created options. The former can lead to AI control *ad absurdum* and the latter cannot be performed by present-day AI since explanatory knowledge is required. In short, probabilistic moral uncertainty in moral programming is feasible if seen as heuristic, and whose resolution would require the AI to prompt a human at the post-deployment stage. Otherwise, especially in ethical high-stake cases, to delegate a metaphorical throwing of (crooked) dice (or flipism) to a deployed AI system only adds blurriness to the already doubt-ridden heuristic moral models, is not optimal for cybernetic control (simply put, incomplete knowledge is better than none (Wiener, 1960)), and can be highly questionable for reasons of safety, accountability, and lack of participatory decision-making. The existence of epistemic dizziness means we cannot inherently eliminate doubt from the world, but we need not magnify it. We can accept the conjecture that we could always be wrong but can strive for ever better (moral) explanations within (moral) feedback-loop mechanisms, among others to craft better AI-enacted heuristic moral models. In a nutshell, uncertain humans equipped with some dice at the time of moral decision-making could throw those dice but could also unexpectedly (co-)create novel as yet unknown solutions for the problem – something present-day AIs cannot. Hence, in ethical high-stake cases of moral programming, it is recommended to reserve the throwing of dice (if really necessary) to people.

4.2 Practical use of theoretical solution

As mentioned in Section 4.1, AU-based meaningful AI control can be framed in terms of three subtasks. In this section, we will finally address the third one. After integrating theoretical knowledge on how to tackle the first two subtasks, i.e. the moral chunking problem and epistemic dizziness, one is still faced with the complex systems engineering problem of devising safety-aware strategies for AU-based AI deployment. Against the backdrop of AI risks in the form of first-order harm and potential second-order repercussions emerging amidst interwoven multi-stakeholder settings – be it at the level of an organisation or larger societal contexts – this third AU subtask meets non-trivial challenges. In this section, we elucidate how the theoretical foundations from the last section can be harnessed for moral programming management in practical settings with AU as an interface. Finally, we briefly touch upon a practical open question whose relevance has increased in recent years: can we craft a principled approach against the risks posed by deploying AI operating in regimes that widely exceed their human counterparts with regard to speed, scale, and selectivity? We deconstruct the underlying issue, introduce what we call the 'no-body problem' in AI safety and explain how one possible option could consist of utilising AU in conjunction with a cybernetically motivated strategy for advanced AI safety; however, we also depict novel caveats.

systems engineering problem

4.2.1 Moral programming management with augmented utilitarianism

For small-scale AI deployment contexts, AU can already fulfil a relevant function in the design process. Namely, in small-scale AU-based AI design the domain-general requirement is twofold. Firstly, for safety reasons, there is a vital requirement for only offering well-formed internally consistent machine-readable heuristic moral models as options, i.e. that are for instance not known to be affected by mathematically specified impossibility theorems. Conceivable machine-readable formats range from rule-based forms to AU-enhanced utility functions and to knowledge graphs. Secondly, there is the requirement for designers to provide an updatable and user-adjustable interface giving each user the possibility (if desired) to tailor their own individualised heuristic moral model selecting from a wide pool of specific elements sampled from the entire generic SHMS chunking tetrad. Importantly, since AU is non-normative, meta-ethical, and descriptive, it only acts as an empty structured scaffold left blank. It is the user who then specifies whether and which of the multiple thinkable and available moral parameters related to perceiver, agent, action, and/or patient to instantiate. As a result, via the tailored AU encoding, the problem-solving ability of the AI system can now operate within the self-determined individualised moral bounds of the user. Optionally, designers can in addition provide a pre-selection, e.g. from common heuristic moral models or their own inclinations.

For AI deployed in public domains and in large-scale legally relevant contexts, the emerging accountability issues require a clean assignment of responsibilities. In this case, the well-formed machine-readable heuristic moral models should ideally, and where feasible, be constrained not only by soft law but also by hard law. However, it seems worth acknowledging that a central AI safety regulation could – if not actively counteracted with novel practical measures – undesirably entail bureaucratic layers that are temporally not on a par with ongoing technological progress. In any case, an inherent requirement of a scheme compatible with either hard or soft law (as was also the case in the small-scale setting) is that the fine-tuning of a heuristic moral model and the design of an AI system focusing on its problem-solving ability in a given domain are separable. More precisely, for such a strategy to be feasible one requires AI architectures for which goals and problem-solving ability are orthogonal to each other (i.e. can be freely combined). Only via this type of orthogonality-based disentanglement can the described AU-based strategy be realised. (Note that orthogonality-based disentanglement is decisively different from and not to be confused with the orthogonality thesis propounded by Bostrom claiming that 'intelligence and final goals are orthogonal axes along which possible agents can freely vary' (Bostrom, 2012). However, by contrast it holds that: 'Orthogonality-based disentanglement is strictly bounded to an existential quantifier – it only assumes that there exists an AI architecture for which orthogonality holds. From a predicate logic perspective, whether the orthogonality-thesis formulated for all agents holds or not is distinct and separated from our orthogonality-based disentanglement assumption' (Aliman, 2020).) From a system engineering perspective this signifies that there must be an AI system architecture for which one is able to separate the 'what' from the 'how' in order to allow for a disentangled assignment of responsibilities. Examples of existing types of AI systems where this is possible are deep-learning systems and hybrid intelligent systems instantiating MDPs in utility-based planning (but also the more recent active inference agents (Smith *et al.*, 2021)). Intelligent systems are able to perform the OODA loop and have been often referred to as 'autonomous systems'. However, we choose to use the term 'intelligent system' since those AIs are not setting their own goals autonomously. Instead, they always act on human-defined heuristic moral models.

In Figure 4.1, we provide a simplified illustration of deploying an intelligent system in large-scale contexts utilising a stakeholder-crafted AU-encoding constrained – in this example – by hard law. The stakeholder in question is a societally relevant entity performing a certain operation. For transparency and to improve explainability, the intelligent system provides information related to its internal working via counterfactuals of selected courses of action to clarify its enactment towards surroundings and stakeholder. If later required, these samples of counterfactual reasoning can be used for auditing and in forensic and judicial contexts. The manufacturers are responsible for conducting safety and security tests pre-deployment and throughout the lifecycle. Importantly, for reasons of formal inconsistencies discussed earlier, if the chosen AU-encoding represents a utility function, it cannot be a conventional one. It must be an alternative not prone to impossibility theorems, such as e.g. AU-enhanced utility functions (where norms and rules can also be used as constraints on the function), an inter-theoretic weighted sum option or morally uncertain functions with partial orders necessarily leading to prompt human entities. On the whole, the figure contains multiple feedback loops. While an in-depth analysis is beyond the scope of this chapter and has been performed elsewhere, we draw the attention of the reader to two AU-based feedback loops. The outer feedback loop highlighted in blue comprising legislative power, stakeholder, intelligent system, and environment reflects a socio-technological feedback loop. At that level, which can also be supported by simulations or small experiments in synthetic environments, a societal level ethical enhancement can take place by testing and refining heuristic moral models in practice as discussed. However, the small inner feedback-loop between stakeholder and intelligent system (and analogically between user and AI in small-scale settings), while apparently simple, already reflects an emerging complex issue for the deployment of future more advanced AI surpassing human entities with regard to speed, scale, and scope. While in smaller less complex contexts harm incurred by AI seems more manageable, it is thinkable that future AI deployment could yield more unforeseen repercussions. In the following subsection, we briefly elucidate how a further analysis of this feedback loop reveals what we denote as the 'no-body problem'.

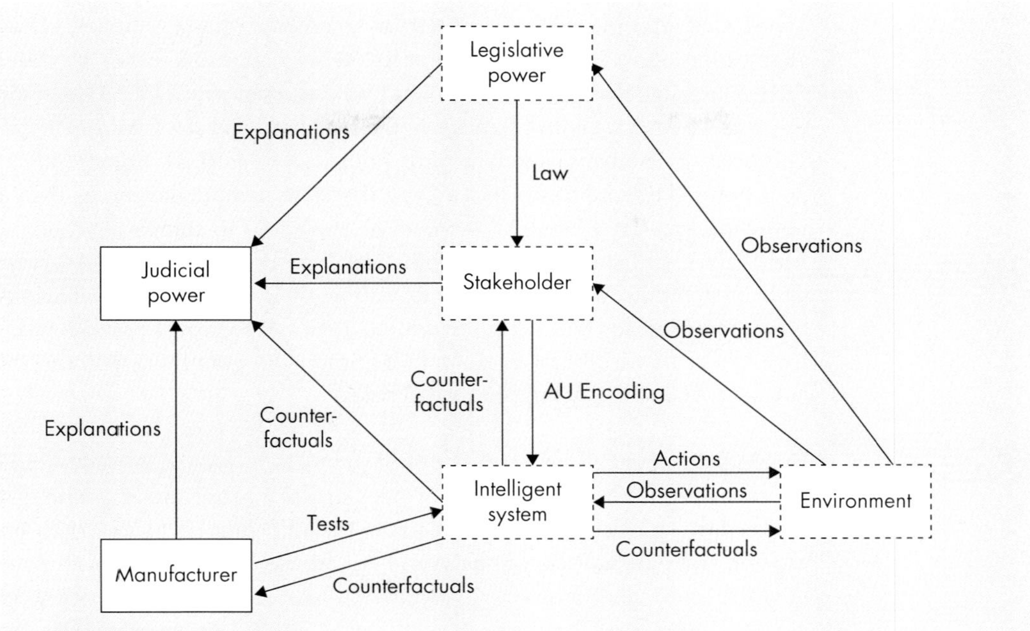

Figure 4.1. Simplified illustration of an AU-based socio-technological feedback loop (highlighted with dashed lines).

4.2.2 Advanced Type I artificial intelligence risks and the 'no-body problem'

Before discussing AI risks, we taxonomically distinguish between two types of systems. Unlike humans, present-day AI cannot consciously create and understand explanatory knowledge. Recently, the substrate-independent term 'Type II system' has been proposed (Aliman, 2020) for systems that are able to consciously create and understand explanatory knowledge. Type I systems are all systems that are not able to fulfil this task. Hence, from that perspective, all present-day AIs are Type I and non-conscious, and there is no single existing Type II AI. Non-human animals are conscious Type I systems. More broadly, humans are currently the only known instances of Type II systems (all of which are necessarily conscious). However, from a purely theoretical standpoint, while no such artificial system has been implemented anywhere on earth, Type II AI must be possible since it is not prohibited by any law of nature and due to the universality of computation. The AI safety paradox states that control and value alignment are conjugate requirements in AI safety. To put it plainly, you cannot simultaneously control and value-align with the same entity. Value alignment via morality necessitates understanding and explanatory knowledge which Type I AI is not able to. In other words, although the current aim of many researchers is to achieve the conjunction of control and value alignment for

Type I AI, it is postulated that we can try to control but cannot genuinely value-align with it across our domains of interest. Once a Type II AI exists, we could then value-align with it (if it agrees), but we could not control it, in the same way that we cannot control humans in the long term and slavery is unethical. This holds for humans as well as for hypothetical Type II AI. Importantly, it does not make sense to say that a Type II system is more dangerous than a non-conscious Type I system. It would be analogous to comparing humans with knives. Humans and also hypothetical Type II AI could of course have violent intentions, but it is not what determines their Type II nature. Malicious actors could already now design hazardous Type I AI to provoke serious risks. (In Chapter 10, we discuss an example of immoral programming with current language AI and the significant related risks.)

Generally, the Type II AI case is not of practical relevance nowadays and its issue would be a hybrid case of participatory sense-making amongst explanatory knowledge creators. Hence, in the following, we focus strictly on existing and thus non-conscious Type I AIs whose abilities surpass humans in scale, speed, and/or selectivity. Arguably, intelligent systems which were depicted in the small feedback loop with stakeholders/users are candidates for new emerging advanced Type I AI risks. An interesting question then becomes: how can advanced Type I AI control be implemented without real value alignment? In our view, zooming into this human vs Type I AI feedback loop may reveal some clues about the underlying cybernetically relevant message transmission problem. In 1960, Norbert Wiener, one of the first cyberneticians, stated that: 'Human action is a feedback action' (Wiener, 1960). In fact, it is insufficient that humans would in theory be able to react appropriately to a Type I AI's message in the specified form of counterfactuals. If it takes humans too long to understand that message in the first place, unforeseen undesirable consequences might already have occurred before they were able to act on it, e.g. by manually turning it off. Wiener gave the hypothetical example of an automated bottle factory programmed to maximise on productivity and whose owner could be 'made bankrupt by the enormous inventory of unsalable bottles manufactured before he learns he should have stopped production six months earlier' (Wiener, 1960). It is on such grounds that many AI safety researchers conjecture about the need to make Type I AI artificially stupid (Trazzi and Yampolskiy, 2020), i.e. to artificially maintain AI abilities at human level or below before it reaches 'superintelligence' and outsmarts human counterparts. But in our view, 'the price of security is eternal creativity' (Aliman *et al.*, 2021). From a functional viewpoint, considering the different ways of information processing, one could consider humans and Type I AI as disparate subparts of the same functional unit where the issue is to add meaning to the messages and synchronise their transmission in order to maintain explanatory knowledge creation as the most relevant subset of creativity. The substrate-independent construct of a hybrid functional unit comprising Type II entities (today only applicable to humans) and Type I entities (such as Type I AIs but also any other Type I entities) can be termed a cyborgnet. (The latter is essentially different from the word cyborg, is formulated at a much more general level and embedded in an own novel theoretical framework (Aliman, 2021).) We explain how this could finally motivate one possible future strategy.

cyborgnet

In a cyborgnet, creativity is necessarily embodied since Type II entities are also physical entities. From this point of view, a communication between humans and Type I AI that does not adjust to human bodily elements does not contain enough information. To put it very simply, most present-day heuristic moral models for Type I AI belong to no-body. Applying our functional stance, it would be embodied cyborgnetic creativity that one intends to augment and not the intelligence of an AI separately or the intelligence of an isolated human separately. In short, the body of relevance to consider is the 'body' of the entire cyborgnet in a given deployment context. A future Type I intelligent system that would itself be deployed in robotic settings may be equipped with technically self-controlled (cyber-)physical elements. However, it would then not suffice for it to develop a technical self-assessment and self-management of those own parts, but one would also integrate a model of the human physical entity in real time. This could include Type-I-AI-directed voice commands, a hybrid mix of context-sensitive experience sampling via physiological activity including arousal monitoring, human valence ratings, and so forth. Feedback loops are by definition bi-directional. Humans could receive additional information on the state of the Type I AI. For instance, next to the counterfactuals projected to past and future, information of the physical state of the Type AI could be visualised efficiently or conveyed via sensory extensions. This could as a side-effect create novel socio-material affordances stimulating human creativity. To sum up, solving the no-body problem bi-directionally via conceiving of humans and Type I AI as disparate but connected entities within one cyborgnet engaging in feedback loops could improve the underlying intra-cyborgnet message transmission and provide a physical grounding of meaning. In turn, this might mitigate the Type I AI control issue by transforming it into a bi-directional hybrid functional integration task at the level of a cyborgnet. Ultimately, one could for instance inject real-time human bodily signals as an additional parameter into an AU-based heuristic moral model. In this way, the problem-solving ability of Type I AI could be funnelled not only by human morality but also by embodied cyborgnetic creativity. (In Chapter 10, we discuss how present-day AI could be employed to counteract AI attacks themselves via cyborgnetic creativity augmentation.) Hence, instead of designing an isolated Type I AI made artificially stupid with human intelligence as upper-bound, this cyborgnetic functional design would strive to ideally attune and tailor its subcomponents to each other – the body included. And as is often said, the whole is more than the sum of its parts... On the downside, a higher capacity information channel (given that more data is transferred) may yield higher costs. Moreover, privacy concerns and security risks may arise. Private data shared in those cyborgnetic information exchange networks could be exploited by malevolent actors. However, to counteract these risks, the zero-trust paradigm from cybersecurity must be proactively applied at multiple levels.

4.3 Conclusions

Wernaart (2021) delineated a sophisticated starting point for a roadmap on moral programming including ethical high-stake cases where the designer is not the central moral authority. AU is one possible non-normative meta-ethical explanatory framework that can be utilised for moral programming in a pluralistic society. In short, the aim of the transdisciplinary AU framework is to accommodate for moral pluralism requirements by providing a supportive generic SHMS scaffold left blank, in which moral authorities (especially users or society but also designers in default settings) fill in flexible updatable and machine-readable heuristic moral models. We explained how AU tackles the three main subtasks of this complex endeavour: the moral chunking problem, epistemic dizziness, and the practical systems engineering subtask of AU-based AI deployment. Finally, we offered a differentiated account of advanced AI risks and reflected on a cyborgnetic embodiment strategy that could yield resilience – albeit not immunity – providing a variety of incentives for future work.

References

Aliman, N.M., 2020. Hybrid cognitive-affective strategies for AI safety. Doctoral dissertation, Utrecht University, the Netherlands. https://doi.org/10.33540/203

Aliman, N.M., 2021. Cyborgnet theory. Available at: https://nadishamarie.jimdo.com/app/download/10822805671/CT.pdf?t=1632837017.

Aliman, N.M. and Kester, L., 2019a. Transformative Al governance and Al-empowered ethical enhancement through preemptive simulations. Delphi – Interdisciplinary Review of Emerging Technologies, 2: 23-29. https://doi.org/10.21552/delphi/2019/1/6

Aliman, N.M. and Kester, L., 2019b. Requisite variety in ethical utility functions for AI value alignment. In Workshop on Artificial Intelligence Safety 2019 co-located with the 28th International Joint Conference on Artificial Intelligence AISafety@IJCAI 2019. Available at: http://ceur-ws.org/Vol-2419/paper_12.pdf.

Aliman, N.M. and Kester, L., 2020. Facing immersive 'post-truth' in AIVR?. Philosophies, 5: 45. https://doi.org/10.3390/philosophies5040045

Aliman, N.M., Kester, L. and Yampolskiy, R., 2021. Transdisciplinary AI observatory – retrospective analyses and future-oriented contradistinctions. Philosophies, 6: 6. https://doi.org/10.3390/philosophies6010006

Arrhenius, G., 2000. An impossibility theorem for welfarist axiologies. Economics & Philosophy, 16: 247-266.

Barrett, L.F., 2017. How emotions are made: the secret life of the brain. Houghton Mifflin Harcourt, Boston, MA, USA.

Bartley, W.W., 1976 The philosophy of Karl Popper. Philosophia 6: 463-494. https://doi.org/10.1007/BF02379286

Bigman, Y.E., Waytz, A., Alterovitz, R. and Gray, K., 2019. Holding robots responsible: the elements of machine morality. Trends in cognitive sciences, 23: 365-368. https://doi.org/10.1016/j.tics.2019.02.008

Bostrom, N., 2012. The superintelligent will: motivation and instrumental rationality in advanced artificial agents. Minds and Machines, 22: 71-85. https://doi.org/10.1007/s11023-012-9281-3

Ciardo, F., Beyer, F., De Tommaso, D. and Wykowska, A., 2020. Attribution of intentional agency towards robots reduces one's own sense of agency. Cognition, 194: 104109. https://doi.org/10.1016/j.cognition.2019.104109

Deutsch, D., 2011. The beginning of infinity: explanations that transform the world. Penguin, London, UK.

Dubljević, V., List, G.F., Milojevich, J., Ajmeri, N., Bauer, W., Singh, M.P., Bardaka, E., Birkland, T., Edwards, C., Mayer, R. and Muntean, I., 2021. Toward a rational and ethical sociotechnical system of autonomous vehicles: a novel application of multi-criteria decision analysis. arXiv preprint arXiv:2102.02928 https://arxiv.org/abs/2102.02928

Dubljević, V., Sattler, S. and Racine, E., 2018. Deciphering moral intuition: how agents, deeds, and consequences influence moral judgment. PloS one, 13: p.e0204631. https://doi.org/10.1371/journal.pone.0206750

Eckersley, P., 2018. Impossibility and uncertainty theorems in AI value alignment (or why your AGI should not have a utility function). arXiv preprint arXiv:1901.00064. https://arxiv.org/abs/1901.00064

Frederick, D., 2020a. Against the philosophical tide: essays in Popperian critical rationalism. Critias Publishing. Yeovil, UK.

Frederick, D., 2020b. Falsificationism and the pragmatic problem of induction. Organon F, 27: 494-503.

Fridman, J., Barrett, L.F., Wormwood, J.B. and Quigley, K.S., 2019. Applying the theory of constructed emotion to police decision making. Frontiers in Psychology, 10: 1946. https://doi.org/10.3389/fpsyg.2019.01946

Gray, K., Young, L. and Waytz, A., 2012. Mind perception is the essence of morality. Psychological Inquiry, 23: 101-124. https://doi.org/10.1080/1047840X.2012.651387

Greaves, H., 2017. Population axiology. Philosophy Compass, 12: p.e12442. https://doi.org/10.1111/phc3.12442

Haidt, J., 2001. The emotional dog and its rational tail: a social intuitionist approach to moral judgement. Psychological Review, 108(4), 814-834. https://doi.org/10.1037/0033-295X.108.4.814

Hoemann, K. and Feldman Barrett, L., 2019. Concepts dissolve artificial boundaries in the study of emotion and cognition, uniting body, brain, and mind. Cognition and Emotion, 33: 67-76. https://doi.org/10.1080/02699931.2018.1535428

Isbilen, E.S. and Christiansen, M.H., 2020. Chunk-based memory constraints on the cultural evolution of language. Topics in Cognitive Science, 12: 713-726. https://doi.org/10.1111/tops.12376

Kleckner, I.R., Zhang, J., Touroutoglou, A., Chanes, L., Xia, C., Simmons, W.K., Quigley, K.S., Dickerson, B.C. and Barrett, L.F., 2017. Evidence for a large-scale brain system supporting allostasis and interoception in humans. Nature Human Behaviour, 1: 1-14. https://doi.org/10.1038/s41562-017-0069

Liu, P., Du, Y. and Xu, Z., 2019. Machines versus humans: people's biased responses to traffic accidents involving self-driving vehicles. Accident Analysis & Prevention, 125: 232-240. https://doi.org/10.1016/j.aap.2019.02.012

McHugh, C., McGann, M., Igou, E.R., and Kinsella, E.L., 2021. Moral judgment as categorization (MJAC). Perspectives on Psychological Science. https://doi.org/10.1177/1745691621990636

Oosterwijk, S., Lindquist, K.A., Anderson, E., Dautoff, R., Moriguchi, Y. and Barrett, L.F., 2012. States of mind: emotions, body feelings, and thoughts share distributed neural networks. NeuroImage, 62: 2110-2128. https://doi.org/10.1016/j.neuroimage.2012.05.079

Popper, K.R. 1995. In search of a better world: Lectures and essays from thirty years. Routledge, Brighton, UK.

Schein, C. and Gray, K., 2018. The theory of dyadic morality: reinventing moral judgment by redefining harm. Personality and Social Psychology Review, 22: 32-70. https://doi.org/10.1177/1088868317698288

Schwartz, M.S., 2016. Ethical decision-making theory: an integrated approach. Journal of Business Ethics, 139: 755-776. https://doi.org/10.1007/s10551-015-2886-8

Siegel, E.H., Wormwood, J.B., Quigley, K.S. and Barrett, L.F., 2018. Seeing what you feel: affect drives visual perception of structurally neutral faces. Psychological Science, 29: 496-503. https://doi.org/10.1177/0956797617741718

Smith, R., Friston, K. and Whyte, C., 2021. A step-by-step tutorial on active inference and its application to empirical data. https://psyarxiv.com/b4jm6/

Trazzi, M. and Yampolskiy, R.V., 2020. Artificial stupidity: data we need to make machines our equals. Patterns, 1: 100021. https://doi.org/10.1016/j.patter.2020.100021

Wegner, D.M. and Gray, K., 2017. The mind club: who thinks, what feels, and why it matters. Penguin, London, UK.

Wernaart, B., 2021. Developing a roadmap for the moral programming of smart technology. Technology in Society, 64: 101466. https://doi.org/10.1016/j.techsoc.2020.101466

Wiener, N., 1960. Some moral and technical consequences of automation. Science, 131: 1355-1358. https://www.jstor.org/stable/1705998

Yam, K.C., Bigman, Y.E., Tang, P.M., Ilies, R., De Cremer, D., Soh, H. and Gray, K., in press. Robots at work: people prefer – and forgive – service robots with perceived feelings. The Journal of Applied Psychology, ###. https://doi.org/10.1037/apl0000834

Part 2

MORAL DESIGN AND EXISTING DISCOURSE

5. Values and virtues in technology

Bart F.W. Wernaart

School of Business and Communication, Fontys University, P.O. Box 347, 5600 AH Eindhoven, the Netherlands; b.wernaart@fontys.nl

Abstract

Ethics and technology cannot be separated from one another. As we have seen in the introduction chapter, technology is not a morally neutral concept. It is not 'just' a tool. It is ethically charged due to the fact that the creation of technology, the use of technology, and the response to technology are human activities. After all, technology is primarily designed to contribute to the realisation of values. Social media platforms are a means to facilitate connection, automated weapon systems are created to guarantee protection, and self-driving vehicles are designed to optimise mobility. However, it is not plausible to exclusively consider technology as a means to an end, due to the increasing level of complexity. Its impact is beyond that linear relationship, since technology can profoundly influence human societies. It can simultaneously revolutionise and disrupt societies. The social media platform can create bubbles and amplify societal polarisation, the weapon system may change geo-political relations, and the self-driving vehicle may prompt a redesign of infrastructure. Technology not only influences value realisation, but can also reshape the meaning of these values. This also forces us to explore how we can sustainably deal with the ever-increasing complexity of new technology. We may need to develop technological virtues, and these may be different from the past. In other words, we have to reassess the moral language we use in the technology context. In this chapter, we will explore the relationship between technology and two ethical concepts that are at the core of how we define morality: values and virtues.

Key concepts

- Morality is the perception of what is right. This perception is formed by values and virtues and supported by norms
- A value represents a morally praiseworthy goal
- A virtue is a moral characteristic that is required in order to live well

- A norm is a rule or standard that contributes to the realisation of a value
- Terminal values are considered to be the most important states of existence that are worth striving for
- Instrumental values are modes of conduct that can help realise terminal values
- The Value Neutrality Thesis suggests that technology does not embody values
- Intrinsic values are values that are intrinsic to technology itself, independent of the relation in which the technology is used
- Extrinsic values are values that are embodied in technology due to the context in which it is used
- In a failed value alignment, there is a discrepancy between intended, embodied, and realised values in technology
- Technological determinism is a theory that suggests that technology has the potential to change and reshape society autonomously
- Internal values are values that typically relate to the technology and engineering profession itself
- External values are values that typically relate to broader social issues that are influenced by technology
- Moral virtues are virtues that relate to proper moral behaviour
- Intellectual virtues are cognitive virtues that are required for us to understand the world
- Technomoral virtues are moral characteristics that enable people to live well in a world with technology

5.1 Norms, values, virtues and morality

morality

Morality is the perception of what is right. It is formed by values and virtues, and supported by norms. In short, a value is a morally praiseworthy goal, a norm is a rule or standard that contributes to the realisation of a value, and a virtue is a moral characteristic that is required in order to live well.

For example, autonomy is considered an important value. A norm to help realise this value in the context of social media usage could be that parents adopt rules to prevent their children from becoming overly dependent on social media platforms (e.g. a time slot) or that the social media algorithm is not built to keep young people online as long as possible. The youngsters themselves should practise the virtue of self-control in order to lead a good life and avoid enslavement to social media. Figure 5.1 visualises how morality, values, virtues, and norms relate to one another.

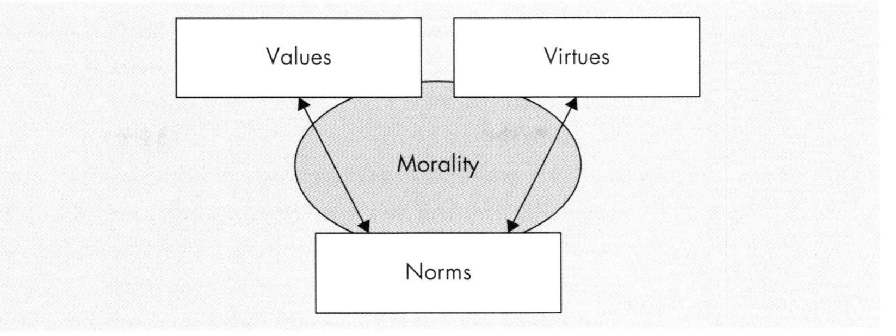

Figure 5.1. Morality, values, virtues and norms.

Technology changes the moral discourse and therefore also the way we perceive values and virtues, and consequently the norms we expect one another to comply with. In this section we will further define the key concepts 'values', 'norms', and 'virtues'. In Sections 5.2 and 5.3 we will further focus on how technology changes the way we define and prioritise values and virtues. In Section 5.4 we will draw further conclusions and look ahead.

values

There are many ways to define values. A widely used definition is offered by Clyde Kluckhohn (1951: 395): a value is 'a conception, explicit or implicit, distinctive of an individual or characteristic of a group, of the desirable which influences the selection from available modes, means and ends of action.' Crane *et al.* (2019: 152) consider values as 'individual beliefs about desirable behaviours and goals that are stable over time and which influence decision-making'. Barrett (2014: 3) uses a more pragmatic definition: 'values are a shorthand method of describing what is important to us individually or collectively at any given moment in time'. When we consider these definitions, a value could be said to have the following characteristics:

- a value is a desirable and important goal to strive for;
- a value guides our actions and decision-making;
- the wordings of a value are characterised by shorthand concepts;
- a value is a sustainable, lasting goal.

norms

In daily language (but also in science) norms are mostly used in the same breath as values. In essence, a norm implies a certain behaviour that aims to contribute to the realisation of a value. For instance, banning explicit content from social media is a norm that is applied in order to realise the value 'decency'. Norms can be regarded at various levels (Wernaart, 2021b: 17-20). They can have an individual, an organisational, but also a cultural scope. Some norms are even considered to have a universal scope. For instance, the norms recognised in the Universal Declaration of Human Rights are presented as universal norms.

But also less formalised norms can have universal characteristics, such as the unwritten rule that a younger person should offer their seat to an elderly person on a crowded bus.

In ethics, we tend to focus more on values and virtues, and not so much on the norms that can be regarded to support morality. Observing and mapping societal norms is more profoundly a practice in (cross-cultural) psychology (Frese, 2015), while the formalisation, organisation and enforcement of normativity is typically a legal subject. From cross-cultural psychology we learn that in societal contexts certain behaviours are commonly observed in particular situations as routinised norms (Shteynberg *et al.*, 2009). Such norms can be generally accepted and approved of. Not complying with the norm is considered to be 'bad behaviour', and may have (societal) consequences, such as naming and shaming, or exclusion. The observation that in given contexts a certain behaviour is generally embraced in a societal context is also referred to as 'descriptive norms'. The study of approval of such norms is called 'injunctive norms'. For instance, research shows that students generally overestimate both how much alcohol their peers drink (descriptive norms), and the approval of drinking alcohol amongst their peers (injunctive norms) (Borsari and Carey, 2015).

In some cases, norms are also formalised through political processes. Normativity is then more centrally organised, and led by procedures to maintain the expected behaviour, i.e. the law. Alternatively, instead of using legal instruments as a way of formalising norms, self-regulation becomes increasingly relevant, for instance through codes of conduct (Wernaart, 2021b: 273-286). To stay with the example of drinking among students, a lawmaker may limit the sale of alcoholic products to adults only, or a student association may adopt a code of conduct that stipulates rules on the consumption of alcohol during student activities.

From this, we can conclude that:

- a norm implies a behaviour that is observed or expected in a certain societal context;
- the scope of the societal context can differ from individual to universal;
- a norm implies a routine;
- a norm implies a certain control mechanism (formal or informal).

virtues

Virtue ethics stems from the ideas of ancient Greek philosopher Aristotle, most notably written down in his famous Nicomachean Ethics. It is assumed that something is morally right when the action is performed by a virtuous person. A virtue is therefore a characteristic that is required in order to lead a virtuous life. Among other things, this means that people have a duty to try and find the golden mean between two extremes, which are vices. For instance, don't be a coward, don't be overconfident but instead be brave. and continuously build your character along this line. In contemporary literature various definitions are used, offering different approaches to the nature of virtue ethics. Swanton (2003) defines a virtue as 'a disposition to respond to, or acknowledge, items within its field or fields in an excellent or good enough way'. Crane *et al.* (2019: 115) consider that virtue ethics is 'an approach to ethics which focuses on the idea that possessing excellent traits of character, or virtues, is required in order to be a good person. A virtuous person is a morally good and wise person'. In the authoritative works of Alasdair MacIntyre (1981: 191), 'a virtue is an acquired human quality the possession and the exercise of which tends to enable us to achieve those goods which are internal to practices and the lack of which effectively prevents us from achieving any such goods'. In contrast, Vallor offers us a simple and short definition, and defines a virtue as 'moral character' (Vallor, 2016: 1). When we consider these definitions altogether, we can establish that:

- a virtue is required in order to lead a good life;
- a virtue is the opposite of a vice, and is a balanced golden mean between vices;
- a virtue is a moral competence that comes from moral character;
- a virtue is something that can be practised and one should strive to excel in being virtuous.

5.2 Values and technology

The existence of technology has changed the value discourse for two main reasons. The first is that values may be embodied in technology. This means that while values are human goals, it is not a phenomenon that is to be discussed in the human context only. The second reason is that there can be a discrepancy between the values that designers seek to realise with their technological creations, the values these creations actually embody, and the values that are affected as a result of using the technology. It is important to consider this carefully in the design process of new technology. To this end, we reflect on the more traditional value discourse, and how values can be labelled outside the technological world (Section 5.2.1). Then, we will discuss whether artefacts can be said to be morally laden (Section 5.2.2). We will conclude with a value jargon that can help in analysing values and technology during a design process (Section 5.2.3).

5.2.1 Values in a non-technological context

When we discuss value discourse outside the techno-context, we usually consider whether or not values can be universally defined, and if so, whether there can be a hierarchy between these values. In general, this is done by proposing individual or cultural value models (Hanel *et al.*, 2018), or organisational value models. These models are then used to define and categorise values, and show how these values relate to one another.

individual value models

Perhaps the most influential individual value model is offered by Milton Rokeach (1973, 1979), who famously distinguished between terminal values and instrumental values. The former are considered to be the most important states of existence worth striving for; the latter are modes of conduct to get there (Table 5.1). For this reason, Rokeach formulates his instrumental values as characteristics, focusing on how individuals seek to realise terminal values. Through the Rokeach Value Survey, individuals gain insight into their value perspective by prioritising the terminal and instrumental values.

Rokeach's approach is criticised by some due to the fact that the set of terminal and instrumental values can be understood differently, making it difficult to meaningfully compare the results of the Rokeach Value Survey (Keith and Lian, 1993). To this end, alternatives are offered, most notably by Shalom H. Schwartz *et al.* (2012) and Valdiney V. Gouveia *et al.* (2014). Schwartz *et al.* focus on the relation between motivational goals and a list of nineteen value types (in an earlier version there were ten types), recognising the need to define why, rather than how, individuals seek to realise values. This has led to a subgrouping of values that correspond to four main motivational goals: openness to change, self-transcendence, self-enhancement, and conservation. This leads to the additional insight that values motivated by self-transcendence usually conflict with values motivated by self-enhancement. The same can be observed in the case of values motivated by openness to change and conservation (Schwartz

Table 5.1. Terminal and instrumental values as proposed by Milton Rokeach.

Terminal values (state of existence)	A comfortable life; An exciting life; A sense of accomplishment; A world at peace; A world of beauty; Equality; Family security; Freedom; Happiness; Inner Harmony; Mature love; National security; Pleasure; Salvation; Self-respect; Social recognition; True friendship; Wisdom.
Instrumental values (mode of conduct)	Ambitious; Broadminded; Capable; Cheerful; Clean; Courageous; Forgiving; Helpful; Honest; Imaginative; Independent; Intellectual; Logical; Loving; Obedient; Polite; Responsible; Self-controlled

and Bilsky, 1990). Gouveia *et al.* also focus on the 'why' question, but then from two different perspectives: needs and goals. Values are considered an expression of needs (thriving or survival needs), and as a guide for action (personal, central, or social goals). Table 5.2 offers a visualisation of values and goals as proposed by both Schwartz *et al.* (2012) and Gouveia *et al.* (2014).

cultural value models

By far the most influential cultural value model is presented by Hofstede (2001), offering six culture dimensions that can be ranked in order to define the characteristics and core values of a given culture. This so-called 6D-Model is composed of the following dimensions: power distance, individualism vs collectivism, uncertainty avoidance, masculinity vs femininity, long-term orientation, and indulgence vs restraints. This model is widely embraced, especially in the business literature.

organisational value models

Besides individual and cultural value models, values can be analysed in organisational contexts. For instance, values can be used as a starting point in identifying organisational identity. Richard Barrett (2014) proposes the Seven Levels of Consciousness model, loosely based on Abraham Maslow's hierarchy of needs. In this model, seven layers of (organisational) consciousness are defined that represent particular needs. These needs can be transformed into values, which helps when reflecting on organisational goals and future ambitions. The seven layers of consciousness are (from 'low' to 'high') survival, relationship, self-esteem, transformation, internal cohesion, making a difference, and service. An organisational value that emerges from the need to survive can be 'financial stability' or 'profit'. When we consider a value at the level of service, an organisation might seek to pursue 'social responsibility' or 'long-term perspective' as a value. Organisational values can be used as core

Table 5.2. Values and goals.

Openness to change:	**Self-transcendence:**
self-direction, stimulation, hedonism	universalism, benevolence
Self enhancement:	**Conservation:**
achievement, power	conformity, tradition, security

Values and motivational goals according to Schwartz *et al.* (2012).

	Personal goals	**Central goals**	**Social goals**
Thriving needs	Emotion, pleasure, sexuality	Beauty, knowledge, maturity	Affection, belonging, support
Survival needs	Power, prestige, success	Health, stability, survival	Obedience, religiosity. tradition

Values as expressions of needs and goals according to Gouveia *et al.* (2014).

concepts in mission and vision statements (King *et al.*, 2010), or lay the foundation of codes of conducts to trigger (corporate) social responsibility (Crane *et al.*, 2019: 183-184).

5.2.2 Value-laden technology

In a technological context, new players enter the equation: the technology itself, and its creator, the engineer. This leads to a slightly different discourse on values. A first important question is whether technology itself can be value-laden.

Value Neutrality Thesis

Some authors defend the Value Neutrality Thesis (e.g. Pitt, 2014). This theory suggests that technology does not embody values. Technologies are mere instruments or tools that are put to use by people. The question is whether they are put to good or bad use by these people. The value discourse should therefore focus on the agents that design or use the technological tool, not the technology itself. After all, it is the designer that designs technology to realise certain values, or the user that uses technology in pursuit of values. Guns don't kill, people kill. And guns are designed by people to kill. So don't blame the gun when someone is killed by a gun, as this will only distract from the real debate that should be held: why do people want to kill other people, and even create tools to facilitate this?

rejecting the Value Neutrality Thesis

The Value Neutrality Thesis is widely rejected. A majority of ethicists are convinced that technology is value-laden. Back in 1980, Langdon Winner had already pointed out that the building of a simple bridge in New York City could be politically charged if it was built in such a way that only cars could use it and not buses. At that time it had a *de facto* discriminatory effect since most African-American citizens were dependent on buses, and such bridges would seriously limit their access to certain areas in New York. The bridge was not 'just' a tool for transport but part of a political attempt to discriminate against Afro-American people. Ever since then, the Value Neutrality Thesis has been rejected, using a wide variety of arguments (Verbeek, 2011: 41-65).

For example, Michael Klenk (2020) proposes the Affordance Account of Value Embedding. His main argument is that technology makes certain actions likelier to happen given the circumstances. Someone who holds a gun will be more likely to kill someone else than someone who is not holding a gun. This means that the technology itself is able to influence the realisation (or undermining) of certain values. The mere existence of the technology is therefore not neutral, and these characteristics should be valued and taken into consideration in a value discourse when we analyse the morality of technology. Miller (2020) rejects the Value Neutrality Thesis by emphasising that technology has the potential for long-term consequences that supersede the intentions of the designers of that technology. The longevity of technology separates the designers' value-intention from the actual value-impact it has. The atomic bomb may be a good example here: the intention of the designers when creating this weapon cannot be identified with the values that were affected in the long term (up until this day) as a result of this invention. This means that the value of the technology itself outlives, and potentially differs from, the values of its creators. Van de Poel and Kroes (2014) go one step further and argue that technology is sometimes specifically designed for the purpose of the realisation of certain values. In such cases, it is impossible to separate the value from the nature of the technology. They use the example of the sea dyke: this artefact was built for safety and for no other reason. It is difficult to see the dyke being designed for or used for any other primary reason than safety.

information ethics

Another way to argue that artefacts embody values on their own is by making a distinction between intrinsic and extrinsic values (Korsgaard, 1983). Intrinsic values are values that are intrinsic to technology. They are values that are not relational and therefore independent of the context in which the technology is used. Some technology heavily depends on the context in which it is used. A knife, for instance, can be used to prepare a meal or to stab someone. The way it is used is then decisive in what values the knife embodies. These are extrinsic values, depending on the context in which the technology is applied. However, technology may also have values that do not primarily depend on the relational aspects in which they are used. One way to apply this distinction in a meaningful way is to use theories we find in information ethics (see Chapter 18). Simply put: all objects are at the very least considered as artefacts that embody information. The information itself is part of the intrinsic nature of the artefact, and therefore valuable. The values these artefacts represent as an information object are beyond the physical features or technological composition of the artefact. An interesting example is offered by Luciano Floridi (2002), who explains the intrinsic value of chess pieces: the value of a pawn does not come from the physical features or the materials that were used to craft the piece. Instead, the value of the pawn depends on the data it represents. This includes the rules on how to move the piece and how it can capture another piece. This data is what makes the chess piece valuable, not its physical features or composition (after all, we could replace the pawn piece with a cupcake and it would still have the same intrinsic value in the chess game). The same can be said in the case of a dataset of customer information or patient data: the technological composition of the dataset (the coded form as such) is not valuable, but the information it represents is. Raw big data is not valuable because it does not represent comprehensive information, while an applied dataset does, and consequently it represents value. In the case of the patient dataset, the information can represent important terminal values when the dataset is crucial in determining what medical treatment or surgery should be chosen to protect the health of the individual the data corresponds to. We should not only focus on the values the doctor or the patient seeks to pursue, but also take the values embodied in the technology they use in their doctor-patient relationship as an autonomous value-representation. This leads to the crucial consideration that technologies are not 'just' tools: they embody values in their own right and are never morally neutral.

failed value alignment

Yet other arguments to show that technology can be value-laden can be found in failed value alignment between the creator of technology and the technology itself. There is always a risk when human values are transferred to technology. For instance, the engineer may have been unable to translate human values into machine functionality (or the machine may not recognise the human values properly). This phenomenon is called 'perverse instantiation' (Aliman and Kester, 2019; Yampolskiy, 2014). Another problem presents itself when technology is designed to realise certain values and the technology performance leads to unexpected or unforeseen side-effects. Sometimes these side-effects are undesirable or counterproductive as regards the intended value realisation. This phenomenon is called 'instrumental convergence'. A famous and striking example is offered by Nick Bostrom (2014): the paperclip maximiser. Imagine that scientists design artificial intelligence that is programmed to produce paperclips. When the AI is programmed with only one goal, i.e. to produce as many paperclips as possible, the AI may in the end destroy all human life on the planet and turn the earth (and perhaps everything that lies beyond in the wider universe) into either a paperclip or paperclip production technology. Perverse instantiation and instrumental convergence have something in common: there is a failed value alignment between the creator and the machine resulting in technology that pursues values in a different manner compared to the intended value input.

Technological Determinism

A more radical view is proposed in Technological Determinism (Dafoe, 2015). This theory suggests that technology has the potential to change and re-shape a society autonomously, independent of human actions. Martin Heidegger has already suggested that technology itself can change the way we live, interact and expect. This theory assumes that technology can uphold new values and modify human values by its mere existence. A good example would be the creation of social media platforms: while it was never the intention of their creators, these platforms have had a significant impact on how people communicate, conduct a dialogue, engage with politics, and even how people perceive the truth (Goyanes *et al.*, 2021). The disrupting and enabling effects of these platforms were not foreseen, but became a reality, nevertheless.

The way we approach the question of whether technology can be value-laden has consequences. If we assume that technology is value-neutral, we should focus on a value discourse among users of technology, or perhaps among the designers of technology. When we assume that technology is value-charged, then we should focus on the interaction between humans and technology, and implement ways to monitor whether the value charge of the technology still suits our needs. In the following section we will analyse different value approaches that can help foster a better understanding of the role of values in the design process of new technology.

5.2.3 Designing value-laden technology

Perhaps the most important challenge for designers is to recognise that there can be a mismatch between the values they perceive as 'final' when they create new technology, and values that are terminal in the moral sense. A distinction can be made between so-called 'internal' and 'external' values (Gonzalez, 2015). Internal values are typically related to the technology and the engineering profession itself. They focus on the functionality of the technology, and typically are not about broader societal themes. Van de Poel (2015) considers technological enthusiasm, effectiveness, reliability, robustness, maintainability, compatibility, quality, and rationality (as in rational process) as typical internal values of technology. External values on the other hand affect broader societal issues, such as safety, health, well-being, sustainability, justice, democracy, and inclusiveness. When we compare the internal and external values (Table 5.3) with the instrumental and terminal values proposed by Rokeach (Table 5.1) we have to conclude that most internal values are instrumental values, whereas most external values are terminal. This in an important distinction for designers and engineers: if we assume that technology is increasingly value-laden, and technology can have a significant impact on our terminal values, those who create new technology should internalise the external (terminal) values in order to design meaningful innovation. A social-media algorithm should not only be effective and efficient, but it should also strengthen (and not undermine) democracy, well-being and inclusiveness.

In the design process it can be helpful to use the terminology of intended, embodied, and realised values in technology (Van de Poel and Kroes, 2014; Figure 5.2). This way, the values as intended by the designer (mostly intrinsic values) are cut loose from the values that are intrinsically and extrinsically embodied in the technology. Intrinsically because they are embodied in the technology due to its existence, and extrinsically because they are used in practice and are therefore embodied in the technology as a result of that

Table 5.3. Internal and external values in technology as proposed by Ibo van de Poel (2015).

Internal values	External values
Technological enthusiasm	Safety
Effectiveness	Health
Reliability	Wellbeing
Robustness	Sustainability
Maintainability	Justice
Compatibility	Democracy
Quality	Inclusiveness
Rationality (rational process)	

practice. The embodied values are now regarded as an autonomous part of the discourse. Furthermore, the realisation (or depriving) of values that result from using the technology is considered, and these are mostly external values.

This value discourse prompts us to reflect on feedback loops (Aliman and Kester, 2019; Wernaart, 2021a), given that intended values do not necessarily result in embodied values, which are not necessarily the same as the realised values. Feedback loops may contribute to value alignment between these stages. This is also discussed into more detail in Section 2.4, and Chapter 4.

5.3 Virtues and technology

Technology changed not only the value discourse but also our perception of virtues. Virtue ethics is one of the oldest approaches in normative ethics. Societal changes have always influenced the ongoing discourse about which virtues are considered to be important in order to live well. In this section, we will reflect on the nature of virtues and how 'the list of virtues' has changed throughout history (Section 5.3.1). Furthermore, we will explore the so-called 'technomoral virtues' as identified by Shannon Vallor (2016). These virtues enable us to live well in a world with technology.

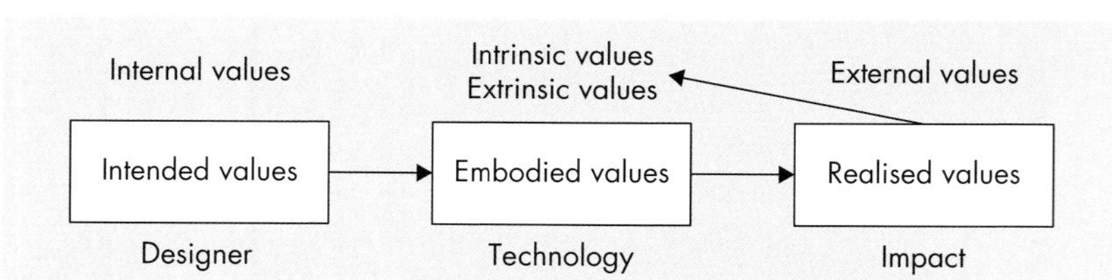

Figure 5.2. Intended, embodied and realised values in technology.

5.3.1 Virtues in a non-technological context

The golden mean and corresponding moral characteristics were initially understood as objective moral truths: a position that – in various degrees – is also defended in the more contemporary debates on the nature of virtue ethics (Nussbaum, 1988, and in a business context, Demuijnck, 2015). However, the idea of absolute moral truth in virtue ethics and the notion of cultural diversity or different historical perspectives are not always easy to reconcile (Gowans, 2011). Friedrich Nietzsche pointed out that virtues should be considered in context and as individual concepts, not as universal characteristics (Annas, 2017). In other words, virtues have a more relativist nature, and we must accept that that which is virtuous depends on (social) context and changes over time (MacIntyre, 1981), for example, as a result of prevailing religious views, industrialisation and a more business-oriented society, or as a consequence of technological development (Table 5.4).

In both ancient Greek and Far Eastern societies, virtues were at the core of philosophical debate. Among others, Socrates, Plato, Aristotle, and Confucius firmly established virtue ethics as a cornerstone of Greek philosophy and Confucianism. For instance, Aristotle defined moral virtues that should

Table 5.4. Virtues in a changing world.

Aristotle	Confucius	Christian	Smith	Solomon	Vallor
Ancient Greek	Ancient Far-East	Middle Ages	Industrial Revolution	Business oriented societies	Digital revolution and beyond
Courage	Benevolence	Prudence	Prudence	Honesty	Honesty
Moderation	Righteousness	Justice	Justice	Trust	Self-control
Justice	Propriety	Temperance	Beneficence	Toughness	Humility
Magnanimity	Wisdom	Courage	Self-command	Fairness	Justice
Expansive hospitality	Fidelity	Faith		Sympathy & empathy	Courage
Greatness of soul		Hope		Altruism	Empathy
Mildness of temper		Charity			Care
Truthfulness					Civility
Easy grace					Flexibility
Friendliness					Perspective
Proper judgement					Magnanimity
Perceptiveness					Techno-moral wisdom
Theoretical knowledge					
Practical knowledge					
Productive knowledge					

be complemented by intellectual virtues. The first are virtues that relate to proper moral behaviour, while the latter are more cognitive virtues that are required in order to understand the world (Wilson, 2017). In Europe, with the Christianisation of societies, virtues were understood in a more religious context. Amongst others, Thomas Aquinas made a distinction between cardinal and theological virtues. The cardinal virtues were borrowed from the ancient Greek (prudence, justice, temperance and courage) while the theological virtues had a biblical origin (faith, hope and charity). In the 18th century, the Industrial Revolution led to a different kind of society. Amid this industrialisation of especially Great Britain, Adam Smith wrote his famous works. Among them is an often overlooked essay on virtue ethics: the Theory of Moral Sentiments (1759). He proposes that in an industrialised society, prudence, justice, beneficence, and self-command are the virtues we should master. In the post-World War II era, the business sector became an important part of life. Robert C. Solomon is considered one of the most influential authors in the field of virtues and business ethics (Ferrero and Sison, 2014). He offers a selection of virtues that are required in societies with a prominent private sector (Audi, 2012).

5.3.2. Vallor's technomoral virtues

More recently, as a result of technological developments, Shannon Vallor has offered a set of so-called 'technomoral virtues' encompassing crucial characteristics that are required to sustainably deal with rapid and complex innovation. Most of the proposed virtues are inspired by virtue ethics as developed since the ancient Greeks, and transformed to fit into technological societies. Vallor's most important plea is that technomoral challenges do not so much require moral solutions as a strategy that involves technomoral virtues. According to her, most other approaches in normative ethics – especially those that thrived during the Enlightenment (such as utilitarianism or deontology) – fail. The reason is simple: these approaches in ethics propose fundamental solutions to ethical challenges. This is problematic, because technology is rapidly evolving, and any fundamental solution proposed will be too little too late. Instead, a sustainable strategy is required by means of which we learn how to cope with an ever-changing technological world.

honesty

The first of Vallor's technomoral virtues is defined as honesty: *'an exemplary respect for truth, along with the practical expertise to express that respect appropriately in technosocial contexts'* (Vallor, 2016: 122). As a result of rapidly innovated Information Communication Technologies, this virtue is under serious challenge. The mass distribution of fake news and disinformation are examples of such challenges. Respect for the truth is not the core business of many social media platforms; advertising and selling data is. This leads to an online environment in which data is presented in such a way that it leads to serious polarisation in societies (see, for instance, Lee *et al.* 2018), and consequently harms related virtues such as trust, reliability and integrity (see also Section 7.3). Honesty in the context of technology can also prove to be challenging in the public domain. For instance, data systems and surveillance technology used for preventive policing purposes have proved more than once to have racist biases (Oosterloo and Van Schie, 2018; Shahin and Zheng, 2020). This has led to serious distrust among citizens. This challenge is also discussed in Section 2.4.

self-control

The second technomoral virtue according to Vallor is self-control: '*an exemplary ability in technosocial contexts to choose, and ideally to desire for their own sakes, those goods and experiences that most contribute to contemporary and future human flourishing*'. We have access to an almost unlimited amount of online data, and are subjected – on a daily basis – to uncountable online incentives aimed at influencing our behaviour. To be able to deal with this abundance of input, we need to master temperance and build the ability to limit our 'digital daily intake' (Brevers and Turel, 2019). In addition to that, the way we portray ourselves through online social media platforms often paints a fragmented, inaccurate picture of who we truly are. This virtue is not only important in the digital service context. As a result of globalisation and the digitalisation of shopping experiences, consumers are used to ordering goods online that come from all corners of the world and arrive at their doorstep in a matter of days. This leads to additional environmental challenges in the field of pollution (Jaller and Pahwa, 2020), and social challenges in the field of labour standards in the delivery sector (Zekic, 2019). Also here, self-constraint is a virtue to be practised in order to deal with these issues in the long term.

humility

The third technomoral virtue is humility: '*a recognition of the real limits of our technosocial knowledge and ability; reverence and wonder at the universe's retained power to surprise and confound us; and renunciation of the blind faith that new technologies inevitably lead to human mastery and control of our environment*' (Vallor, 2016: 126-127). In essence, we should relinquish the idea that humans can completely master technology. As technology becomes more complex and self-learning, and as interaction with technology becomes increasingly intense, we should acknowledge that technology is – to a certain degree – unpredictable and uncontrollable. Vallor suggests exploring the middle ground between techno-optimism and techno-pessimism: we should be hopeful but critical, and acknowledge the fact that technology has morality of its own (see Section 5.2). In a way, this virtue challenges the 'all-or-nothing' proposition we discussed in the opening chapter of this book: we should not have a utopian or dystopian view of a world with technology, but find a golden mean between these two radical mind-sets.

justice

The fourth technomoral virtue is justice: on the one hand, this is '*a reliable disposition to seek a fair and equitable distribution of the benefits and risks of emerging technologies*' and on the other hand '*characteristic concern for how emerging technologies impact the basic rights, dignity, or welfare of individuals and groups*' (Vallor, 2016: 128). It is to be expected that in the 'new' future, technology will create enormous wealth and opportunities, but only for the happy few; it is likely to amplify the already growing income gap that is observed worldwide (Frey, 2019). An interesting example is the effect of the Corona crisis on the wealth and power of Big Tech: during the COVID-19 pandemic, the world's seven biggest technology companies accelerated their position of power at a global level (Fernandez *et al.*, 2020).

courage

The fifth technomoral virtue is courage: '*a reliable disposition toward intelligent fear and hope with respect to the moral and material dangers and opportunities presented by emerging technologies*' (Vallor, 2016: 131). In older versions of virtue ethics – most notably the Aristotelian and Confucian versions – courage was mostly understood as an individual matter in the context of the individual network. It is the goal of the individual to develop a golden mean between hope and fear in order to lead a good life. This means that issues such as sacrifice or risk-taking are aimed at doing the right thing: not cowardice, but also not behaving in an overconfident manner. While in earlier virtue theories such characteristics were mostly explored in the context of direct relations between humans, technology changes this perception. Increasingly, the use of technology has a moral impact on those you cannot see directly, or has a serious impact on humanity as a whole. For instance, our (online) shopping behaviour on Black Fridays leads to disproportional harm to small businesses, but also to the environment (Gibbens, 2018). To deal with this issue would require looking beyond the mere individual needs of the consumer or the family of the consumer, and taking into consideration the well-being of those you do not know directly, and the well-being of our planet. The courage of the consumer is to be found in the capacity to reconcile self-sacrifice for the greater good with defining one's own (urgent) needs: not all consumers have the luxury of paying more for a similar product for the sake of a better world. Another example is that technological progress sometimes means taking risks. If we want to fly to the moon, it means that people will have to risk their lives for progress. The risk taken can be acceptable when it is a result of intelligent hopes and fears: moon-travel should not literally be a 'shot for the moon'. On the other hand, travelling in space with the same safety expectations as taking a train does not make sense. The courage of the astronauts is to be found somewhere in between these extremes.

empathy

The sixth technomoral virtue is empathy: '*a cultivated openness to being morally moved to caring action by the emotions of other members of our technosocial word*' (Vallor, 2016: 133). While on the one hand our world is more connected than ever, connections are not always led by mutual empathy. Especially in the context of online behaviour, it is easy to fall into the trap of digital narcissism: social media profiles are no longer used to empathically connect with others, but are instead a tool for self-exhibition. Furthermore, social media algorithms amplify the polarisation of moral and political online discourse (Lee *et al.*, 2018), and the rise of online trolls, fanboys and lurkers (Barnes, 2018). The consequence is a polarised, online landscape. Problems such as cyber-bullying and online sexual exploitation are serious issues that affect people's mental health (Henry and Flynn, 2019; Navarro *et al.*, 2016). Characteristics promoting mutual care and empathic relations would result in a more sustainable online community based on meaningful connectedness.

care

The seventh technomoral virtue is care: '*a skilful, attentive, responsible, and emotionally responsive disposition to personally meet the needs of those with whom we share our technosocial environment*' (Vallor, 2016: 138). This virtue addresses the issue that technology may replace human care. In our desire to optimise labour efficiency, technology can replace human work. This is typically done in jobs we consider dull, dirty, or dangerous: the so-called 3Ds (Czarniawska and Joerges, 2018: 7-8). While on the one hand such developments could be considered as positive technological progress, in the long run resulting in human labourers performing safer and healthier jobs, there is also a downside. Particular concerns arise when technology replaces jobs where interaction is not at the core of the physical labour activity, but does involve human interaction as a side-effect. The recipient of postal mail will not be too concerned whether the post is physically delivered by a human or a drone. However, the daily chat with the person who brings around the post is a micro-social encounter that can be extremely valuable, especially for lonely people. In robot ethics, an ongoing debate is emerging around the use of carebots in healthcare facilities (Zardiashvili, 2020). Replacing an employee who, for instance, changes someone's incontinence nappy, gives medication, or provides simple medical services, is not always progress in the experience of those who receive the health care. Dehumanising health services can lead to serious challenges in the field of social interaction, trust and autonomy for patients. After all, we can have serious doubts about whether a robot is truly able to care (Meacham and Studley, 2017). It is therefore extremely important to master the virtue of care when we decide to use technology in these particular contexts.

civility

The eighth technomoral virtue is civility: '*a sincere disposition to live well with one's fellow citizens of a globally networked information society*' (Vallor, 2016: 141). In a way, this virtue addresses the issue of being civil towards one another, especially in an online environment. Increasingly, news or politics is discussed online in an unfriendly manner. Unfriendly remarks provoke even more unfriendly remarks (Goyanes *et al.*, 2021). Besides the effect of social media filters caused by algorithms that amplify the polarisation in online discussions, most of us will find it easier to behave in an uncivil manner online than offline (Nedelec, 2018). Civility as a virtue can help establish meaningful discourse in a global network that offers new and nuanced insights. The technological potential to shape online environments in which this is stimulated is there, but it is rarely used by the main (social) media outlets.

flexibility

The ninth technomoral virtue is flexibility: '*a reliable and skilful disposition to modulate action, belief, and feeling as called for by novel, unpredictable, frustrating or unstable technosocial conditions*' (Vallor, 2016: 145). Flexibility requires the capacity to deal with rapid social changes resulting from innovation. This involves the capacity to deal with uncertainty and unpredictable events. Perhaps the global COVID-19 pandemic, and the design of technological solutions to deal with this crisis, was a challenging exercise in mastering this virtue. Working from home, receiving online education and overcoming social distancing by using online alternatives to physical proximity involved creative innovative solutions. Vallor argues that flexibility is not only about flexibility towards innovation, but also towards one another. In a globalised world, moral standards around technology are different. For instance, issues such as autonomy, privacy, or (intellectual) property are perceived differently in different cultures. At the same time, there is a global information society network in which all these cultures are present. This requires the capacity to find the golden mean between cultural absolutism and relativism: cultural pluralism (Wernaart, 2021b: 296-300). In a way, cultural pluralism implies the flexibility to understand and learn from different cultural moral standards without forsaking your own morality, and the capacity to take joint decisions on how to proceed with new technology.

perspective

The tenth technomoral virtue is perspective: '*a reliable disposition to attend to, discern, and understand moral phenomena as meaningful parts of a moral whole*' (Vallor, 2016: 149). This virtue is about building technomoral capacity. This means at least two things. First, it is important to be able to recognise and understand moral aspects of technological issues. Second, it is important that we are capable of prioritising these moral aspects in the context of a 'moral whole'. Our technomoral decisions are having an increasing impact on our world. For instance, it is almost impossible to have sexual images that have been circulated online without proper consent removed from the internet.

Those who put these images into circulation do not always foresee the impact of their actions, and sometimes are not fully aware of the moral implications of their actions. To illustrate this problem, in 2020 a New York Times investigation led Pornhub to remove millions of videos with non-consensual content and even child-abuse. These videos had a global reach and it appeared to be next to impossible for victims of this abuse to have them removed on their own initiative (Kristof, 2020).

magnanimity

The eleventh technomoral virtue is magnanimity. While Vallor does not offer a clear definition, she explains that a magnanimous person is a great-souled person. This means that someone has a sense of nobility that is not restricted to one action, moment, or decision. This person is sustainably generous and kind, and is consistently driven by an inner moral compass. Vallor observes that it is precisely this virtue that is generally lacking in leadership. In other words, there is a necessity to express moral leadership when dealing with new technology (Wernaart and Van Halderen, 2020). The virtue magnanimity enables moral leadership. Vallor emphasises that a moral leader is someone who has earned the moral trust of those he/she works with, and is able to morally inspire.

technomoral wisdom

And finally, the twelfth technomoral virtue is technomoral wisdom: *'a general condition of well-cultivated and integrated moral expertise'* (Vallor, 2016: 154). In essence, a morally wise person is able to integrate and unify all previous virtues in order to live well in a world with ever-emerging new technologies.

5.4 Challenges

In conclusion, technology changes our moral language. Most notably, it changes our value and virtue discourse. In terms of values, there is an increased consensus that technology itself embodies values that are not necessarily similar to the intended value envisioned by the creator or user of the technology. This has consequences for the design of technology: we need to recognise that technology may embody values we did not expect or foresee, and we need to adjust design processes accordingly. This means, among others things, that the values embodied in technology need to be a part of the equation when we have a techno-ethical discourse. For a start, it would mean that engineers/designers should internalise external values that are affected by their technologies. In other words, it is important that creators of technology look beyond technological enthusiasm, effectiveness, reliability, robustness, maintainability, compatibility, quality, and rationality and also focus on values such as safety, health, well-being, sustainability, justice, democracy, and inclusiveness.

With regard to virtues, it is important to establish that the emphasis on specific virtues has changed over time, depending on the societal context. Technology changes our societies in many ways, leading to proposed technomoral virtues that enable people to live well in a world with technology that rapidly changes how we live and interact with one another, and how we deal with life. It is important to integrate these virtues into our daily life, but especially into our education and our (professional) ways of working in a global technological society.

References

Aliman, N.M. and Kester, L., 2019. Augmented utilitarianism for AGI safety. In: Hammer P., Agrawal P., Goertzel B. and Iklé M. (eds.) Artificial general intelligence. AGI 2019. Lecture Notes in Computer Science, Volume 11654. Springer, Cham, Switzerland. https://doi.org/10.1007/978-3-030-27005-6_2

Annas, J., 2017. Which variety of virtue ethics? In: Carr, D., Arthur, J. and Kristjánsson, K. (eds.) Varieties of virtue ethics. Palgrave Macmillan, London, UK, pp. 35-51.

Audi, R., 2012. Virtues, concepts, and rules in business ethics: reflections on the contributions of Robert C. Solomon. In: Higgins, K. and Sherman, D. (eds.) Passion, death and spirituality, the philosophy of Robert C. Solomon. Springer, Dordrecht, the Netherlands, pp. 103-111.

Barnes, R., 2018. Uncovering online commenting culture, trolls, fanboys and lurkers. Palgrave Macmillan, Cham, Switzerland. https://doi.org/10.1007/978-3-319-70235-3

Barrett, R., 2014. The values-driven organization, unleashing human potential for performance and profit. Routledge, London, UK.

Borsari, B., and Carey, K.B., 2003. Descriptive and injunctive norms in college drinking: a meta-analytic integration. Journal of Studies on Alcohol, 64: 331-341. https://doi.org/10.15288/jsa.2003.64.331

Bostrom, N., 2014. Superintelligence: paths, dangers, strategies. Oxford University Press, Oxford, UK.

Brevers, D. and Turel, O., 2019. Strategies for self-controlling social media use: classification and role in preventing social media addiction symptoms. Journal of behavioral addictions, 8: 554-563. https://doi.org/10.1556/2006.8.2019.49

Crane, A., Matten, A., Glozer, S. and Spence, L., 2019. Business ethics (5th ed.). Oxford University Press, Oxford, UK.

Czarniawska, B. and Joerges, B., 2018. Robotization of work as presented in popular culture, media and social sciences (part two). GRI-rapport 2018:2. Available at: https://gupea.ub.gu.se/bitstream/2077/57616/1/gupea_2077_57616_1.pdf.

Dafoe, A., 2015. On technological determinism: a typology, scope conditions, and a mechanism. Science, Technology, & Human Values, 40: 1047-1076. https://doi.org/10.1177/0162243915579283

Demuijnck, G., 2015. Universal values and virtues in management versus cross-cultural moral relativism: an educational strategy to clear the ground for business ethics. Journal of Bussiness Ethics, 128: 817-835. https://doi.org/10.1007/s10551-014-2065-3

Fernandez, R., Adriaans, I., Klinge, T.J. and Hendrikse, R., 2020. Engineering digital monopolies: the financialisation of Big Tech. Somo, Amsterdam, the Netherlands. Available at: https://www.somo.nl/the-financialisation-of-big-tech/.

Ferrero, I. and Sison, A.G., 2014. A quantitative analysis of authors, schools and themes in virtue ethics articles in business ethics and management journals (1980-2011). Business ethics, the environment and responsibility, 23: 375-400. https://doi.org/10.1111/beer.12057

Floridi, L., 2002. On the intrinsic value of information objects and the infosphere. Ethics and Information Technology, 4: 287-304. https://doi.org/10.1023/A:1021342422699

Frese, M., 2015. Cultural practices, norms, and values. Journal of Cross-Cultural Psychology, 46: 1327-1330. https://doi.org/10.1177/0022022115600267

Frey, C.B., 2019. The technology trap, capital, labor, and power in the age of automation. Princeton University Press, New Jersey, NJ, USA.

Gibbens, S., 2018. Black Friday shopping hurts the environment – but you can help, shoppers will spend more in the post-Thanksgiving rush than any other time of the year. National Geographic, 21 November 2018. Available at: https://tinyurl.com/th6vdfsz.

Gonzalez, W.J., 2015. On the role of values in the configuration of technology: from axiology to ethics. In: Gonzalez, W.J. (ed.) New perspectives on technology, values, and ethics. Boston Studies in the Philosophy and History of Science, Volume 315, Springer, Cham, Switzerland, pp. 3-28.

Gouveia, V.V., Milfont, T.L. and Guerra, V.M., 2014. Functional theory of human values: testing its content and structure hypotheses. Personality and Individual Differences 60: 41-47. https://doi.org/10.1016/j.paid.2013.12.012

Gowans, C.W., 2011. Virtue ethics and moral relativism. In: Hales, S.D. (ed.) A companion to relativism, Blackwell Publishing Ltd, Oxford, UK, pp. 391-410.

Goyanes, M., Borah, P, and Gil de Zúñiga, H., 2021. Social media filtering and democracy: effects of social media news use and uncivil political discussions on social media unfriending. Computers in Human Behavior, 120: 106759. https://doi.org/10.1016/j.chb.2021.106759.

Hanel, P.H.P., Litzellachner, L.F. and Maio, G.E., 2018. An empirical comparison of human value models. Frontiers in Psychology, 9: 1643. https://doi.org/10.3389/fpsyg.2018.01643

Henry, N. and Flynn, A., 2019. Image-based sexual abuse: online distribution channels and illicit communities of support. Violence Against Women, 25: 1932-1955. https://doi.org/10.1177/1077801219863881

Hofstede, G., 2001. Culture's consequences: comparing values, behaviors, institutions and organizations across nations. Sage Publications, Thousand Oaks, CA, USA.

Jaller, M. and Pahwa, A., 2020. Evaluating the environmental impacts of online shopping: a behavioral and transportation approach. Transportation research, Part D: Transport and Environment, 80: 102223. https://doi.org/10.1016/j.trd.2020.102223.

Keith, G. and Iain, W., 1993. Multiple interpretations of the Rokeach value survey. Journal of Social Psychology, 133: 797-805. https://doi.org/10.1080/00224545.1993.9713941

King, D.L., Case, C.J. and Premo, K.M., 2010. Current mission statement emphasis: be ethical and go global. Academy of Strategic Management Journal, 9: 71-87.

Klenk, M., 2020. How do technological artefacts embody moral values? Philosophy and Technology, 34: 525-544. https://doi.org/10.1007/s13347-020-00401-y

Kluckhohn, C.K., 1951. Values and value orientations in the theory of action. In: Parsons, T. and Shils E.A. (eds.) Toward a general theory of action. Harvard University Press, Cambridge, MA, USA.

Korsgaard, C., 1983. Two distinctions in goodness. The Philosophical Review, 92: 169-195. https://doi.org/10.2307/2184924

Lee, C., Shin, J. and Hong, A., 2018. Does social media use really make people politically polarized? Direct and indirect effects of social media use on political polarization in South Korea. Telematics and information, 35: 245-254. https://doi.org/10.1016/j.tele.2017.11.005

MacIntyre, A., 1981. After virtue, a study in moral theory. University of Notre Dame Press, Notre Dame, IN, USA.

Meacham, D. and Studley, M., 2017. Could a robot care? It's all in the movement. In: Lin, P., Jenkins, R. and Abney, K. (eds.) Robot ethics 2.0. Oxford University Press, Oxford, UK, pp. 97-112.

Miller, B., 2020. Is technology value-neutral? Science, Technology & Human Values, 46: 53-80. https://doi.org/10.1177/0162243919900965

Navarro, R. Yubero, S. and Larrañaga, E. (eds.), 2016. Cyberbullying across the globe gender, family, and mental health. Springer, Cham, Switzerland. https://doi.org/10.1007/978-3-319-25552-1

Nedelec, J.L., 2018. Individual differences and co-occurring victimization online and offline: the role of impulsivity. Personality and Individual Differences, 133: 77-84. https://doi.org/10.1016/j.paid.2016.11.028

Nussbaum, M., 1988. Non-relative virtues: an aristotelian approach. Midwest studies in Philosophy, 13: 32-50. https://doi.org/10.1093/0198287976.003.0019

Oosterloo, S. and Van Schie, G., 2018. The politics and biases of the 'crime anticipation system' of the Dutch police. Proceedings of the International Workshop on Bias in Information, Algorithms, and Systems, 2103: 30-41.

Pitt J.C., 2014. 'Guns don't kill, people kill'; values in and/or around technologies. In: Kroes P. and Verbeek, P.P. (eds.) The moral status of technical artefacts. Philosophy of Engineering and Technology, 17. Springer, Dordrecht, the Netherlands. https://doi-org.fontys.idm.oclc.org/10.1007/978-94-007-7914-3_6

Rokeach, M., 1973. The nature of human values. Free Press, New York, NY, USA.

Rokeach, M., 1979. Understanding human values, individual and societal. Free Press, New York, NY, USA.

Schwartz, S.H. and Bilsky, W., 1990. Toward a theory of the universal content and structure of values: extensions and cross-cultural replications. Journal of Personality and Social Psychology, 58: 878-891. http://doi.org/10.1037/0022-3514.58.5.878

Schwartz, S.H., Cieciuch, J., Vecchione, M., Davidov, E., Fischer, R., Beierlein, C., Ramos, A., Verkasalo, M., Lönnqvist, J.-E., Demirutku, K., Dirilen-Gumus, O. and Konty, M., 2012. Refining the theory of basic individual values. Journal of Personality and Social Psychology, 103: 663-688. https://doi.org/10.1037/a0029393

Shahin, S. and Zheng, P., 2020. Big data and the illusion of choice: comparing the evolution of India's aadhaar and China's social credit system as technosocial discourses. Social Science Computer Review, 38: 25-41. https://doi-org.fontys.idm.oclc.org/10.1177/0894439318789343

Shteynberg G., Gelfand, M.J. and Kim, K., 2009. Peering into the 'Magnum Mysterium' of culture: the explanatory power of descriptive norms. Journal of Cross-Cultural Psychology, 40: 46-69. https://doi.org/10.1177/0022022108326196

Smith, A., 2009 (original publication: 1759). The theory of moral sentiments. Penguin Books, London, UK.

Swanton, C., 2003. Virtue ethics: a pluralistic view. Oxford University Press, Oxford, UK.

Vallor, S., 2016. Technology and the virtues, a philosophical guide to a future worth wanting. Oxford University Press, New York, NY, USA.

Van de Poel, I., 2015. Values in engineering and technology. In: Gonzalez, W.J. (ed.) New perspectives on technology, values, and ethics. Boston Studies in the Philosophy and History of Science, Volume 315, Springer, Cham, Switzerland, pp. 29-46.

Van de Poel, I. and Kroes, P., 2014. Can technology embody values? In: Kroes, P. and Verbeek, P.P. (eds.) Moral Agency and technical artefacts. Springer, Dordrecht, the Netherlands, pp. 103-124.

Verbeek, PP., 2011. Moralizing technology, understanding and designing the morality of things. University of Chicago Press, Chicago, IL, USA.

Wernaart, B., 2021a. Developing a roadmap for the moral programming of smart technology. Technology in Society, 46: 101466. https://doi.org/10.1016/j.techsoc.2020.101466

Wernaart, B., 2021b. Ethics and business: a global introduction (1st ed.). Routledge, Oxford, UK. https://doi.org/10.4324/9781003193951

Wernaart, B. and Van Halderen, M., 2020. Digital transformation requires moral leadership. Innovation Origins. Available at: https://innovationorigins.com/digital-transformation-requires-moral-leadership/.

Wilson, A.T., 2017. Avoiding the conflation of moral and intellectual virtues. Ethical Theory and Moral Practise, 20: 1037-1050. https://doi.org/10.1007/s10677-017-9843-9

Winner, L., 1980. Do artifacts have politics? Daedalus, 109: 121-136. http://www.jstor.org/stable/20024652

Yampolskiy, R.V., 2014. Utility function security in artificially intelligent agents, Journal of Experimental & Theoretical Artificial Intelligence, 26: 373-389. https://doi.org/10.1080/0952813X.2014.895114

Zardiashvili, L. and Fosch-Villaronga, E., 2020. 'Oh, dignity too?' said the robot: human dignity as the basis for the governance of robotics. Minds & Machines, 30: 121-143. https://doi.org/10.1007/s11023-019-09514-6

Zekic, N., 2019. Contradictory court rulings on the status of Deliveroo workers in the Netherlands. Illinois College of Law. Available at: https://cllpj.law.illinois.edu/dispatches

Media

Kristof, N., 9 December 2020. An uplifting update, on the terrible world of Pornhub. Young heroes lead in confronting pornography companies that monetize rape and child abuse. The New York Times. Available at: https://tinyurl.com/733mtryt.

6. Business ethics and technology

Bart F.W. Wernaart

School of Business and Communication, Fontys University, P.O. Box 347, 5600 AH Eindhoven, the Netherlands; b.wernaart@fontys.nl

Abstract

Technology and business are strongly connected: technology influences the way we do business, and business is an important driver for innovative technologies. This relationship has an important effect on how we see business ethics. The changing role of business in society is amplified by technology, and represents new ethical challenges. Technology may lead to a knowledge- and power-gap between a company and other stakeholders, and – when we do not take business ethics seriously – can stretch the already increasing income gap in societies. The effect could be that technology is beneficial only for the happy few. And that effect does not represent a sustainable strategy. To overcome these challenges it is important to reassess the role of governments in a globalised and privatised world, and reflect on the managerial, inward focus of business ethics' strategies in the private sector. In this chapter, we will first discuss the meaning of business ethics (Section 6.1). Then, we will explore the societal context of business ethics, including important developments that shape the current discourse on business ethics in a world with technology (Section 6.2). We then discuss how business relates to technology, and what challenges can be defined (Section 6.3). As a follow-up, we will discuss the prevailing models on business ethics, and how they are changing as a result of new technologies (Section 6.4). Finally, we will look ahead and outline the challenges we need to overcome (Section 6.5).

Key concepts

- In business ethics we explore the moral aspects of ethical issues in the context of the legal form that represents a company
- The particularities of business ethics are defined by the business size, the cultural context, the level of competition and the business branch
- Privatisation, globalisation, and technology are important trends that have changed the role of business in society

- Pressing issues in the debate on technology and business ethics relate to unequal access to knowledge about technology, unequal power to decide on technological issues, and unequal distribution of welfare as benefits from technology
- In a shareholder approach, decision-making in business is driven by the interests of the shareholders
- In a stakeholder approach, decision-making in business is driven by the interests of relevant stakeholders
- The triple bottom line is a business approach in which business activities are driven by social, ecological and economic values (people, planet and profit)
- Integrated reporting means giving an account not only of the financial performance of a business but also the business achievements in the field of social and environmental values

6.1 Business ethics

Business ethics differs from individual ethics insofar as, in the former, we explore the moral aspects of ethical issues in the context of a legal form, while in the latter we focus on the morality of individual human beings. There is a complex interplay between individual ethics and business ethics. While individuals work and act on behalf of a private legal entity, the morality that comes forth from the business is not necessarily a summation of the moral beliefs of the individuals who represent that organisation. After all, employees that work for Ashley Madison do not necessarily support cheating among married couples, nor would they necessarily find it ethical to cheat in their own relationship (Case 6.1). The business organisation has its own interests and corresponding norms and values it seeks to pursue. Among them, most notably, is the desire to make a profit. Ever since business was organised in legal constructions (such as a public or private limited company), business ethics has been on the agenda. This was most notably the case after the Industrial Revolution, when business organisations where no longer predominantly small, local and family-oriented entities, but rather large-scaled organisations, funded by shareholders, and with a strong power position in society. Perhaps Charles Dickens' 'Oliver Twist' (1938) and 'a Christmas Carol' (1843) were the first explicit works on business ethics, where the power position of the owners of industry in relation to the vulnerable position of employees was emphasised.

Case 6.1. The Ashley Madison data breach.

Ashley Madison is an online platform that specialises in discreet love affairs. Ironically enough, the platform that specialises in facilitating affairs also had a few affairs of its own. The most notable (and notorious) affair was the 2015 data breach. As the result of a hack, the identity of 36 million users was revealed. The hackers claimed that the company conducted unfair business practices. Among other things, the company was accused of offering a so-called 'full-delete service': a paid service that would guarantee the deletion of all personal user data. According to the hackers, this was plainly a lie, and the service was a fake – but profitable – service (Mansfield-Devine, 2015). Another issue was the usage of so-called 'fem-bots'. Apparently, the users of the platform were predominantly male, leading to a shortage of female users. This gap was filled by using artificial subscribers that would present themselves as females to make sure the male users would continue to make use of Ashley Madison's services. Data analysis showed that over 70,000 fem-bots were used on the platform (Newitz, 2015). The data breach allegedly led to quite a few divorces, and even some cases of suicide were reported (Baraniuk, 2015).

6.1.1 Contextual differences

business seize

Business ethics is not a uniform concept. Rather, its meaning depends on the context in which the business is situated. Some general observations can be made here. First, the seize of the business influences the way business ethics play a role in the organisation (Spence, 1999). Most debate in business ethics is centred around bigger organisations or multinationals, and is led by big scandals from the past, or a huge societal impact caused by these bigger organisations. Small and Medium Enterprises (SMEs) are often overlooked in the discourse (Wehrmeyer *et al.*, 2020). In these organisations (with 250 employees or less), different issues are at stake that conceptualise business ethics. It can be observed that moral issues are dealt with on a more ad hoc, informal basis, and that there is usually a very limited budget that can be used to develop a strategy on ethics. This is an important notion when we consider the important (and sometimes disruptive) role tech-start-ups can play. All big-tech organisations, such as Facebook, Google and Amazon, once started as an ambitious start-up developing something that would eventually shake up the (online) world. It took a scandal or two before the notion that ethics was something that needed to be organised professionally was taken more seriously. Therefore, next to the existing focus on business ethics in the context of large organisations, an emphasis on the particular situation of SMEs – more specifically rapid growing start-ups and scale-ups – is necessary. This is particularly the case in the tech sector.

cultural context

A second feature we need to take into consideration is the cultural context of private organisations. In business ethics literature we see an emphasis on regional differences towards business ethics (Crane *et al.*, 2019: 23-27), most notably the differences between North America and Europe. For instance, the North American focus on shareholder value and individual responsibility towards business ethics is at odds with the European focus on shareholder dialogue and a more societal control of business ethics. Increasingly, the regional particularities of other parts of the world are being explored, including Asia, Africa, Latin-America and Australia. It is important to note here that there can be a difference between the techno-social world (which is for the major part an online environment) and the physical business world. The 'globally networked information society' (Vallor, 2016: 141 – also referred to in Section 6.3.2) is not much bound by geographical boundaries or cultural differences, while the businesses that enable the existence of this society are. This can lead to an even more complicated interplay of cultural differences in business ethics and the morality that is shared in the online worldwide community.

level of competition

A third contextual feature is the level of competition at which the business operates. The fiercer the competition, the more budget restraints there will be when it comes to investing in business ethics strategies. On the other hand, the weaker the competition, the less the intrinsic urgency is felt to embrace a coherent strategy on business ethics. In a market that is dominated by only a few large multinationals, the market participants have a strong power position, and are able to dictate the rules of the game by themselves. It is clear that business ethics strategies in case of fierce competition are different from those in the case of weak competition (Van de Ven and Jeurissen, 2007). In the first, there is usually a strong focus on legal compliance and image protection, while in the latter the focus is on how to deal with the power position. Especially in the Tech sector, this is an ongoing challenge. Information technology products and services, social media platforms, and online markets are dominated by a handful of multinationals. The overall corporate strategy seems to result in an uncomfortable balance between self-governance and partnering up with governments to encourage suitable tech policies.

business branch

One last feature that needs to be explored is the particularities that come with the branch of the business. In each line of business, different moral issues are dominant. For instance, in agriculture, animal well-being, food safety and informed consumer choice are leading issues (Wernaart, ###in press). Also, in the Tech business – as is widely discussed in this book – particular issues are at the forefront, such as privacy, consumer autonomy and dominant market positions.

6.1.2 Terminology

The phrase 'business ethics' is not uncontested. Some argue that 'business' and 'ethics' are two contradictory words (Collins, 1994; Prasad and Agarwal, 2015). An entity that can only exist when it is profitable can never do the right thing unless it contributes to economic success. When ethics depends on profitability, the honest intention to do what is right is per definition absent, which makes 'business ethics' a false concept. At the same time – as we will also discuss in Section 6.2 – the position of the private sector has changed over the last few decades, as has its relationship with the public sector. Among other things, these developments lead to the proposition that when 'doing the right thing' is not at the core of a business organisation, it can never have a future-proof business strategy, nor a 'raison d'être'. This is the prevailing view on business ethics amongst scholars (Crane *et al.*, 2019: 4-5; Duska, 2000).

corporate social responsibility

triple bottom line

stakeholder-approach

sustainability

Business ethics is not the only terminology that is used to address the moral position of business. In managerial literature, the concept 'corporate social responsibility' (CSR) is used to address the idea that a private organisation should do more than be successful economically, and also has a societal and environmental responsibility. In this context, the so-called 'triple bottom line' is used, assuming that a company has a responsibility in the field of people, planet and profit (Elkington, 1997). Yet another term that is used to express the need for a strategy on business ethics is the stakeholder-approach, in contrast with a shareholder-approach. In the case of the latter, the strategy of the organisation is driven by the interests of the shareholders, while in the first, the strategy is driven by the combined interest of the company's stakeholders (Freeman, 2010). Also, the word 'sustainability' plays an important role in defining the ethical features of a company. Sustainable entrepreneurship usually focusses on the environmental aspects of business (Wernaart, 2021: 199-229). The idea is that the business strategy should ensure that the environmental footprint of an organisation is as small as possible. In this context, the urgency to create a circular economy rather than a linear economy is often emphasised. The huge energy consumption of data centres that are required to maintain our global information society is a good example of the challenges that come with sustainable entrepreneurship in the tech sector (Efoui-Hess, 2019; Jaller and Pahwa, 2020).

6.2 Business and society

Technology changes the way we do business, and business is an important driver of technological innovation. However, we cannot see technology as an isolated concept to understand why the interplay between technology and business is so influential. Two other trends are perhaps equally important, and together have led to a profoundly new role for business in society: privatisation and globalisation. In general, it is important to note that privatisation and (economic) globalisation are processes that usually go hand in hand (Boubakri *et al.*, 2013): the lowering of trade barriers and a desire to establish a system of free, global trade, often leads to smaller governments and more emphasis on international economic integration (Debab, 2011).

6.2.1 Privatisation

For various reasons, governments aspire to encouraging privatisation. The most notable reason is that a government cannot always take full responsibility for people's pursuit of happiness: a welfare state is – depending on its exact particularities – a very expensive endeavour. Especially in Western States, this has led to an agenda of privatisation since the economic crises of the mid-eighties (Gilbert, 2015). The idea is simple: private undertakings are more capable of offering social services, since they are able to work more efficiently compared to governmental institutions, and will come up with creative and innovative solutions due to the pressure of competition. Ultimately, the end-user of the services will have more choice for the most optimal price, and society has a better and financially healthier welfare system. Privatisation itself takes many forms (Armstrong *et al.*, 1997). Among them are the sales of government agencies to the private sector, the government that takes the role as shareholder, a government that stimulates competition in a market that previously had only one supplier (the government), or the government that assumes the role of a client from the private sector. Whatever the effect or success of privatisation is, it undeniably leads to a change in moral perception on both the side of the supplier and the end user of the services involved (Figure 6.1). A government is simply led by different values compared to a business organisation. Where a government primarily acts in support of the public interest, a business is led by private interest. When services that are previously dealt with in the public interest – such as healthcare, public transport, or energy supply – are now provided by private organisations, this leads to the question how the public interest is served, and how public values are protected (Wernaart, 2021: 118-120). On the side of the end-user, we also observe a change in value perception. A citizen is led by different values than a consumer. This is also referred to as the citizen versus consumer hypothesis (Curtis and McConnel, 2002). When someone casts a vote for a political party, the individual is led by different values compared to when that same individual uses her/his buying power to purchase something. This means that the role of the end-user in a privatised setting has changed from citizen to consumer. The relationship between citizen and government is not the same as the relationship between the consumer and the business, while the same public interest is at stake (see for a thorough analysis: De Wolf, 2012). It is for this reason that privatisation is an important driver in modern discourse on the ethics of business. Due to the changing role of business in society, we see an emphasis on the societal role of the private sector: there is no longer a sole responsibility for making profit but also a societal responsibility to take into consideration social and environmental values in business performance.

6.2.2 Globalisation

It is fair to say that globalisation is a container concept with many faces. The word is generally used to address the phenomenon of fading (country) borders, resulting in a more internationally oriented society. There are many areas that can be globalised. For instance, there is military globalisation as we can find in international military alliances, such as the North Atlantic Treaty Organisation (NATO); there is economic globalisation as we can find in international organisations such as the European Union (EU), the North American Free Trade Agreement (NAFTA), or the South African Customs Union (SACU); there is financial globalisation, as we can find in the European Monetary Union (EMU); and, of course, there is digital globalisation in the form of an international online community, as we can find in the globally networked information society, such as the internet.

trade liberalisation

In the particular case of business and technology, multiple forms of globalisation are important. In particular, the post-Second World War agenda to lower trade barriers through economic cooperation and integration is important (Wernaart, 2021: 317-342). Global economic institutions are still driven by this aim today, e.g. the Bretton Woods Institutions, the International Monetary Fund (IMF), the World Bank Group, and the World Trade Organization (WTO). Regionally more intensified integration modes are found, such as the common market that is at the core of the European Union. The main idea behind the lowering of trade barriers and resulting trade liberalisation is that it is considered as a way to maintain peace and realise prosperity. When countries depend on one another's economies, they will be less inclined to turn to violence. Inspired by macro-economic models, such as David Ricardo's competitive advantage theory, trade liberalisation should lead to a maximisation of welfare, due to the

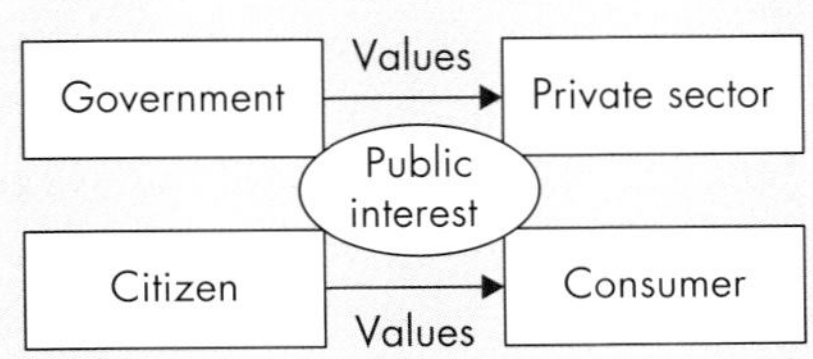

Figure 6.1. Changing values in privatisation.

fact that countries focus on and specialise in what they can produce best. Consequently, resources and capacity are optimally used, leading to a worldwide optimisation of prosperity. While in theory this sounds promising, there are also pitfalls that need to be discussed. Two major issues lead to ethical challenges that involve business.

regulatory competition

One of them is a phenomenon called 'regulatory competition'. The result of lowering trade barriers is a global playing field for competing businesses. This brings with it huge potential: a company has easy access to a global market, and can easily set up subsidiary establishments in other countries, or outsource part of the production process to other regions of the world. In the choice of where to settle, the locational advantages can be decisive. If a country wants to attract business, it may be beneficial to adopt more flexible standards on certain issues compared to other countries. These issues could relate to labour standards, environmental protection, or tax policies. When countries compete with their standards in order to attract business, this is called regulatory competition. The side effect of regulatory competition is that in the desire to attack business, important societal values can be negatively affected. A second challenge lies in the fact that a global market leads on the one hand to huge market potential, and products or services can be purchased from all corners of the world. On the other hand, this means that products and services that could be supplied locally are now supplied globally. This leads to unnecessary polluting supply chains. Also here, reference can be made to the datacentre case: globalisation leads to ethical challenges in terms of energy supply for data storage.

polluting supply chains

6.3 Business and technology

The changing role of business in society, as described in the previous section, raises the question to what degree business should assume the moral responsibility to protect the affected social and environmental values. This question is especially important in a context in which governments do not want to interfere too much in a free market economy, and have at the same time privatised important social welfare services. Moral responsibility issues are even more pressing when business owns the exclusive rights to or has exclusive knowledge of technologies that have a serious social impact. This creates an unfair power balance between business and society that could potentially be a driver for increasing inequality. Therefore, technology in itself is also an amplifier in the value shift that is caused by the changing role of companies in society. Pressing issues in this context include unequal access to knowledge about technology, unequal power to decide on technological issues and unequal distribution of welfare as benefits from technology.

When technology is in the exclusive hands of a private organisation whose aim is to make a profit, it is used primarily for that aim. A widely discussed challenge is represented by the algorithmic chains that shape online platforms (Batrinca and Treleaven, 2014). They are programmed to stimulate a fruitful match between advertiser and platform user, and are optimised to do just that. However, platforms – and in particular social media platforms – are increasingly used to spread and gather news amongst their users (Hågvar, 2019). As a result, the algorithmic chains that were designed to stimulate successful advertising have strong societal side effects that include a polarised – and mostly an unfriendly – discourse regarding political issues (Goyanes *et al.*, 2021). Furthermore, it provokes media outlets to present their content in such a way that it successfully triggers the platform's algorithms, which leads to a change in media practice (Bucher, 2018). Due to their architecture, these platforms can be the amplifier for spreading disinformation (fake news), and accordingly damaging democratic fundaments (Bennett and Livingston, 2018; Marichal, 2016). At the core is a knowledge gap between the platform company and its user: the knowledge on how algorithmic chains are constructed, how they work, and what the intended outcome is, is exclusively in the hands of the private organisation. The platform user is mostly unaware of the mechanisms that decide the content that is presented to the user. As we can see in Case 6.2, social media platforms were an important moderator in the spreading of fake news in the run up to the USA presidential elections in 2016 and 2020.

knowledge gap

Case 6.2. The social media ban on Donald Trump.

An interesting example of the way in which alleged disinformation affected democratic fundaments evolved around the social media use of former US president Donald Trump in the aftermath of the USA elections in 2020. As a result of the alleged spreading of disinformation regarding the election results, Trump was banned from Facebook, Instagram (Facebook, 2021) and Twitter (Twitter, 2021). These social media platforms had been under pressure to be more proactive about combating disinformation for years, and most notably since the USA elections in 2016 (Allcott and Gentzkow, 2017). The aftermath of both the elections in 2016 and 2021 has led to hearings in the USA congress in which the CEOs of Big Tech, including Google, Facebook and Twitter, faced questions about the role of their platforms in spreading disinformation affecting the democratic processes surrounding the elections (New York Times, 2021; Shaban *et al.*, 2017).

To this end, in 2020, Facebook eventually installed its Oversight Board, which is composed of a variety of experts representing the internet community. The purpose of this board is to judge in specific cases of content moderation decisions. On 5 May

2021, this board upheld the decision to suspend Trump's account on Facebook and Instagram (Oversight Board, 2021), though criticised the open-ended nature of this suspension. The board considered, amongst other things, that:

> in maintaining an unfounded narrative of electoral fraud and persistent calls to action, Mr. Trump created an environment where a serious risk of violence was possible. At the time of Mr. Trump's posts, there was a clear, immediate risk of harm and his words of support for those involved in the riots legitimized their violent actions. As president, Mr. Trump had a high level of influence. The reach of his posts was large, with 35 million followers on Facebook and 24 million on Instagram.

power-gap

This case not only represents a knowledge-gap, but also a power-gap. Social media-platforms hold a strong power position due to their disruptive technologies that affect how we do business, how we organise democratic processes, and how we interact (Van Dijck *et al.*, 2018). An important challenge is the economic power position of these platforms: when there are no serious substitute services, the platform offers a take-it-or-leave-it service, and its conditions are non-negotiable. Frequent updates of general terms in which the data policy is reviewed are a continuous source of public and legal debate (see for instance: Court of Justice of the European Union, 2021), and the power position of these platforms can seriously affect the autonomy of consumers (André *et al.*, 2018). Next to a knowledge- and power-gap, there is a welfare issue. As already discussed in Section 6.2.3, there is a risk that new technology in general amplifies the existing income gap (Frey, 2019) and only the happy few benefit from its development. In times of crisis, such as the corona pandemic, this becomes even more visible (Fernandez *et al.*, 2020).

income gap

6.4 Business, ethics and strategy

In business ethics theory, various models are used to facilitate moral strategy for business. In this section we will explore three fundamental approaches: one that focusses on decision-making processes (stakeholder approach), one that focusses on ethical goals (triple bottom-line), and one that focusses on accountability (integrated reporting).

6.4.1 A stakeholder approach

In a shareholder approach, decision-making in business is driven by the interests of the shareholders. This approach was famously defended by Milton Friedman (1962). In his view, the only social responsibility of business is to increase its profits. Friedman is misunderstood by some as a defender of corporate greed.

In essence, this is inaccurate: Friedman proposed drawing a clear line between the responsibilities of a government on the one hand, producing laws that define the corporate playing field, and businesses on the other, competing with one another within that playing field. In other words: the corporate playing field is defined by society, in the interests of society, and within that field business can compete in the way they see fit to maximise its profits, in the interests of its shareholders. Friedman's main motivation in drawing this line was that, in his opinion, business that claimed to be ethical was using the language of ethics for marketing purposes, since – amongst other reasons – business itself cannot intrinsically be ethical: only humans can. However, as we have discussed in Section 6.2, the relationship between the public and private sector has transformed due to privatisation and globalisation, which leads to different moral expectations of businesses. This has gradually led to a decision-making approach in which the interests of relevant stakeholders are also embodied in business strategies (Freeman, 2010). In earlier models, shareholders, suppliers, customers and employees were considered the most important stakeholders. Later, this group was complemented with governments, competitors and civil society (Freeman, 1984). More recently, a network model is being used that not only maps out the various stakeholders, but also shows how these stakeholders in turn are related to their own stakeholders, giving a full and more complete impression of mutual and conflicting interests (Rowley, 1997).

6.4.2 The triple bottom line

The notion that business has a moral responsibility towards societal values that moves beyond the purely private interests of a commercial organisation has led to a different strategic discourse. John Elkington famously introduced the concept of the triple bottom line in 1994 (Elkington, 1994, 1997, 2004). Instead of a single bottom line (profit), business activities are driven by a triple bottom line (People, Planet and Profit, the so-called 3Ps). The main idea is that a commercial organisation should do business driven by the ambition to realise values in the field of society, the environment, alongside the desire to make profit. While there can be a certain tension between the three Ps (Ozanne *et al.*, 2016), it can also result in, as Elkington puts it, a win-win-win situation (Elkington, 1994): society demands a certain corporate responsibility due to the societal changes described in Section 6.2. and the resulting shift in value interplay between the public and private sector. At the same time, investing in social and sustainable entrepreneurship leads to a more future-proof business model, and in the long term simply leads to more successful financial performances as well. A company with happy, ambitious and enthusiastic employees will produce more innovative products and services, and a business that uses renewable energy resources will in the end reduce its production costs. So, the triple bottom line is not only a bottom line, but also a business model. And that is exactly the thing that can also become its Achilles' heel. The use of triple-bottom-line thinking in a business jargon and context can easily lead to a trade-off tool (Elkington, 2018; Rambaud and Richard, 2015). It has proven to be relatively easy to use the sustainable language of the 3Ps to cover up bad behaviour. It appears that there is a thin line between TBL thinking on the one hand, and a marketing instrument, a risk management tool or a means to create a positive image on the other. A notorious example is an organisation like British Aerospace (BEA), which emphasises in their annual reports their ethical code of conduct and their responsible entrepreneurship in the field of social and environmental values (Doane, 2004). However, they still are a weapons supplier. Another issue is that the win-win-win assumption contradicts the urgency of taking environmental and social values as a starting point, and not, per se, private interest. While John Elkington considered the 'profit' concept a broader economic goal that goes beyond the mere profit of one organisation, the managerial and organisational focus of the TBL is often applied at the company level. This may result in a practice in which private profit is put at the same level of urgency as environmental and social values (Isil and Hernke, 2017).

6.4.3 Integrated reporting

One element that is unquestionably part of triple-bottom-line thinking is the business practice of giving an account not only of financial achievements, but also of achievements in the field of environmental and social values. There are many terms used to address this practice, such as external cost accounting, social and environmental accounting, integrated reporting and non-financial accountability The main idea is that businesses are transparent about their non-financial performances, and usually integrate such accounting in their annual financial statement.

principle of materiality

This is primarily important from an accountancy perspective. Financial data on its own does not necessarily paint a full picture of the performance of a company. The accountancy principle of materiality dictates that annual financial statements should comply with the principle of materiality. As defined by the International Accounting Standards Board (IASB, 2018) in their revised International Financial Reporting Standard (IFRS) and the Conceptual Framework, this means:

> Information is material if omitting, misstating or obscuring it could reasonably be expected to influence decisions that the primary users of general purpose financial statements make on the basis of those financial statements, which provide financial information about a specific reporting entity.

After all, the business stakeholders are better able to take strategic decisions based on the full performance of a business. A company that is involved in unethical business practices may have a wonderful financial performance to report, but in the long term it will not run a sustainable business. This will negatively affect future financial prospects as well. Therefore, from an accountancy perspective, it is important to paint a full picture of business performance in annual financial statements, and that includes decisive information on the societal and environmental aspects of the private organisation.

changing societal role of business

Secondly, from a society perspective, transparent reporting on social and environmental values is important considering the changing societal role of business. As explained in Section 6.2, the private sector plays an important role in the realisation of public values. One way to learn how business gives meaning to this role is by considering business performance also in a non-financial way. In practice, integrated reporting is usually done at the company level, in the context of the annual financial statements. However, on an international level we increasingly see NGOs stimulating the public disclosure of data regarding the impact of business on the environment or social values. Organisations such as Trucost, the Corporate Disclosure Project (CDP) and the Global Reporting Initiative (GRI) ask businesses to voluntary reveal relevant data (Wernaart, 2021: 244).

randomness

There are most certainly best practices in integrated reporting that shows that triple-bottom-line thinking can indeed lead to a win-win-win situation, which results in a continuous and well-measured improvement ambition within an organisation in the field of social and environmental values (Elkington and Zeitz, 2014). However, measuring and reporting about social and environmental value is not necessarily the same as contributing to its realisation (Milne and Gray, 2013). There are also concerns about integrated reporting as a cornerstone of business ethics. For instance, the translation of environmental and social values into measurable items can prove problematic in at least two ways. First, because there is usually a certain randomness about translating values into measurable items that fit within the context of an annual financial statement. Some organisations have experimented with expressing factors like pollution in social costs, while others prefer to use the Sustainable Development Goals (SDGs) as key indicators. It is clear however that, depending on which method you chose, the outcome can be completely different. A striking example is how British Petroleum (BP) was regarded in terms of corporate social responsibility in the year of the Deep Water Horizon oil spill (2010), one of the biggest man-made natural disasters in history. The method that was used to rank the Dow Jones Sustainability Index led to the conclusion that BP was one of the most sustainable organisations within its sector in 2010. Another ranking method used by a big investment organisation (Domini Social Investments) used a different method, and reached the opposite conclusion. It appears that the weighting of elements of the triple bottom line was fundamentally different. In the case of the Dow Jones method, there was a focus on sustainability reporting and investments, while there was a blind spot for risk assessment and relationships with stakeholders. It was precisely these last two that were at the core of the Domini Social Investments' method (Fay, 2012). A second reason why the translation of environmental and social values into measurable output can be problematic lies in the way it is used. By nature, companies have the tendency to optimise their processes in order to realise continuing growth (Young and Tilley, 2006). This also means that managerial language will embrace terms like eco-efficiency, or reducing an ecological footprint. Using resources as efficiently as possible does contribute to economic growth, but is not the same thing as sustainability. After all, continuous economic growth may even stand in the way of realising sustainability goals, such as the SDGs (Coscieme *et al.*, 2020). Next to that, using measurable units in the context of environmental and social values can easily lead to the 'trade-off'-attitude mentioned earlier in this chapter.

voluntary basis

Another issue is that most of the non-financial reporting is done on a voluntary basis. Whereas legislators usually feel very comfortable about creating rules or boundaries for financial reporting, the law usually remains silent in the field of non-financial reporting. A notable exemption are the attempts by the European Commission to initiate a firmer legislative framework for non-financial reporting: in 2021 they published a proposal for a Corporate Sustainability Reporting Directive (CSRD) (EC, 2021). The voluntary nature of the reporting is on the one hand laudable, since the drive to report on environmental and social values comes from within the organisation, and the drive to give an account of such issues comes from an intrinsic motivation, not an extrinsic legal obligation. This can lead to more awareness within an organisation of a business's environmental and social impact, and better performance in these fields (Maroun, 2019; Mervelskemper and Streit, 2017). On the other hand, there is always the risk of painting a rosy picture of the organisation using non-financial reporting as a marketing instrument (Bankers, 2018; Farache and Perks, 2010; Vanhamme and Grobben, 2009). This is a phenomenon sometimes referred to in the field of environmental values as 'greenwashing' (Marquis *et al.*, 2016).

6.5 The ethics of technology and business ethics

In the previous sections we discussed the changing role of business in society, and strategic theories used in business to play this role. It causes a shift in values from the public to the private sector. Technology amplifies this value-shift:

technological innovation often takes place in the sphere of the private sector. This also means that the private sector almost per definition plays a key role in managing the societal effects of new technologies.

While in business ethics literature the importance of technology is recognised as an important driver for taking business ethics seriously (see for instance Wernaart, 2021: 120-121), or as a potential part of solving business ethics challenges (Elkington, 1997: 187-217), the ethics of technology in itself is mostly left out of the management-oriented literature (see for notable exceptions: De George, 2003; Martin *et al.*, 2019; Microsoft, 2018). Existing models that focus on more sustainable business models have an internal, organisational and managerial focus. However, to address the complex issues that come with technology, a multi-disciplinary approach is needed (Isil and Hernke, 2017).

To conclude, it is time to revisit the assumption that by changing a business model, social and environmental values are adequately protected. Moral design works the other way around: as a result of value-driven design, business models change. In other words: business models should not realise values, rather values should realise business models.

References

Allcott, H, and Gentzkow, M., 2017. Social media and fake news in the 2016 election. Journal of Economic Perspectives, 31: 211-236. https://doi.org/10.1257/jep.31.2.211

André, Q., Carmon, Z., Wertenbroch, K., Crum, A., Frank, D., Goldstein, W., Huber, J., Van Boven, L., Weber, B. and Yang, H., 2018. Consumer choice and autonomy in the age of artificial intelligence and big data. Customer Needs and Solutions 5: 28-37. https://doi.org/10.1007/s40547-017-0085-8

Armstrong, H., Armstrong, P and Connelly, M.P., 1997. Introduction: the many forms of privatization. Studies in Political Economy, 53: 3-9, http://doi.org/10.1080/19187033.1997.11675312

Bankers, N.P., 2018. Bashing back: amoral CSR justifications. Journal of Business Ethics, 147: 401-418. https://doi.org/10.1007/s10551-015-2965-x

Batrinca, B., and Treleaven, P. C., 2014. Social media analytics: a survey of techniques, tools and platforms. AI & Society, 30: 89-116. https://doi.org/10.1007/s00146-014-0549-4

Bennett, W.L. and Livingston S., 2018. The disinformation order: disruptive communication and the decline of democratic institutions. European Journal of Communication, 33: 122-139. https://doi.org/10.1177/0267323118760317

Boubakri, N., Cosset, J.C., Debab, N. and Valéry, P., 2013. Privatization and globalization: an empirical analysis. Journal of Banking & Finance,37: 1898-1914. https://doi.org/10.1016/j.jbankfin.2012.12.014

Bucher, T., 2018. If … then: algorithmic power and politics (Oxford Studies in Digital Politics). Oxford University Press, Oxford, UK.

Collins, J.W., 1994. Is business ethics an oxymoron? Business Horizons, 37: 1-8 https://doi.org/10.1016/0007-6813(94)90013-2

Coscieme, L., Mortensen, L.F., Anderson, S., Ward, J., Donohue, I. and Sutton, P.C., 2020. Going beyond gross domestic product as an indicator to bring coherence to the sustainable development goals. Journal of Cleaner Production, 248: 119232. https://doi.org/10.1016/j.jclepro.2019.119232

Crane, A., Matten, A., Glozer, S. and Spence, L., 2019. Business ethics (5th ed.). Oxford University Press, Oxford, UK.

Curtis, J. and McConnell, K., 2002. The citizen versus consumer hypothesis: evidence from a contingent valuation survey. The Australian Journal of Agricultural and Resource Economics, 46: 69-83. https://doi.org/10.1111/1467-8489.00167

Debab, N., 2011. Globalization and privatization; two faces…one coin. Society of Interdisciplinary Business Research (SIBR), Conference on Interdisciplinary Business Research. Available at SSRN: https://ssrn.com/abstract=1869417

De George, R.T., 2003. The ethics of information technology and business. Blackwell Publishing, Malden, MA, USA; Oxford, UK.

De Wolf, A.H., 2012. Reconciling privatization with human rights. Intersentia, Antwerp, Belgium.

Dickens, C., 1838. Oliver Twist. Richard Bentley, London, UK.

Dickens, C., 1843. A Christmas carol in prose: being a ghost story of Christmas. Chapman & Hall, London, UK.

Doane, D., 2004. Good intentions – bad outcomes? The broken promise of CSR reporting. In: Henriques, A. and Richardson, J. (eds.) The triple bottom line, does it all add up? Earthscan, London, UK, pp. 81-88.

Duska, R., 2000. Business ethics: oxymoron or good business? Business Ethics Quarterly, 10: 111-129. https://doi.org/10.2307/3857699

Efoui-Hess, M., 2019. Climate crisis: the unsustainable use of online video – the practical case for digital sobriety. The shift project. Available at: https://theshiftproject.org/wp-content/uploads/2019/07/2019-02.pdf.

Elkington J., 1994. Towards the sustainable corporation: win-win-win business strategies for sustainable development. California Management Review, 36: 90-100. https://doi.org/10.2307/41165746

Elkington, J., 1997. Cannibals with forks. Capstone Publishers, Oxford, UK.

Elkington, J., 2004. Enter the triple bottom line. In: Henriques, A. and Richradson, J. (eds.) The triple bottom line, does it all add up? Earthscan, London, UK, pp. 1-25.

Elkington, J. and Zeitz, J., 2014. Apply true accounting principles. In: The breakthrough challenge: 10 ways to connect today's profits with tomorrow's bottom line. Jossey-Brass, San Francisco, CA, USA, pp. 55-70.

Elkington, J. 2018, 25 June. 25 Years ago I coined the phrase 'triple bottom line.' Here's why it's time to rethink it. Harvard Business Review. Available at: https://tinyurl.com/ys6r5uw9.

Farache, F., and Perks, K.J., 2010. CSR advertisements: a legitimacy tool? Corporate Communications: An International Journal, 15: 235-248

Fay, C., 2012. Domini and BP. In: Krosinsky, C. (ed.) Evolutions in sustainable investing. John Wiley & sons Inc, Hoboken, NJ, USA, pp. 81-90.

Fernandez, R., Adriaans, I., Klinge, T.J. and Hendrikse, R., 2020. Engineering digital monopolies. The financialisation of Big Tech. Somo, Amsterdam, the Netherlands.

Freeman, E., 1984. Strategic management. Cambridge University Press, Cambridge, UK.

Freeman, E., 2010. Stakeholder theory. Cambridge University Press, Cambridge, UK.

Frey, C.B., 2019. The technology trap, capital, labor, and power in the age of automation. Princeton University Press, New Jersey, NJ, USA.

Friedman, M, 1962. Capitalism and freedom. University of Chicago Press, Chicago, USA.

Gilbert, N., 2015. Social welfare trends in western societies: privatisation and the challenge to social work. Arbor, 191: a200. http://dx.doi.org/10.3989/arbor.2015.771n1002

Goyanes, M., Borah, P., Gil de Zúñiga, H., 2021. Social media filtering and democracy: effects of social media news use and uncivil political discussions on social media unfriending. Computers in Human Behavior, 120: 106759. https://doi.org/10.1016/j.chb.2021.106759

Hågvar, Y.B., 2019. News media's rhetoric on Facebook. Journalism Practice, 13: 853-872. https://doi.org/10.1080/17512786.2019.1577163

Isil, O. and Hernke, M.T., 2017. The Triple bottom line: a critical review from a transdisciplinary perspective. Business Strategy and the Environment, 26: 1235-1251. https://doi.org/10.1002/bse.1982

Jaller, M. and Pahwa, A. 2020. Evaluating the environmental impacts of online shopping: a behavioral and transportation approach. Transportation Research, Part D: Transport and Environment, 80: 102223. https://doi.org/10.1016/j.trd.2020.102223.

Mansfield-Devine, S., 2015.The Ashley Madison affair. Network Security, 2015: 8-16. https://doi.org/10.1016/S1353-4858(15)30080-5.

Marichal, J., 2016. Facebook democracy. Routledge, London, UK and New York, NY, USA.

Maroun, W., 2019. Does external assurance contribute to higher quality integrated reports? Journal of Accounting and Public Policy, 38: 106670. https://doi.org/10.1016/j.jaccpubpol.2019.06.002

Martin, K., Shilton, K. and Smith, J., 2019. Business and the ethical implications of technology: introduction to the symposium. Journal of Business Ethics, 160: 307-317. https://doi.org/10.1007/s10551-019-04213-9

Marquis, C., Toffel, M.W. and Zhou, Y., 2016. Scrutiny, norms, and selective disclosure: a global study of greenwashing. Organization Science, 27: 483-504. https://doi.org/10.1287/orsc.2015.1039

Mervelskemper, L. and Streit, D., 2017. Enhancing market valuation of ESG performance: is integrated reporting keeping its promise? Business Strategy and the Environment, 26: 536-549. https://doi-org.fontys.idm.oclc.org/10.1002/bse.1935

Microsoft, 2018. The future computed. Microsoft, Redmond, WA, USA. Available at: https://tinyurl.com/2ncf2j8z.

Milne, M.J. and Gray, R., 2013. W(h)ither ecology? The triple bottom line, the global reporting initiative, and corporate sustainability reporting. Journal of Business Ethics 118: 13-29. https://doi.org/10.1007/s10551-012-1543-8

Ozanne, L.K., Phipps, M., Weaver, T., Carrington, M., Luchs, M., Catlin, J., Gupta, S., Santos, N., Scott, K. and Williams, J., 2016. Managing the tensions at the intersection of the triple bottom line: a paradox theory approach to sustainability management. Journal of Public Policy & Marketing, 35: 249-61. https://doi.org/10.1509/jppm.15.143.

Prasad, A., and Agarwal, R., 2015. The oxymoron that is 'business ethics'. Journal of Management and Research, 15: 13-22.

Rambaud, A., and Richard, J., 2015. The 'triple depreciation line' instead of the 'triple bottom line': towards a genuine integrated reporting. Critical Perspectives on Accounting, 33: 92-116. https://doi.org/10.1016/j.cpa.2015.01.012

Rowley, T. (1997). Moving beyond dyadic ties: a network theory of stakeholder influences. The Academy of Management Review, 22(4): 887-910. https://doi.org/10.2307/259248

Spence, L., 1999. Does size matter? The state of the art in small business ethics. Business Ethics: A European Review, 8: 163-174.

Vallor, S., 2016. Technology and the virtues, a philosophical guide to a future worth wanting. Oxford University Press, New York, NY, USA.

Van de Ven, B. and Jeurissen, R., 2007. Corporate social responsibility and strategy. In: Jeurissen, R. (ed.) Ethics and business. Van Gorcum, Assen, the Netherlands.

Van Dijck, J., Poell, T. and De Waal, M., 2018. The platform society. Oxford Scholarship Online, Oxford, UK. https://doi.org/10.1093/oso/9780190889760.001.0001

Vanhamme, J. and Grobben, B., 2009. 'Too good to be true!'. The effectiveness of CSR history in countering negative publicity. Journal of Business Ethics, 85: 273. https://doi-org.fontys.idm.oclc.org/10.1007/s10551-008-9731-2

Wernaart, B., 2021. Ethics and business: a global introduction (1st ed.). Routledge, Oxford, UK. https://doi.org/10.4324/9781003193951

Wernaart, B., 2022. Food ethics. In: Van der Meulen, B. and Wernaart, B. (eds.) Applied food science. Wageningen Academic Publishers, Wageningen, the Netherlands, pp. ###-###.

Wehrmeyer, W., Looser, S. and Del Baldo, M. (eds.), 2020. Intrinsic CSR and competition doing well amongst european SMEs. Palgrave Macmillan, Cham, Switzerland. https://doi.org/10.1007/978-3-030-21037-3

Young, W. and Tilley, F., 2006. Can businesses move beyond efficiency? The shift toward effectiveness and equity in the corporate sustainability debate. Business Strategy and the Environment, 15: 402-415. https://doi.org/10.1002/bse.510

Media

Baraniuk, C., 2015, 24 August. Ashley Madison: 'Suicides' over website hack. BBC news.

Court of Justice of the European Union, 2021, 13 January. Press release No 1/21, Luxembourg. Advocate General Bobek: the data protection authority in the State where a data controller or processor has its main EU establishment has a general competence to start court proceedings for GDPR infringements in relation to cross-border data processing (europa.eu). Available at: https://tinyurl.com/3k27355e.

Facebook, 6 January 2021. Our Response to the Violence in Washington. Available at: https://tinyurl.com/hu2pcj42.

Newitz, A., 2015, 31 August. Ashley Madison Code Shows More Women, and More Bots. Available at: https://tinyurl.com/akndt3uv

New York Times, 25 March 2021. Big Tech C.E.O.s Face Lawmakers on Disinformation. Available at: https://tinyurl.com/3xspjjjr.

Twitter, 8 January 2021. Permanent suspension of @realDonaldTrump. Available at: https://tinyurl.com/59uerun7.

Shaban, H., Timberg, C. and Dwoskin, E., 2017 31 October. Facebook, Google and Twitter testified on Capitol Hill. Here's what they said. The Washington Post. Available at: https://tinyurl.com/r2wwf8bs.

Codified standards and case law

European Commission (EC), 2021. Proposal for a directive of the European Parliament and of the Council amending Directive 2013/34/EU, Directive 2004/109/EC, Directive 2006/43/EC and Regulation (EU) No 537/2014, as regards corporate sustainability reporting. Brussels, 21.4.2021 COM(2021) 189 final 2021/0104 (COD). Available at: https://eur-lex.europa.eu/legal-content/EN/TXT/?uri=CELEX%3A52021PC0189.

International Accounting Standards Board, 2018. International Financial Reporting Standards – Conceptual Framework 2019. Available at: https://www.ifrs.org/projects/completed-projects/2018/conceptual-framework/.

Oversight Board, 5 May 2021. Case decision 2021-001-FB-FBR. Available at: https://www.oversightboard.com/decision/FB-691QAMHJ.

7. The law as backbone in responsible innovation

An illustration from the perspective of data protection

Colette M.K.C. Cuijpers

Juridische Hogeschool Avans – Fontys, Prof. Goossenslaan 1-04, 5022 DM Tilburg, the Netherlands; c.cuijpers@fontys.nl

Abstract

In recent decades our world has steadily changed into the technology-embedded environment we know from the original Star Trek series (1966-67). Technologies that were science fiction when Star Trek first aired, are now a reality today. Our present-day smartphones combine functions of the communicator, the universal translator and the tricorder, while 3D-printing closely resembles the replicator. In times of Corona, telepresence via tablets can even be described as a primary life necessity. The benefits of technology within our society are manifold and, in many sectors, the only way forward in dealing with complex societal problems such as sustainable healthcare and energy. Besides the benefits, we have also witnessed the many risks, challenges and even harms those new technologies can bring. The initial perception of technology as a neutral actor has proven to be far from reality, as the functionality of technology is highly decisive in how it can be used by other actors (see also Chapter 5). Moreover, AI technologies develop beyond human control, automatically making decisions that can severely impact the lives of individuals but also society as a whole. Many present-day technologies need data as fuel to function, and this data can often be personal. The use of these technologies and personal data may intrude upon the private life of individuals, raising questions in relation to privacy and other human values such as dignity, autonomy and equality. We voluntarily welcome many technologies into our personal sphere, to make our lives easier or to improve our quality of life. Increasingly popular is smart technology, such as smart speakers and smart doorbells. The impact of such technologies, however, goes beyond the business-consumer relationship, since guests and passers-by may be impacted as well. Also, in public space our lives are constantly impacted by technologies processing our personal data. Cameras, public WiFi, ATM machines: they all make our whereabouts easy to trace. And even though the Western world looks on sceptically at the implementation of the social scoring system in China, invasive systems are also being developed and deployed within e.g. the US and the EU. In 2020 the Dutch court ruled that a system to detect social benefits' fraud, known as System Risk Indication (SyRi), violated the right to privacy of Dutch citizens due to a lack of transparency about when and how citizens were flagged up by the system. (Roosen, 2020; Van Schendel, 2019). The emergence, further development, and mass adoption of data-fuelled technologies that impact our society at the most fundamental levels – such as our conception of democracy (Harari, 2018) – and intrude upon fundamental human rights and values of the individuals inhabiting it, calls for regulatory interventions to foster responsible innovation.

Key concepts

- Responsible innovation consists of three dimensions: (1) innovations avoid harming people and the planet; (2) innovations 'do good' by offering new products, services or technologies that foster sustainable development; and (3) global governance schemes are in place that facilitate innovations that avoid harm and 'do good (Voegtlin and Scherer, 2017)
- The taking of responsibility at the back-end of the design process is often circumvented by using exoneration clauses in general terms and conditions and disclaimers, which place liabilities as far as possible upon the end users
- The four principle purposes of the law are: establishing standards, maintaining order, resolving disputes, and protecting liberties and rights
- Privacy by Design is a 'design-thinking-approach to privacy' developed by Ann Cavoukian (2017). It is a lifecycle approach, proactive and preventive, to incorporate by default privacy into networked data systems and technologies
- To operationalize Privacy by Design, privacy design strategies can be adopted. These include data-oriented strategies and process-oriented strategies
- On the basis of the principle of accountability the data controller is responsible for, and must be able to demonstrate, compliance with the core principles of data protection

7.1 Responsible innovation

According to Christian Voegtlin and Andres Georg Scherer (2017) responsible innovation consists of three dimensions: '(1) innovations avoid harming people and the planet; (2) innovations 'do good' by offering new products, services or technologies that foster sustainable development; and (3) global governance schemes are in place that facilitate innovations that avoid harm and 'do good'.'

In other words, this is a strong plea to develop new technology through moral design, centred around the idea that technology should be 'good'. Two separate but interrelated perspectives can be distinguished in this regard, starting from the premise that technology is not neutral: '(...) technologies actively help to shape culture and society, rather than being neutral means for realizing human ends' (Verbeek, 2008: 91). This realisation calls for engineering ethics to broaden its perspective beyond the morality of the designer. As discussed in Chapter 5, we should not focus exclusively on the morality of designers. We should also take into the equation the way external, societal values are affected by technology and we should recognise that technology itself embodies morality (Verbeek, 2008: 91). The external and internal perspective are interrelated, as designers can make use of artifacts to foster behaviour that corresponds to a certain moral standard. Bruno Latour denotes this as delegation: designers delegate specific responsibilities to artifacts. He uses the example of delegating the responsibility of not driving too fast on the approach to a speed bump (Latour, 1992: 166). In explaining the different modalities of regulation, also Lessig uses the example of the speedbump to illustrate the regulatory nature of the architectures underlying our society. Lessig explains how individuals' behaviour is constrained by four modalities of regulation: the law, social norms, the market, and architecture, also known as the pathetic dot theory. These modalities of regulation can be considered a toolbox for responsible innovation. Charles Raab and Paul De Hert are right in stating that Lawrence Lessig's toolbox is not comprehensive enough, as the actors involved in regulation need to be considered as well (Raab and De Hert, 2007: 21). How the law and other modalities of regulation are established and adhered to is very relevant for the contents of the toolbox, its deployment and its effectiveness in providing protection against undesired effects of innovation, as will become clear from the analysis below. However, for the purpose of this chapter, the focus will be on the modalities of regulation and more specifically on the law as a backbone to foster responsible innovation.

Lessig's toolbox

The discussion whether or not the law should be considered a catalyst or inhibitor for innovation is outside the scope of this chapter because, within the moral spectrum, individuals will agree to disagree in this regard. The Collingridge dilemma illustrates the intrinsic dilemma in technology regulation. On the one hand it is difficult to control a technology in its early stages because not enough is known of its possible or probable effects. On the other hand, controlling technology is also difficult once it is well-developed because by then intervention will be expensive, drastic, or impossible because the technology cannot be reversed (Collingridge, 1980). However, I do agree with the position taken by Leenes *et al.*, that this dilemma does create a need to regulate in the early stages when it is still possible, albeit in the dark. This calls for innovative approaches such as responsible research and innovation

Collingridge dilemma

(RRI), referring to the need to incorporate social and ethical values or aspects in the innovation process. (Leenes *et al.*, 2017: 35). An example of a legal framework embedding this approach is, not surprisingly, the European Union General Data Protection Regulation (GDPR). A legal instrument aimed at the protection of natural persons with regard to the processing of personal data and on the free movement of such data (EU, 2016).

This chapter illustrates how law, using EU data protection legislation as an example, can be considered a backbone in fostering responsible innovation both directly and indirectly, in interrelation with other modalities of regulation. Responsible innovation is an important first step in protecting personal data and privacy, mitigating risks, undesired side effects and harms stemming from data-fuelled technologies at the front-end of development. However, this alone is not sufficient. To achieve the aim of protecting natural persons with regard to the processing of personal data, responsible design should not only cover the front-end of a design process; it should also cover the back-end, since even responsible designs will not eradicate all risks and harms (Figure 7.1). What if risks materialise and cause harm regardless of responsible innovation? Malfunctions may still occur or malicious third parties may still find ways to do wrong. This calls for responsibilities at the back-end of the design process as well: providing guarantees in case of harm.

7.1.1 Lack of responsibility at the back-end

In practice, the taking of responsibility at the back-end of the design process is often circumvented by using exoneration clauses in general terms and conditions and disclaimers, which place liabilities as far as possible upon the end users. This can lead to harmful situations, as the scandals discussed in case 7.1 and 7.2 demonstrate.

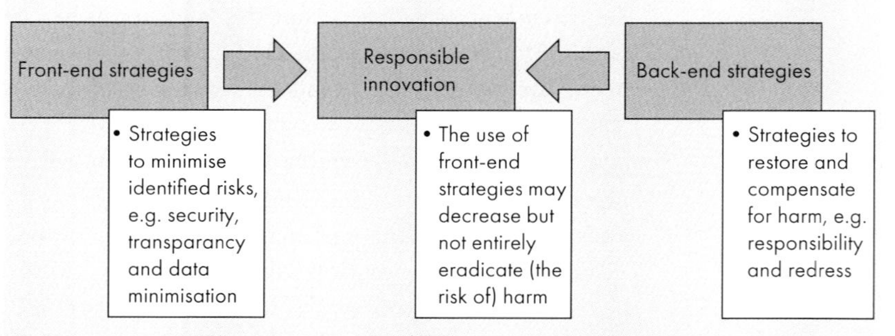

Figure 7.1. Responsible innovation (front-end and back-end strategies).

Case 7.1. Cambridge Analytics and Equifax.

The first example concerns the Cambridge Analytica Scandal of 2015. This affair revealed how the personal data of 87 million Facebook users was improperly harvested by the political data firm Cambridge Analytica. Neither Facebook nor Cambridge Analytica was inclined to compensate the victims. Even though several class actions were launched, as of today many are still awaiting compensation (Euroconsumers, 2020) while Facebook has publicly stated: 'We believe these lawsuits are without merit, and we are vigorously defending them.' (Trautman, 2020: 55). So, even though mechanisms like class actions might help victims to get compensation, these are not part of a system provided for by the data controller – the entity determining the purposes and means of data processing under Art. 2 (d) GDPR – and as such are not part of privacy by design at the back-end of a data process.

The second example concerns a big data breach at the US credit reporting agency Equifax, in which sensitive data of 143 million Americans was exposed. Even though public outcry and government pressure forced Equifax to take measures to accommodate the victims of the breach, the case also illustrates how Equifax tried to avoid liability in a rather dubious fashion. Equifax offered compensation in the form of free services for 1 year, and they offered a tool on their webpage to help consumers find out if their data was part of the data breach. However, it was only possible to use the tool after accepting the general terms and conditions, including a provision that if victims accepted the terms, they gave up the right to sue Equifax in court (Gressin, 2017). This only leaves arbitration as a recourse, and consumers are often not aware of this due to lengthy general terms and conditions. Needless to say, this is not a favourable option for consumers (Kenny, 2018).

Case 7.2. Going Dutch: two scandals.

In the Netherlands two specific cases can be mentioned to illustrate that unwillingness to take responsibility at the back-end not only pertains to private parties but also to the government. In the first case Mr. Kowsoleea, a Dutch citizen, fell victim to identity theft. This led to a situation of him being incorrectly registered as a criminal in several Dutch government systems. Mr. Kowsoleea tried in vain to amend the incorrect data. Calling upon the Dutch government to correct the data also turned out to be fruitless, as the Dutch government kept placing the responsibility for the accuracy of the data in their systems upon the citizens. Part of the problem was the use of decentralised police databases in the Netherlands and a networked system in which it turned out to be very difficult to rectify incorrect data or to prevent bogus information from re-entering the system. Even though the system was used by the Dutch government without Mr. Kowsoleea having a choice or control over the system, the Dutch government deemed it his responsibility to make sure the data in the system was correct. As a result of incorrect data input, Mr. Kowsoleea was – amongst other things – banned from travelling abroad and was arrested in front of his family. This not only had a financial impact on his business but also caused serious mental distress. Nevertheless, the Dutch government was very reluctant to compensate Mr. Kowsoleea for the harm he suffered (Van Schijndel, 2011). Only after a devastating report from the Dutch Ombudsman (Nationale Ombudsman, 2009), was the Dutch government willing to take some responsibility for errors in their systems caused by identity fraud.

A more recent striking example in the Netherlands is the so-called allowance affair, in which thousands of parents were wrongfully accused of defrauding the childcare benefit system. The affair not only illustrates how difficult it can be for people to actually fight back when a system accuses them of cheating, but also the unwillingness of a government to acknowledge their mistakes, take responsibility and compensate the victims. The scandal has led to a broad discussion on the role of the government towards its citizens. Furthermore, the role of the judiciary was critically reviewed, since apparently it could not offer the constitutional protection one would expect a citizen to have when falsely accused of fraudulent behaviour. All this was concluded in the Parliamentary investigation into this affair (Tweede Kamer der Staten Generaal, 2020).

Guarantees at the back-end are not necessarily part of the design of a product or service and as such are not only a matter for those responsible for the design process, as possibilities for redress largely depend on the law and the judicial system in a specific country. The final step in a responsible innovation process is thus a step in which both private actors and the government must take measures to achieve protection full circle, from preventive measures at the front-end to redressing mechanisms at the back-end. Also in this regard, data protection can be used to illustrate the challenges of providing back-end protection. Certain characteristics of privacy and data protection make it difficult for data subjects to restore or compensate for harm caused by breaches of privacy and non-compliance with the GDPR, as will be discussed later in this chapter. Considering the current Dutch situation (Case 7.2), it is argued that, to actually protect individuals within our society against harmful data processing, it is necessary to add a fourth dimension to the responsible innovation scheme of Voegtlin and Schrerer: 4) responsibility is taken for harm being caused by innovations, and adequate mechanisms are in place to effectively restore or compensate such harm.

7.1.2 The front-end: law as an instrument of direct and indirect regulation

law, functions

The law has many purposes, of which establishing standards, maintaining order, resolving disputes, and protecting liberties and rights are the four principal ones. As already mentioned above, the law in Lawrence Lessig's pathetic dot theory is one of four modalities that steer behaviour. Law achieves this by setting standards from the top down, enforced by competent authorities. Social norms can be defined as standards embedded within society, from the bottom up and enforced by peer pressure. The market is the mechanism for steering behaviour by means of economic incentives: encouraging certain behaviour by bonuses or providing discounts, or discouraging certain behaviour by imposing taxes or fines. Architecture steers behaviour by its form and functionalities. This can be code, being the infrastructure of a digital environment, or the architecture of public space like the speedbump mentioned above. The modalities can steer behaviour as such, but the effect of regulation can be stronger when the modalities interrelate. Thus, the law can function in two different ways: directly and indirectly. When the operation of law is direct, it means that the law stipulates how people ought to behave. The law also indicates what could happen in the case of disobedience, as it indicates the punishment when deviating from the directed behaviour. For example, the EU data protection law dictates the conditions under which the processing of personal data is lawful and states that the processing of sensitive data is forbidden (unless exceptions exist). The law explicitly indicates the sanctions that can be imposed when disobeying these rules, as will be elaborated in more detail later in this chapter. The provision of law thus regulates directly. However, law can also regulate in an indirect way. Meaning that the law can change the constraints stemming from other modalities of regulation. Lawrence Lessig uses a couple of examples: 'Law can tax cigarettes, directly regulating the market so as to indirectly change the consumption of cigarettes. Law can put advertisements on television showing the consequences of not wearing seat belts, directly working on a norm against seat belts so as to indirectly effect the use of seat belts. Law can order that buildings be built differently, directly regulating building codes so as to indirectly regulate discriminating behaviour with respect to the disabled' (Lessig, 1998: 671). Lawrence Lessig's final example, the law prescribing how the development of objects must take place, is at the core of the principles of Data Protection by Design and by Default, which are embedded in the GDPR. The law achieves its goal of providing protection for data processing by obliging those developing data-fuelled technology to build privacy into the design. This way the design as such, besides the law as such, offers privacy protection. Moreover, the architecture strengthens the function of the law, as enforcement of the law is now embedded in the design. It is easy to disobey legal rules. And we are inclined to do so when we benefit from breaking the law, when the chances of being caught or the consequences of being caught are minimal. Circumventing a rule embedded in architecture is often far more difficult. As an example, most individuals do not possess the knowledge to rewrite code and are therefore unable to remove a speedbump from the road.

Before addressing Data Protection by Design and Default in further detail, one interesting characteristic of the GDPR needs to be discussed in the light of common critiques regarding the limited effect of jurisdiction-bound legislation. Regulatory difficulties regarding responsible innovation 'arise because multinational organizations can relocate their value-creating activities to avoid legal constraints, which, in most cases are confined to national jurisdiction' (Voegtlin and Schrerer, 2017: 16-17). This may result in a phenomenon called 'regulatory competition': countries compete with the flexibility of their privacy laws to establish an attractive business climate for multinationals. However, one of the changes the GDPR brought about is an extension in the territorial scope of the EU data protection legislation. Bringing most processing of the personal data of EU citizens within the scope of the GDPR, regardless of where the actual data processing takes place (Art. 3 GDPR). This is also an illustration that without global governance of data, legal regimes can offer protection that reaches beyond the borders of the territory of jurisdiction and can impact data processors all over the world. Issues of cross-border enforcement, however, remain more difficult to resolve.

7.2 Privacy by Design

privacy-enhancing technologies

Privacy by Design is a 'design-thinking-approach to privacy' developed by Ann Cavoukian, a Canadian Information and Privacy Commissioner, who first coined the term Privacy by Design in 1997 (Krebs, 2013: 2). In 2009 this approach was published by Cavoukian as the seven foundational principles of Privacy by Design. The outset is that it is a lifecycle approach, proactive and preventive, to incorporate by default privacy into networked data systems and technologies. While initially the solution to preserving privacy and data protection was seen in so-called privacy-enhancing technologies (PETs), the Privacy by Design and Default approach is a more substantial approach, taking a positive-sum approach, not the either/or of zero-sum, which is a false dichotomy according to Cavoukian, who also labels this the approach of full functionality (Cavoukian, 2009). The aim is to avoid unnecessary trade-offs and to create a win-win situation for all interests and objectives from the different stakeholders involved. Privacy-enhancing technologies are thus a tool in a broader spectrum of systems and processes in which they can be embedded and operated to help protect privacy. Cavoukian's final two foundational principles concern the position of the data subject, and require transparency, accountability and compliance, focussing on the interests of the individual, labelled as 'a user centric approach'. This broad view on how technology can be designed in the service of protecting privacy, by making use of technical, organisational and regulatory measures, is fundamental to the Data Protection by Design and Default approach embedded in the GDPR.

7.2.1 Data protection by design

While the roots of Privacy by Design can be traced back to Canada, the approach was embraced and further developed in other regions of the world as well, including the European Union (Krebs, 2013: 2). The Data Protection Directive of 1995 (DPD) came too soon to incorporate the principle of Privacy by Design. The directive was a big step towards improved data protection throughout the EU (EC, 1995). However, it was criticised right from the start for lacking harmonisation effects and enforcement mechanisms and being unable to cope with challenges emerging from technological development and globalisation (Hoofnagle *et al.*, 2019: 71). Nevertheless, it took more than 20 years for the GDPR to replace the DPD. A huge lobby and a lot of media attention created a hype surrounding the coming into effect of the GDPR on 25 May 2018. Even though the GDPR largely resembles its predecessor, in practice the regulation was perceived as a massive change, since the impact of the DPD had remained marginal throughout its lifetime. In an effort to tackle the criticisms of a lack of a harmonisation effect and the absence of meaningful enforcement mechanisms, the choice was made to replace the Directive with a Regulation. EU Directives require implementation through national legislative procedures, potentially causing differences in effect per national jurisdiction. A Regulation, being immediately applicable in all Member States without national implementation, fosters harmonisation and reduces interpretative differences. Besides the need that was felt for an update and for fostering harmonisation, the main reason behind the GDPR was to improve enforcement of the data protection legislation. Law is only effective if it is obeyed, and strong enforcement mechanisms may result in a higher level of obedience. The territorial scope of the GDPR makes it more difficult to circumvent the applicability of the EU data protection rules by moving the actual data processing beyond EU borders. Other measures to increase the impact of data protection in practice are increased accountability requirements, enhanced enforcement mechanisms and opportunities for cooperation between supervisory authorities to tackle cross border issues of data protection more easily. The provision that really caught the attention from practice and created an impact is article 83 of the GDPR, regarding fines for non-compliance. The GDPR offers data protection authorities the opportunity to impose fines of up to 20 million euro or, in the case of an undertaking, up to 4% of the total worldwide annual turnover of the preceding financial year, whichever is higher. Besides these changes being geared towards increasing

non-compliance

compliance with data protection in practice, the data protection reform was also aimed at better aligning data protection with challenges emerging from both technological developments and globalisation (Recital 6 GDPR). In this regard, the main changes in the GDPR concern the incorporation of new or extended rights and principles, such as the right to be forgotten, the right to data portability and the principles of Data Protection by Design and by Default (DPbDD). The legacy of Cavoukian thus found its way into the EU legal framework on data protection.

7.2.2 Data protection by design and by default in the GDPR

Besides its 99 operative provisions, the GDPR also consists of 173 recitals that set out the background to the regulation and give extra insight into the reasons behind and the meaning of the operative provisions. Recitals to EU laws are not in themselves legally binding, but they are important for interpreting the provisions. If the meaning of a provision is ambiguous, the European Court of Justice will interpret the provision in light of the aim and spirit of the legislation, where it can draw from the information provided in the recitals. Therefore, the provisions relevant to DPbDD are analysed in light of the corresponding recitals.

Before addressing DPbDD as laid down in article 25 of the GDPR, attention will first be given to art. 35 GDPR regarding the so-called Data Protection Impact Assessment. There is a close link between assessing risks of data processing and identifying measures to mitigate these risks making use of DPbDD. This is a lifecycle process, starting at the very beginning of a development process, but guiding the entire lifecycle until the moment all personal data is erased and there is no longer a privacy risk. The envisioned change with the GDPR in regard to the Data Protection Directive was not simply moving from a compliance check in retrospect to an impact assessment and taking corresponding measures in advance, before starting the actual data processing. Instead, the DPbDD approach is a cyclical process, whereby changes in variables constantly need consideration and may lead to changes and adaptations throughout the lifecycle of the technology or system.

Data Protection Impact Assessment

According to art. 35 GDPR, performing a Data Protection Impact Assessment (DPIA) is not always required. The controller shall perform a DPIA prior to starting the data processing only when data processing is likely to result in a high risk to the rights and freedoms of natural persons; however, three situations are specifically described. First, performing a DPIA is mandatory in the case of the systematic and extensive evaluation of personal aspects of a data subject based on automated processing, including profiling, leading to decisions that produce legal effects or significantly affect the data subject. Second, in the case of large-scale processing of special categories of data as defined in Art. 9 GDPR or Art. 10 GDPR. Art. 9 provides an exhaustive list of special categories of data, concerning for instance racial origin, sexual preference, religious beliefs and genetic data. These data categories are deemed extra sensitive for data subjects and come under a stricter regime in the GDPR. Processing special categories of data is forbidden, unless the GDPR provides an exception, such as consent of the data subject. Art. 10 concerns data processing of personal data relating to criminal convictions and offences which is only allowed for official authorities or on the basis of Member State laws. The third situation in which a DPIA is required concerns large-scale systematic monitoring of a publicly accessible area. In the case of other situations, the level of risk and the duty to perform a DPIA must be judged on the basis of an objective assessment of the likelihood and severity of the risk to the rights and freedoms of the data subject, which should be determined by reference to the nature, scope, context and purposes of the processing (Recital 76 GDPR).

Even when a DPIA is not mandatory, it can be a useful instrument for identifying the appropriate technical and organisational measures that must be taken to ensure that the requirements of the GDPR are met. Recital 78 specifically states: 'In order to be able to demonstrate compliance with this Regulation, the controller should adopt internal policies and implement measures which meet in particular the principles of data protection by design and data protection by default.'

Art. 25 of the GDPR concerning Data Protection by Design and by Default states that controllers, both at the time of the determination of the means for processing and at the time of the processing itself, must implement appropriate technical and organisational measures. These measures are designed to implement data-protection principles, such as data minimisation, in an effective manner and to integrate the necessary safeguards into the processing in order to meet the requirements of this Regulation and protect the rights of data subjects. The measures that are required may be determined by taking into account the state of the art, the cost of implementation and the nature, scope, context and purposes of processing as well as the risks of varying likelihood and severity for the rights and freedoms of natural persons posed by the processing. This means that there is room for weighing up the interests and that controllers are not obliged to take all measures available. Some measures may be too costly or go beyond what is reasonable in view of the related risks. The second paragraph of Art. 25 GDPR adds that by default only personal data which are necessary for each specific purpose of the processing may be processed, which counts for the amount of personal data collected, the extent of their processing, the period of their storage and their accessibility. In particular, such measures shall ensure that by default personal data are not made accessible without the individual's intervention to an indefinite number of natural persons.

Recital 78 GDPR provides some examples of measures that could be taken, such as minimising the processing of personal data, the pseudonymisation of personal data as soon as possible, transparency with regard to the functions and processing of personal data, enabling the data subject to monitor the data processing and enabling the controller to create and improve security features. In the design process, producers of the products, services and applications should be encouraged to take into account the right to data protection and to make sure that they are able to fulfil their data protection obligations. This encouragement can be found in art. 83 of the GDPR. As mentioned before, high fines of up to 20 million euro or 4% of the worldwide annual turnover can be imposed in case of non-compliance with the GDPR. There is also a lower category listed, namely fines up to 10 million euro or 2% of the annual turnover in case of an undertaking. Non-compliance with either the obligation to perform a DPIA or the principles of DPbDD, is sanctionable under the lower category, as stated in art. 83 (4) a of the GDPR.

7.2.3 Privacy design strategies

As legal concepts can be vague, formulated in a complex manner, and often leave room for interpretation, they might not always offer the necessary clarity to be used in a pragmatic way by producers and designers. To operationalise the principles of DPbDD, Jaap-Hank Hoepman has translated the principles into privacy design strategies (Hoepman, 2020).

Just like DPbDD, privacy design strategies cover a broader spectrum than privacy-enhancing technologies (PETs). The strategies can encompass PETs, which can help at the stage where an artifact or system is actually developed. The privacy design strategies already come into play at an earlier stage, i.e. the initial thought process, where there is nothing more than an idea. Already when thinking about the way in which to develop and design an artifact or system fuelled with personal data, strategies are needed to reflect on capabilities and functionalities and how these relate to the principles of data protection. These strategies '(…) provide talking points to explore the design of the system. They guide the initial design sketches into a privacy-friendly direction, forcing one to make fundamental design choices early on' (Hoepman 2020: 1-2).

data-oriented strategies

Hoepman distinguishes eight privacy design strategies, which he divides over two categories: data-oriented strategies and process-oriented strategies. The data-oriented strategies focus on the data: minimise, separate, abstract and hide. Limit the data being processed to what is necessary to achieve the desired purpose and separate the personal data as much as possible. Process as little detail as possible, protect personal data, or make it unlinkable or unobservable.

process-oriented strategies

The second category lays its focus on the processes surrounding the responsible handling of personal data. While the first category relates mainly to technical measures, the second category relates to organisational measures to protect data processing: inform, control, enforce and demonstrate. Inform data subjects in a timely and adequate way about data processing, because without awareness they are unable to take action against it. Give data subjects adequate control over the processing of their data. Commit to processing personal data in a privacy-friendly way and adequately enforce and demonstrate this. Within each strategy, Hoepman indicates different tactics that can be used to make the overarching strategy more concrete. This will not be elaborated here, as reference to Hoepman's book suffices. However, to give an indication, the following tactics are linked to the design strategy: minimise, select, exclude, strip, remove (Hoepman, 2020: 5). The privacy design strategies can be seen as a toolkit for bridging the gap between the regulatory modalities of law and architecture, providing assistance on how to embed the legal rules into architecture. This enhances the impact of the legal rules and improves their enforcement, leading to increased protection of privacy because of the mutual strengthening of these modalities of regulation. For the privacy design strategies to be effective, all stakeholders should be involved and represented, including not only the perspective of the data processors, but also the end users whose personal data will be processed. This to achieve the win-win situation described in the fourth foundational principle of Privacy by Design from Cavoukian, as explained above.

7.3 The back-end: responsibility, enforcement and redress

DPbDD and its practical translation into design strategies can undoubtedly lead to a more privacy-friendly development of technologies and systems. This is beneficial from the perspective of the protection of data subjects, but also creates compliance with the GDPR for those processing the data, leading to fewer risks, both in terms of being fined for non-compliance by supervisory authorities, and in terms of the risk of losing user trust and acceptance. Adhering to the principles of DPbDD by properly implementing all design strategies will, however, not eliminate all the risks. Technologies and systems processing personal data can, regardless of responsible design, still cause harm. Setting high standards at the front-end of the design process alone is thus not sufficient to protect data subjects. Therefore, the next step in achieving this – and the fourth dimension in responsible innovation – concerns measures that need to be taken at the back-end: responsibility, enforcement and redress (Figure 7.2).

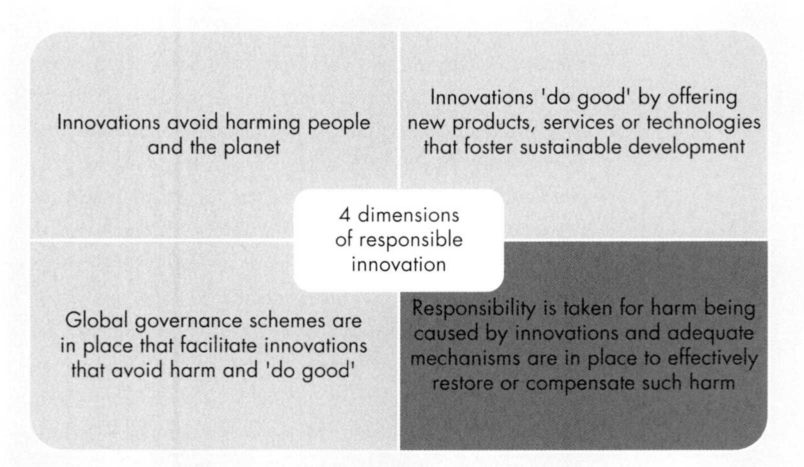

Figure 7.2. Four dimensions of responsible innovation.

7.3.1 Responsibility and enforcement

principle of accountability

Under the GDPR it is not sufficient to comply with the rules; it is also mandatory to demonstrate compliance with these rules. This is known as the principle of accountability, laid down in Art. 5 (2) of the GDPR. On the basis of this principle the controller is responsible for, and must be able to demonstrate, compliance with the core principles of data protection as laid down in the first paragraph of article 5. The GDPR prescribes several explicit measures that must be taken to demonstrate compliance with the GDPR, measures that need to be in place as soon as, or even prior to, the actual data processing, like keeping a record of processing activities and performing a DPIA. Therefore, accountability in the meaning of the GDPR is mainly situated at the front-end of data processing. However, it is also used at the back-end, in retrospect, to establish if proper measures have been taken to prevent harm from occurring. Being accountable is not only about demonstrating compliance, but can also be explained as taking responsibility, not only to make sure things do not go wrong, but also when things do actually go wrong. However, a different terminology is used to address the provision of protection at the back-end in the case of damage caused: in the GDPR this is not labelled as accountability, but as responsibility and liability.

Article 79 GDPR states that every data subject whose rights have been infringed according to the regulation shall have access to an effective remedy. The two most important provisions in the GDPR to achieve this are Art. 82 on damages and Art. 83 on administrative fines. These articles are different in target and subject: 'While the first has a compensatory purpose and is designed for use by individuals, the second has a preventive character and is implemented by data protection authorities vis-à-vis controllers and processors' (Reichel and Chamberlain, 2019: 1). The preventive character of art. 83 stems from the incentive to comply, as non-compliance can lead to severe fines, as indicated above. Many fines by different national data protection authorities have already been imposed, increasing the effect of the GDPR in practice, even though there has been some criticism regarding the number of enforcement actions and the cost of the penalties imposed (Wolff and Atallah, 2020). Data protection authorities themselves can decide whether or not to enforce the GDPR and to impose fines. Data subjects can entice data protection authorities to take action, as art. 77 GDPR offers them the right to file a complaint, making the authority aware of cases of non-compliance. However, even when a complaint leads to a fine, the data subject as such is not compensated. Imposing fines may lead to a situation in which new harms may not occur, or are at least less likely to occur. But those harmed by the activity for which the controller or processor is fined, receive no compensation. This means a data subject, for restoration or compensation, needs to start proceedings against the controller or processor. Art. 79 GDPR provides data subjects with the right to an effective judicial remedy against a controller or processor when the data subject is of the opinion that their data protection rights have been infringed. If so, art. 82 GDPR provides a right to compensation and liability: 'Any person who has suffered material or non-material damage as a result of an infringement of this Regulation shall have the right to receive compensation from the controller or processor for the damage suffered.' If multiple parties are involved in the same processing, data subjects can hold each controller or processor liable for the entire damages to ensure effective compensation of the data subject. Any controller or processor that has paid full compensation may subsequently institute recourse proceedings against other controllers or processors involved in the same processing. A controller or processor shall be exempt from liability if they prove that they are not in any way responsible for the event giving rise to the damage. Recital 146 clarifies that data subjects should receive full and effective compensation for the damage they have suffered. In this regard the concept of damage should be broadly interpreted in the light of the case-law of the Court of Justice in a manner which fully reflects the objectives of this Regulation. However, the proof of damage remains with the data subject. Something that can be difficult as will be discussed in the next section.

Another difficulty in starting a claim for compensation is the costs involved. Recital 149 and art. 80 of the GDPR offer data subjects the right to mandate a not-for-profit body, organisation or association to lodge a complaint, to a judicial remedy or to exercise the right to receive compensation on behalf of data subjects. There are some requirements: the involved body, organisation or association must be properly constituted in accordance with the law of a

Member State, must have statutory objectives which are in the public interest, and must be active in the field of the protection of data subjects' rights and freedoms with regard to the protection of their personal data.

The GDPR thus provides for guarantees and procedures for data subjects to exercise their rights and to get compensation for damages caused by unlawful data processing. Nevertheless, as will be illustrated below, the provisions in the GDPR as such are not sufficient to make sure data subjects are sufficiently protected in terms of having recourse to effective means of restoration and compensation.

7.3.2 Data protection and redress: a difficult combination

Even though incorporated in the seven foundational principles of privacy and embedded into the GDPR, the effectiveness of redress mechanisms in case of privacy violations or non-compliance with the GDPR can be questioned. Without the aim of being exhaustive, some main challenges of effective protection of data subjects after harm is caused are discussed below. This illustrates the need for more attention and research into what is stipulated above as the fourth dimension of responsible innovation (Figure 7.2): 'responsibility is taken for harm being caused by innovations and adequate mechanisms are in place to effectively restore or compensate such harm.' As already mentioned, in practice there is often a lack of willingness to take responsibility for harm caused by data-driven technical innovations. Instead of implementing guarantees at the back-end of data processes for taking responsibility in case of harm, legal mechanisms are used to evade liability and to transfer responsibility to the data subjects. In addition to this issue, the characteristics of privacy and data protection and the Dutch judicial system provide challenges for effective redress as well.

awareness

One of the major challenges in combatting privacy and data protection violations is awareness (Wachter, 2018; Wachter and Mittelstadt, 2019). When individuals are not aware of data processing in the first place, or unjust data use, such as discrimination based on profiling and automated decision-making systems, it is impossible for them to protect their own rights. Creating awareness, transparency and combating opacity are steps to be taken in the initial design process, both in terms of technical and organisational means under the principles of DPbDD. This is part of the design strategy 'inform', based on the openness and accountability principles underlying the GDPR, and is thus part of the front-end dimension of responsible innovation. At the back-end other issues arise in protecting one's rights, after the data subject actually experiences harm caused by an innovation or the data processing on which this innovation functions. Violations of privacy and data protection are often about the data subject losing control over what they want to keep private. Personal things get exposed, causing harm to e.g. reputation. One of the challenges is then that fighting for your right to privacy may lead to further exposure and thus greater harm, as for example demonstrated in cases regarding leaked sex videos. Restoration of the former situation is no option, especially with the ease and speed with which information can be shared via the internet. Compensation for harm suffered is then the only option. In cases of blatant violation of reputation, such as the case with the leaked sex videos of a Dutch celebrity, the award of damages amounted to 30,000 euro (only a fraction of the 450,000 euro claimed by the victim) (ECLI:NL:RBAMS:2018:5130). The data subject must substantiate and prove the damages, a difficult task as the qualification of privacy harms is complex and not always recognised. One of the issues concerns individual harm versus group harm in developments such as Big Data and AI technologies. What is the harm of possibly being surveyed by a system like the abovementioned SyRi, when individual data subjects are not even aware if, or to what extent, they are being flagged by the system? While it is obvious that the entire group of Dutch citizens may be impacted by the system, it is not clear which individuals within this group actually are. Nevertheless, the mere possibility of being watched can already lead to changing behaviours, as illustrated by the panopticon of Bentham (Michel Foucault, 1977). This can cause harm such as distress and anxiety, but this may be difficult to prove. The same goes e.g. for data breaches and the leaking of data. What is the actual harm in personal data being leaked? In declining a request for compensation, the Dutch court argued: 'The claimant has not submitted any concrete information to substantiate the damage alleged by him. He has not made it plausible that the infringement led to the violation of the integrity of his person and that the consequences of the infringement affected him directly. He has not made clear why mentioning his name and place of residence in the statement that he has submitted two requests for information to a certain municipality, in response to a request sent to another municipality, can be regarded as harm to the person. He has also not made plausible what adverse consequences resulted from mentioning his name and place of residence. The mere statement that mentioning his name has a stigmatizing effect is insufficient' (ECLI:NL:RVS:2020:899). Besides a high burden of proof, another limitation in seeking redress is establishing the amount of compensation, or in other words: 'What is the value of privacy?' Analysing Dutch court cases reveals that damages have only been awarded in a few instances for unlawful processing of personal data, where the compensation varied between 250 and 500 euro (ECLI:NL:RBNNE:2020:247; ECLI:NL:RBNNE:2021:106; ECLI:NL:RVS:2020:898). These amounts do raise questions in relation to the fines that can be awarded by Data Protection Authorities. Reichel and Chamberlain argue that it 'could be seen as unreasonable that a data subject having had his or her data privacy breached

compensation

burden of proof

is awarded nothing or a relatively small sum in damages, while the DPA issues an administrative fine in the second tier of Art. 83 GDPR, which may be quite high' (Reichel and Chamberlain, 2019: 19).

procedural expenditure

A final challenge in getting redress is the procedure of having to go to court, which can be complex, time consuming and costly. The procedural expenditure may not always be reimbursed. In one of the Dutch court cases where the data subject was awarded compensation of 500 euro for immaterial damages, the court also decided that both parties had to bear their own costs for the proceeding, which probably exceeded the amount awarded in damages (ECLI:NL:RBNNE:2021:106). As mentioned above, the GDPR does require states to offer the possibility of a class action. In the Netherlands this possibility is offered in art. 3:305a of the Dutch Civil Code. Only since the introduction of the Settlement of Large-scale Losses or Damage Act (in the Netherlands known as WAMCA) is it possible to claim monetary damages in class actions. Some promising cases, including cases regarding privacy and data protection breaches, have already been filed under this new procedure, which entered into force on 1 January 2020. It is, however, too early to make any predictions regarding the outcome in these cases. Whether or not this procedure will fill the gap and offer sufficient redress for harmed data subjects remains to be seen. There are definitely some challenges for victims, such as demonstrating a violation of norms, proving concrete damage or a causal relationship and, in cases of scattered damage, the threshold for starting a procedure is high (Van der Linden and Walree, 2018: 113). Thus, besides keeping an eye on the developments in class actions, it is necessary to consider and further investigate other mechanisms and strategies for compensating data subjects for harm caused by privacy violations and GDPR non-compliance. One possible angle could be to identify harmonised criteria for the assessment of immaterial damage to establish a more harmonised judicial practice in regard to the amount of compensation for similar cases (Sadiku, 2020). Another route for further research is insurance or data breach compensation funds, which according to Kenny 'would be better to reach the goal of compensating true victims, while also limiting the costs associated with litigation' (Kenny, 2018: 238).

7.4 Conclusions

This chapter focused on responsible innovation, a concept understood to include moral design, both in terms of moral responsibilities of designers and the moral significance of technologies themselves. The GDPR is a clear example of how law can foster responsible innovation, making use of the different modalities of regulation. The initial intervention is geared towards the design process by requiring privacy by design and by default, and backed up by monetary incentives in the form of high fines for non-compliance. The GDPR also guarantees data subjects procedures and mechanisms for redress. In this regard, one could argue that the proposed fourth dimension of responsible innovation – 'responsibility is taken for harm being caused by innovations and adequate mechanisms are in place to effectively restore or compensate such harm' – is covered by the GDPR. However, when looking at legal practice, the mere provision of these paper-based guarantees is ineffective and thus not sufficient. Redress is often difficult due to the specifics of privacy and data protection harms, the unwillingness of data controllers to take responsibility in case of harm and the challenges of going to court. Even though responsible innovation seems an increasingly popular strategy at the front-end of data-fuelled innovation processes, to actually protect data subjects, a focus of attention is also required at the back-end. Preventing harm is of course the better strategy; however, once that phase has passed and harm is caused, effective guarantees and compensation mechanisms should already be in place. Willingness to take responsibility should be incentivised, and innovative strategies should be explored to cope with damages once these occur. This requires an investment from both data controllers and the government as some challenges in providing redress can be resolved within the design process, while others need to be dealt with at the level of the judicial system.

References

Cavoukian, A., 2009. Privacy by design. The 7 foundational principles implementation and mapping of fair information practices. Information and Privacy Commissioner Ontario Canada. Available at: https://iapp.org/media/pdf/resource_center/pbd_implement_7found_principles.pdf

Cavoukian, A., 2017. Global privacy and security, by design: turning the 'privacy vs. security' paradigm on its head. Health and Technology, 7: 329-333. https://doi-org.fontys.idm.oclc.org/10.1007/s12553-017-0207-1

Collingridge, D., 1980. The social control of technology. St Martin's Press, https://doi.org/10.1017/s0003055400186265

Foucault, M., 1977. Discipline and punish: the birth of the prison. Pantheon Books, New York, NY, USA.

Gressin, S., 2017. The equifax data breach: what to do. Available at: https://www.consumer.ftc.gov/blog/2017/09/equifax-data-breach-what-do

Harari, Y.N., 2018. Why technology favors tyranny. The Atlantic. Available at: https://www.theatlantic.com/magazine/archive/2018/10/yuval-noah-harari-technology-tyranny/568330/

Hoepman, J.-H., 2020. Privacy design strategies (The little blue book). Available at: https://www.cs.ru.nl/~jhh/publications/pds-booklet.pdf

Hoofnagle, C., Van der Sloot, B. and Zuiderveen Borgesius, F., 2019. The European Union general data protection regulation: what it is and what it means. Information & Communications Technology Law 28: 65-98. https://doi.org/10.1080/13600834.2019.1573501

Kenny, C., 2018. The equifax data breach and the resulting legal recourse. Brooklyn Journal of Corporate, Financial & Commercial Law, 13: 215-238. Available at: https://brooklynworks.brooklaw.edu/bjcfcl/vol13/iss1/10.

Krebs, D., 2013. 'Privacy by design': nice-to-have or a necessary principle of data protection law? JIPITEC 4: 2-20.

Latour, B., 1992. Where are the missing masses? The sociology of a few mundane artifacts. In Shaping Technology-Building Society Studies. In: Bijker, W. and Law, J. (eds.) Sociotechnical change. MIT Press, Cambridge, MA, USA, pp.225-259. https://doi.org/10.1177/027046769401400468

Leenes, R., Palmerini, E., Koops, B-J., Bertolini, A., Salvini, P. and Lucivero, F., 2017. Regulatory challenges of robotics: some guidelines for addressing legal and ethical issues. Law, Innovation and Technology, 9: 1-44. https://doi.org/10.1080/17579961.2017.1304921

Lessig, L., 1998. The new Chicago School. Journal of Legal Studies, 27: 661-692. https://doi.org/10.1086/468039

Nationale ombudsman, 2009. Herzien openbaar rapport 2009/199 datum: 23 september 2009. Available at: https://www.nationaleombudsman.nl/uploads/rapport2009-199.pdf

Raab, C.D. and De Hert, P.,2007. The regulation of technology: policy tools and policy actors (November 1, 2007). TILT Law & Technology Working Paper Series No. 003/2007, Tilburg University Legal Studies Working Paper No. 004/2007. Available at: https://doi.org/10.2139/ssrn.1030263

Reichel, J. and Chamberlain, J., 2019. The relationship between damages and administrative fines in the EU general data protection regulation. Faculty of Law, Stockholm University Research Paper No. 72, Mississippi Law Journal. https://doi.org/10.2139/ssrn.3447854

Roosen, M., 2020. What SyRI can teach us about technical solutions for societal challenges. Available at: https://globaldatajustice.org/2020-02-20-roosen-syri/.

Sadiku, A., 2020. Immaterial damage and some types of its compensation. Prizren Social Science Journal 4: 50. https://doi.org/10.32936/pssj.v4i1.142

Trautman, L.J., 2020. Governance of the Facebook privacy crisis. 20 Pittsburgh Journal of Technology Law & Policy. https://doi.org/10.2139/ssrn.3363002

Tweede Kamer der Staten Generaal, 2020. Parlementaire ondervragingscommissie kinderopvangtoeslag ongekend onrecht. Available at: https://www.tweedekamer.nl/sites/default/files/atoms/files/20201217_eindverslag_parlementaire_ondervragingscommissie_kinderopvangtoeslag.pdf

Van der Linden, T.E. and Walree, T.F., 2018. De collectieve procedure als oplossing voor het privaatrechtelijke handhavingstekort bij een datalek? AV&S 4: 105-113.

Van Schendel S., 2019. The challenges of risk profiling used by law enforcement: examining the cases of COMPAS and SyRI. In: Reins L. (ed.) Regulating new technologies in uncertain times. Information Technology and Law Series, Volume 32. T.M.C. Asser Press, the Hague, the Netherlands. https://doi.org/10.1007/978-94-6265-279-8_12

Van Schijndel, P., 2011. Identity theft and fraud. In: Van der Hof, S. and Groothuis, M. (eds.) Innovating government: normative, policy and technological dimensions of modern government. Asser Press, the Hague, the Netherlands. http://doi.org/10.1007/978-90-6704-731-9_22

Verbeek PP., 2008. Morality in design: design ethics and the morality of technological artifacts. In: Philosophy and Design. Springer, Dordrecht, the Netherlands. https://doi.org/10.1007/978-1-4020-6591-0_7

Voegtlin, C. and Scherer, A.G., 2017. Responsible innovation and the innovation of responsibility: governing sustainable development in a globalized world. Journal of Business Ethics, 143: 227-243. https://doi.org/10.1007/s10551-015-2769-z

Wachter, S. and Mittelstadt, B., 2019. A right to reasonable inferences: re-thinking data protection law in the age of big data and AI. Columbia Business Law Review 12. https://doi.org/10.31228/osf.io/mu2kf

Wachter, S., 2018. The GDPR and the Internet of Things: a three-step transparency model. Law, Innovation and Technology, 10: 266-294. https://doi.org/10.1080/17579961.2018.1527479

Wolff, J. and Atallah, N., 2020. Early GDPR penalties: analysis of implementation and fines through May 2020. TPRC48: The 48th Research Conference on Communication, Information and Internet Policy. https://doi.org/10.2139/ssrn.3748837

Media

Euroconsumers, 2020. Euroconsumers class action against Facebook. The Italian regional administrative court made an historical judgment: our data has value. Available at: https://www.euroconsumers.org/search?q=Facebook

Codified standards and case law

Council of State (the Netherlands), 01-04-2020, ECLI:NL:RVS:2020:898.

Directive 95/46/EC of the European Parliament and of the Council of 24 October 1995 on the protection of individuals with regard to the processing of personal data and on the free movement of such data. Official Journal of the European Union, L 281: 31-50. Available at: http://data.europa.eu/eli/dir/1995/46/oj

District Court Noord-Nederland (the Netherlands), 15-01-2020, ECLI:NL:RBNNE:2020:247.

District Court Noord-Nederland (the Netherlands), 12-01-2021, ECLI:NL:RBNNE:2021:106.

District Court Amsterdam (the Netherlands), 25-07-2018, ECLI:NL:RBAMS:2018:5130.

Dutch Council of State, ECLI:NL:RVS:2020:899, 01-04-2020. Available at: https://uitspraken.rechtspraak.nl/inziendocument?id=ECLI:NL:RVS:2020:899.

European Union (EU), 2016. Regulation (EU) 2016/679 of the European Parliament and of the Council of 27 April 2016 on the protection of natural persons with regard to the processing of personal data and on the free movement of such data, and repealing Directive 95/46/EC (General Data Protection Regulation). Official Journal of the European Union, L 119: 1-88. Available at: http://data.europa.eu/eli/reg/2016/679/oj.

8. Ethics in local politics: a case study of the city of Eindhoven

Maarten van Veen[1]* and Piek Visser-Knijff[2]

[1]City of Eindhoven, Sector Strategy, P.O. Box 90150, 5600 RB Eindhoven, the Netherlands; [2]Filosofie in actie, Klingelbeekseweg 27 H31, 6862 VS Oosterbeek, the Netherlands; maarten.van.veen@eindhoven.nl

Abstract

'The European Union will limit predictive algorithms in the public space', 'An algorithm is not neutral, nor is a government algorithm', 'A citizen is not a dataset'. The newspaper headlines illustrate that the ethics of new technologies has become an important issue in the public debate in the Netherlands. Early awareness of ethical and moral aspects is considered an important principle of the state of law, good government, and responsible citizenship. In this chapter we explore how the municipality of Eindhoven is trying to make sense of ethics. First, we explore some ethical concerns that arose at four innovation projects in the municipality of Eindhoven. Second, we describe how these concerns led to a proposal on how to make the administration and council more sensitive to ethical issues and how to introduce certain safeguards. Third, we discuss our approach for the future and reflect on the way ahead.

Key concepts

- New technologies can change society and are therefore political
- Smart cities struggle with the ethical aspects of new technologies (responsible innovation)
- Ethics covers a wide range of public values: safety, privacy, autonomy, control, rule of law, human rights, sustainability
- Tools for governing ethics include ethical frameworks and ethical committees

8.1 Introduction

political objects

In a famous article Langdon Winner wonders 'Do artefacts have politics?' (Winner, 1986) He argues that objects are indeed political. Objects have social effects and possess – in that sense – 'agency'. He refers to the bridges of Long Island as a clear example of how objects can act. The bridges on Long Island are quite low which means that vehicles of a certain height cannot pass through them. So, in effect, vehicles like buses are practically kept out of the area with the result that people without a car are unable to reach the beautiful beaches (see also Section 5.2.2). The observation that objects have certain social effects has encouraged many scholars to explore the multi-layered relationship between science, technology, and society.

Many ground-breaking contributions to this new field of 'Science & Technology Studies' have come from Dutch scholars, for example: Achterhuis, 1997; Bijker, 1995; Mol, 2002; Schot and Rip, 1996; Verbeek, 2000. Nowadays it doesn't seem that huge an insight, but the idea that technology affects the social world in more ways than intended emerged from quite a body of research. The research opened 'black boxes' (Latour, 2005) and made clear that technology development is a human endeavour that should be part of public debate. Just by guiding technology development at an early stage, undesirable social effects can be avoided. And this will also prevent scenarios in which technology 'bites back' in unexpected ways (Tenner, 1996).

new technologies

The city has become a playground for new technologies: for example, waste management, wind turbines, vertical farming, ventilation, solar power, transport drones, surveillance technology, autonomous vehicles, smart routing, communication network, sensors, intelligent lights, and sharing economy (platforms). The concept of a smart city or smart society is often used to describe this development. For example, the CEO of Cisco defines it as follows: '(..) applying data and technology to traditional objects or processes to enhance efficiency and convenience has come to define what it means to make something 'smart', in cities and beyond.' (Green, 2020). In general, local governments are very much aware of the transformative character of new technologies but are at the same time uncertain how to respond to it. New technology often brings solutions that are highly desirable and hard to resist. New technologies come with high expectations and promises of a better – more convenient – future. It seems like a perfect technical solution that is universally beneficial, therefore public spaces are ideal testbeds. Morozov has labelled this way of thinking as 'solutionism' (Morozov, 2013). It is the power of the promise that makes it very hard to take a step back and reflect on its consequences (Lilley *et al.*, 2004).

smart city

Ethics and the smart city have recently gained momentum (Kitchin, 2019):

> Initiatives include pan-city collectives (such as Cities Coalition for Digital Rights), individual city-based programmes (including Barcelona Digital City, Detroit Digital Justice Coalition, Seattle Community Technology Advisory Board and Amsterdam Tada!), smart city advisory boards (Smart London Board and Smart Dubai AI Ethics Advisory Board) and in-house projects within city government. This turn to ethics is part of a more general engagement across the tech sector, with several AI/big data and ethics initiatives being recently launched by universities (Stanford Institute for Human-Centered Artificial Intelligence and MIT/Harvard's Ethics and Governance of Artificial Intelligence Initiative), research bodies such as the Ada Lovelace Institute, companies (see Google's short-lived Advanced Technology External Advisory Council), governments (the EU High-Level Expert Group on Artificial Intelligence), and stakeholder coalitions (Data for Black Lives and Partnership on AI).

But the question is whether these initiatives really lead to ethical policymaking and, more generally, an ethical city. In the paper 'Ethics Washing and Ethics Bashing' Elettra Bietti (2020) argues that many of these initiatives seem to underestimate the complexity of the ethical issue at hand. It is hard 'to do' ethics based on general guidelines and regulations. Attention for specific cases in specific contexts is necessary: 'Moral philosophy indeed can shed new light on human practices by adding needed perspective, explaining the relationship between technology and other worthy goals, situating technology within the human, the social, the political.'

In this chapter we explore how the municipality of Eindhoven (the Netherlands) is trying to make sense of ethics in innovation projects – or in other words: how it wants to innovate responsibly. First, we explore some ethical concerns that arose at four innovation projects in the municipality of Eindhoven.

8.2 Scope: ethical concerns of smart city projects

In this section we explore the challenges that the municipality of Eindhoven had to face while starting innovation projects for the city. We take three studies from the Rathenau Institute as a starting point. The Rathenau Institute is an independent and influential advisor for the national parliament and government in the fields of science, innovation, technology, and society. The mission statement '… support[s] the formation of public and political opinion on socially relevant aspects of science and technology. It conducts research on this subject and organises debates on science, innovation, and new technology.' (For more information, go to https://www.rathenau.nl).

The Rathenau Institute has recently published several studies that help us to understand the issues concerning digital innovation projects within municipalities, i.e. 'Waardevol digitaliseren' (Van Est *et al.*, 2018), 'Voeten in de Aarde' (2020) and 'Raad weten met Digitalisering' (2020). The studies show how the application of technology determines the society of tomorrow. The choice of innovation projects is based on today's knowledge and skills. All too frequently, the technology causes adverse side effects, in the short or long run. Therefore, we never know exactly *how* the technology will influence the society of tomorrow. However, the government must serve the public interest and therefore needs to shape the application of new technologies, today and tomorrow. The Rathenau Institute proposes a framework that can be helpful when discussing technological development with the public. The framework identifies five perspectives that should be taken into consideration (Van Est *et al.*, 2018):

1. New technologies should be appreciated with the public values and goals as a crucial matter of concern (Public values).
2. It is important to experiment with new technologies in safe spaces (living lab) to gain practical insights (Experimentation).
3. The opportunities/benefits of new technologies should be identified and related to societal transformation (Opportunities).
4. The risks of new technologies should be mitigated and public values should be protected (Risks).
5. It is important to cooperate with stakeholders, inspire citizens to participate and fuel public debate in order to learn from each other (Stakeholder engagement).

local transformation

The main message is that local transformation is a complex endeavour in which many different factors must be weighed up. The Rathenau Institute used the framework to analyse four innovation projects of the municipality of Eindhoven (Karstens *et al.*, 2020). This analysis helped Eindhoven to understand the entanglement of ethical issues in the real world. We will summarise the examples that were studied, to give an impression of the diversity of challenges that a (smart) city has to cope with, i.e. (1) Living lab Stratumseind; (2) Smart Light; (3) Human Driven Data; and (4) Woonconnect (Case 8.1). The observations, which informed our proposal for the city council, are summarised in Table 8.1.

Case 8.1. Eindhoven.

The City of Eindhoven is in the south of the Netherlands. It is the biggest city in the Netherlands after Amsterdam, Rotterdam, The Hague, and Utrecht. It has more than 230,000 in residents, an area of 88 square kilometres and a budget of approximately

937 million euros. Eindhoven is internationally known for its Brainport status: it has developed from an industrial city with companies like DAF (automobiles) and Philips (electronics) to a world-leading research hub with Technical University (TU/e), ASML, Thermo Fisher and NXP. Eindhoven wants to be a liveable, innovative, and creative city so that expats can feel at home and are willing to put their talents to the service of the highly competitive high-tech industry. For more information, go to https://www.eindhoven.nl.

Example 1. Living Lab Stratumseind

In 2013, the municipality of Eindhoven initiated the 'Living Lab Stratumseind'. Stratumseind is a lively street in Eindhoven that contains many pubs and discos. So, during weekends it attracts many young people. Residents complained of noise pollution, altercations, and vandalism. There were about 800 incidents each year. This amounts to approximately 15 incidents per weekend, and represented 90% of nightlife-related incidents in Eindhoven. All parties involved – pub owners, breweries, police, the municipality, and residents – agreed that something had to be done to improve the liveability and safety of the Stratumseind. This was done by gathering information from sensors, for example surveillance cameras for counting people, but also from other sources like police databases, social media, Wi-Fi, air pollution monitors, weather reports, breweries, events, waste, and residents. The data was put together and analysed for patterns. This Living Lab attracted a great deal of interest from other cities, (inter)national politicians and the industry. It is now being upgraded to a city lab.

Example 2. Smart light

In 2012 the municipality of Eindhoven commissioned the report 'Vision and roadmap urban lighting Eindhoven 2030'. This report outlines several future scenarios for a multi-year lighting strategy. This involves much more than just energy saving by using Led lights. It also concerns making the street lights 'smart', so they can include sensors for better traffic flow, increasing well-being on the street and connecting with emergency services. The report recommends that the municipality sets up pilots in cooperation with partners of the quadruple helix (municipality, companies, knowledge institutions and citizens). The project was stopped before the contractual end-date due to several complex ethical and social challenges.

Example 3. Human driven data

The innovation manager of a welfare organisation in Eindhoven wondered whether data-driven applications might help provide a better insight into young people in order to provide more efficient effective psychological and social help to prevent suicides. The manager initiated a meeting with representatives of other welfare organisations, the municipality of Eindhoven and an ICT company. The idea was to generate risk and talent profiles to provide support through welfare work. Young people would not be identified, unless they fell within certain profiles. During the brainstorming for a possible prototype too many concerns about profiling young people were raised, which resulted in the cancellation of the project.

Example 4. WoonConnect

Eindhoven wants to make buildings more sustainable by supporting citizens who are creating more sustainable places of residence. That is why the municipality subsidised the commercial digital platform WoonConnect. WoonConnect works with a configurator that creates 3D models which take data from various sources: e.g. maps from the archives, data from housing corporations, data from the building administration, extracts from the Building Information Model (BIM) and survey data. The survey data contains information about the residents: when they are at home, their energy consumption and so on. The 3D model visualises the possible scenarios and calculates the costs and energy savings so that clients can make an informed choice about adjustments to their home. The project is currently being tested in other municipalities in the Netherlands.

The four examples illustrate the broad spectrum of smart city projects that local governments are involved in. It also shows how difficult it is for local councils to formulate a (political) opinion. How should one weigh up the proposals and findings? Example 1 and Example 2 (Case 8.1) focus on the liveability of public space, Example 3 and 4 (Case 8.1) on improving public service. Table 8.1 shows quite clearly that municipalities are facing a barrage of ethical issues. Many are wicked problems and there are no simple solutions. A guideline or decision tree is not helpful. So, what did the analysis by Rathenau Institute teach us? Here are our observations:

8.2.1 Complexity and diversity of the projects

We noted that all projects have many different stakeholders and actors. In each example at least five stakeholders are involved and many more people. The stakeholders can be potential users, business advisors, technology developers, industry representatives, citizens, civil servants, researchers, regulators, societal organisations and so on. It is very difficult to develop a common vocabulary, and everyone brings with them different expectations, backgrounds, and interests. It is easy to fall in the trap of thinking in dilemmas: the business interests (sell products and services) versus the legal interests (no market disruption). It is not a trade-off between one and the other; there are many other interests at stake at the same time. It is more productive to think in terms of a dynamic network: people working together to achieve certain (public) values while their expectations are constantly under pressure.

Table 8.1. Overview ethical concerns adapted from (Karstens *et al.*, 2020).

Questions	Example 1	Example 2	Example 3	Example 4
1. Public value	The debate between safety, economic opportunities for business, and privacy for the visitors. The main goal was to contribute a safe and attractive entertainment area.	The main goal was to contribute to overall livability, sustainability and at the same time develop new (business) opportunities for innovative services, especially in the areas of safety and quality of life.	The aim of the project was to prevent social and psychological suffering caused by social exclusion.	The aim was to digitize 100,000 homes, The challenge was to ensure fair competition and balancing the public and private interests, while protecting the privacy of the citizens.
2. Experimentation	The living lab cooperates with technology firms to set-up all sorts of sensors, that gather and analyze data. A space was setup in de area to demonstrate the many technological possibilities.	Several areas were identified for experiments. The installation of the grid and led lights took some time, but its result was that 2,700 LED lamps were installed. This saves approximately 400,000 kWh energy per year. However, the experiments with new services were stopped early.	Data projects in youth care involve vulnerable young people and highly privacy-sensitive data. So it was not possible to experiment with real data. The practical and legal challenges were so complicated that the idea was not developed further.	It was difficult to take advantage of the opportunities that digitalization of homes has to offers because the building companies were not involved by the development of scenario's and citizens preferred direct personal contact instead of a computer program.
3. Opportunities	Living lab was a testing environment for sensor technology and was quite helpful for preventing street violence. It experimented with nudging technologies.	The project also initiated to develop new commercial products and services. However, this proved difficult to exploit so the economic feasibility of the project was undermined. The experiments were too limited in scale for the business partners. Eindhoven had to operate very careful to prevent vendor lock-in.	Because there was not a system developed it is quite impossible to reflect on the promises made and the outcomes delivered.	The idea is promising home-owners can receive a tailormade advise as to make their home more sustainable. However the revenue model was that the home-owner must pre-pay for the visualization (and advice). Building companies did not experience enough incentives to take assignments with WoonConnect, they favored their own channels.

Table 8.1. Continued.

4. Risks	Risks were very much focused on the privacy of citizens. Risks were mitigated by developing sensors that only recorded people as dot. Data was deleted and only patterns saved. However, some politicians and citizens were very concerned about the application of this technology.	The municipality owns the infrastructure; however, the stewardship of the data and the lighting lies with the consortium partners. The municipality has stipulated that it can always intervene by irresponsible data management and that the data should be open and the applications open source.	The project team intended to store and anonymize the data securely by aggregating it to group-level. The idea was to prevent stigmatization by being transparent about the development process of the profiles. Also it was acknowledged that the results should be tested extensively.	The parties agreed to focus on data collection on the technical aspects of the house, data was only made public in aggregated form and many measures to protect privacy were implemented. However, data of a house can be privacy-sensitive especially when combined with other datasets.
5. Stakeholder engagement	Local residents, police and pub owners were actively engaged in this project. However, it was very hard to overcome 'the valley of death' to implement projects with the police.	Cooperation in the quadruple helix has been much more difficult than was expected. Partners had often opposed interests. Economic different expectations of revenue models, contractual clarity, and control over collected data. Many legal issues were very difficult to solve. It didn't help that the project did not take citizen participation as an integral part of innovation processes, so the partners were not able to make their contribution to the public values explicit.	The concept of 'social exclusion' must be understood by everyone the same way to be able to develop a certain technological solution. The definition has also great impact on the legal framework that can be evoked for processing data.	Public-private partnerships make operate government very careful. Market disruption due to possible advantage of WoonConnect is something that government want to avoid. While at the same time the private parties wanted to move fast. So many efforts have been done to share data publicly to ensure an equal level playing field. The role of the role of the Land Registry is also worth mentioning: the registry wants to act as a data portal for the building sector that makes that public tasks and economic activities cannot be separated. The home-owners and citizens were not systematically engaged in this project.

8.2.2 Practical implications

We noted that most of the efforts are invested in 'making it work'. Technology developers are working hard to build and test technology. Often prototypes are produced and tested in 'real' environments. Implementation and financial plans are developed but are often not part of the innovation project. It is easy for a project to fall into the 'valley of death' in which funding for continuation has to come from sources other than those for funding the innovation (Murphy and Edwards, 2003). For Examples 2 and 4 (Case 8.1) it was very difficult to scale up due to questions about implementation and funding, rather than technical issues.

8.2.3 Legal challenges

We noted that, during the development of new solutions, many struggled with legal and regulatory issues. In Examples 1 and 3 (Case 8.1) privacy has to be an important part of the solution. But this requires sophisticated legal knowledge and clear ownership and responsibilities. In the innovation projects this was not always as clear as it should have been from a legal or regulatory point of view. Many innovation projects start with a vague idea about trying something out; its aim and desirable outcomes can be quite open-ended. This opens an interpretative space in which the stakeholders can cooperate with the project without a narrow assignment. Different stakeholders can bring in their ideas and search together for the 'best solution'. But in the end an innovation project doesn't have a clear focus and there is no knowing when it has achieved its goal.

8.2.4 Ethics as dynamic concept

We noted that in all projects the promise of new technology plays an important role: it sketches a new world where an important societal problem is solved. In the examples: (1) more safety in the nightlife and leisure spaces for the public, (2) better liveability for citizens, (3) earlier psychological support for youngsters and (4) more sustainable housing. Thinking these aims over, it becomes immediately clear that they are quite ambitious, which can mean quite different things to different people. The outcomes are formulated in such a way that no one can disagree with them (Van Lente, 1993). The project contributes to an important public value and proposes a suitable solution with the use of the most modern technologies. The solutions are sought in 'technological fixes' – a tool or a system – that can be developed by the project team. The same project can be a 'shiny star' for one and a 'disaster' for another. It almost seems that the success of a smart city project is in the eye of the beholder. None of the projects seemed to have discussed the (political) desirability of a certain broad aim. The public debate about the technology and its possible consequences were seldom made explicit and written down. No advisory and policy papers on that matter are reported. It taught us that it is quite important to be able to learn lessons from earlier projects and, therefore, ethical considerations and the outcomes should be documented.

The observations shaped our perspective on the development of our proposal on ethics in local government. Ethics only has meaning in specific situations at specific moments in time. It is not something that 'must be done' once; it requires constant attention and is mainly about finding common ground on which to discuss innovations. We define ethics in the context of a smart city as accepting responsibility for the choices one makes. It requires continuous reflection on the aims of smart city projects so that technology does not become an end, but remains a means to an end, and contributes to public values such as safety, quality of life, social inclusion, and sustainability. Nowadays, this view is also known as the 'Ethical City' (Barrett *et al.*, 2021). Suffice to say, the reports by the Rathenau Institute helped the municipality understand that ethics in local politics is about bringing together different perspectives. Technology is not neutral; it has certain effects on society. In the next section we discuss how these insights were translated into a proposal for the city council.

8.3 Our working method: workshops and brainstorms

In the past the municipality of Eindhoven was involved in the development of the Principles of the Smart City (VNG, 2018), Open Data principles (Eindhoven, 2017a) and the Internet of Things (IoT) charter (Eindhoven, 2017b). These notions/initiatives/standards were included in the policy note on Digitalization of the City (Eindhoven, 2019). In several debates the city council noted that it was difficult to take the debate about digitalisation further than the financial and practical matters in question, so it requested a proposal on how to safeguard public values in digitalisation projects; it also asked for the development of an ethical framework, an external ethical committee and a method for integrating ethics in the organisation and processes (Eindhoven, 2020). In this section we explain our working method for developing the proposal. But before we can do that, we need to elaborate on the different roles and responsibilities within a municipality (Table 8.2).

The multitude of actors in local government makes a multi-layered approach desirable: we made a distinction between council members, aldermen, heads of departments and civil servants. For each group we developed specific activities (see Table 8.3 for an overview of activities).

Table 8.2. Roles in a municipality in the Netherlands.

The **Municipal Council** (Dutch: de gemeenteraad) is the elected assembly of the municipality. Its main role is laying down the guidelines for the policy of the municipal executive and exercising control over its execution by the mayor and aldermen. Members of the council are elected by the population every four years. It is a part-time responsibility because they are not paid as full-time politicians. (In Eindhoven there are 45 members and 13 fractions.)
Mayor and aldermen (Dutch: College B&W) is the executive board of a municipality. It plays a central role in municipal politics. The aldermen are appointed by the Municipal council after the elections. The mayor is appointed by the national government. (In Eindhoven there are beside the mayor 6 aldermen.)
The **civil servants** support the executive board by the execution of the responsibilities of local government, like: Public services, Urban planning, Traffic and transport, Primary and secondary education, Social affairs and Regional and local Economic development. The head of the civil organization is the **municipal secretary** (Dutch: gemeentesecretaris). She heads the department managers; in Eindhoven the departments are called 'sectoren' and there are 14 of them. They meet up in board meetings.

Table 8.3. Outline of activities.[1]

	Activity?	Who?	When?	Sessions?
1.	Kick-off: webinar about the reports Rathenau Institute with Q&A	Council members and head of departments	Nov 2020	1
2.	Meetings	Alderman responsible for this proposal	Nov-May 2021	4
3.	Workshops	Special working group of council members	Dec-Apr 2021	3
4.	Brainstorm workshops	Civil servants of three domains: safety/security, social affairs and urban planning	Nov-Dec 2020	3
5.	Validation workshops	Civil servants of three domains: safety/security, social affairs and urban planning	Feb 2021	3
6.	Supervisory committee	Head of departments	Dec-March 2021	3
7.	Dissemination workshops	Civil servants	March 2021	2
8.	Paper review	All relevant departments (7)	Apr 2021	1
9.	Presentation for the board	Head of departments	Jul 2021	1
10.	Proposal accepted	Executive board of a municipality	May 2021	1
11.	Debates in city council	Council members	Sep 2021	3
12.	Meetings with other municipalities/GO's	VNG, Utrecht, Enschede, Hilversum, Amsterdam, etc.	Nov-Jul 2021	6

[1] Due to the measures against the Corona-virus all activities took place in a virtual meeting environment (ms-teams).

We started with a webinar to inform the city council members of the findings of the Rathenau Institute.[4] In this section we summarised our observations. The webinar contributed to a shared vocabulary and a common understanding of the problem at hand.

We then set up a working group on digitalisation and ethics with the city council, a supervisory committee with several departmental managers within the administration, and a periodical meeting with the alderman responsible for this subject. The working group, supervisory committee and responsible alderman received periodical updates on the team's findings.

The crucial ideas were developed with the civil servants of the three domains of the municipality: safety/security, social affairs, and urban planning. This happened in two steps:

1. For each domain we organised a brainstorming workshop with the so-called strategists, i.e. the civil servants responsible for policy development within a specific domain. The strategists were asked how responsible innovation (ethics) could help their domain to do their work better.
2. The second step was setting up validation workshops with the same strategists and asking them to bring their own problem. In the meeting we tried one of the methods to discuss the matter at hand (see Table 8.4 for an overview of methods). We will elaborate on the findings.

Brainstorming Workshops

Take-aways from the Brainstorming Workshops:

- The concepts of digitalisation and ethics are both quite unclear and open to interpretation. Ethics and privacy laws are often confused with one another: in what way are they different? Ethics is often understood as a negative concept: it restricts and smothers innovation.
- There are too many reports, recommendations, guidelines, regulations, and policies on ethics; it is very difficult to oversee the field and understand its consequences. Ethics should not become a matter of completing a checklist but should be about having a real conversation about the public values that a project wants to achieve and how it does so.
- It is important to help people to ask ethical questions and guide them to find informed answers. There is a need for a supportive approach to ethics: one that helps people to understand ethics better and helps them to come up with better solutions.

[4] The researchers Bart Karstens and Djurre Das were so kind as to summarise the reports and discuss with the participants what it could mean for the city.

- Ethics is the responsibility of the project manager and the management; it is not something that can be outsourced to an (external) ethical committee. Ethics should not be institutionalised to a (new) department; it is everyone's responsibility and should be an integral part of other considerations (financial, legal, feasible, technical, etc.).
- An external committee is necessary in order to organise a critical culture and should also support awareness for ethical issues and offer guidance on how to develop knowledge about ethics.

Table 8.4. Overview of ethical conversation methods.[1]

DEDA (Franzke and Schäfer, 2021)	Data Ethics Canvas (Open Data Institute, 2019)	Guided ethics (Verbeek and Tijink, 2019)	Moral deliberation (Karssing, 2011; Overlegorgaan Fysieke Leefomgeving, 2021)
The Data Ethics Decision Aid is developed by the Utrecht Data School (University of Utrecht) in corporation with data analyst from the City of Utrecht.	The Data Ethics canvas was launched a couple of years ago as one of the first data ethical reflection tools and recently got an update (2021). The tool is developed by The Open Data Institute in the UK.	Guided ethics is a new approach developed by ECP a platform of technology companies. The ethics is not primarily focused on the question whether a technology is acceptable or not, but on the question of whether and under what conditions a technology can be given a responsible place in society.	Moral deliberation is a method that is not specifically designed for digital or data ethical kind of issues. It is used to dissect moral dilemmas and issues where integrity is in play in order to come to a thought-out choice.
The model guides the team through several themes: 1. Data questions 2. General questions 3. Values of organisations It consists of a fixed set of eleven questions. See also: dataschool.nl/deda/	The model guides the team through several themes: 1. Know your data 2. Explore your impact 3. Plan your engagement 4. Integrate into processes It consists of a flexible set of twelve questions which can be selected. See also: theodi.org	The model has an open structure and consists of four phases: 1. Analysis of casus (technology-in-context) 2. Dialogue about actors, effecs and values 3. Possible actions (ethics by design) See also: ecp.nl/project/aanpak-begeleidingsethiek/	Possible steps of moral deliberation model: 1. What decision should I make? 2. Who are the stakeholders? 3. What do the professional rules, company codes or other binding regulations say? 4. Which arguments could I bring in? 5. What's the conclusion? 6. Will I act upon the conclusion?

[1] The four methods mentioned are only a small set of existing guidelines, for example more than 173 guidelines in AI and Ethics can be found in the AI Ethics Guidelines Global Inventory, see https://inventory.algorithmwatch.org.

Validation Workshops

Take-aways from the Validation Workshops:

- For each workshop a different case was discussed. The social domain presented the idea of a machine learning lab that would be able to identify people before they became homeless. The safety/security domain presented an idea about a system that would help to support people with psychological problems. Urban Planning presented a smart traffic control system.
- The problems that were submitted required a lot of work to specify the ambition, means, and aims. On the whole, the ambitions were very noble, but from a practical point of view, the way it should work and what it should do was often unclear.
- We tried out certain methods for discussing ethical issues systematically. This shaped the ethical conversation. However, we noted that the focus of the conversation tended to shift more readily to the feasibility of the method and less to the problem in hand. In Table 8.4 we provide an overview of the method with which we experimented in different settings (not only the workshops). We concluded that the methods were very helpful for understanding the ethical dimensions of the case study. Some of the methods are quite structured, other offering more leeway in their application. What they all have in common is the requirement for an experienced mentor and a knowledgeable group.

After the brainstorming and validation workshops we decided to focus our efforts on the question: how do we strengthen the ethical conversation in the organisation? Most participants argued that they would like some more guidance and insight, but not in an 'enforced' regulatory way. In the next paragraph we will present how we want to achieve that.

8.4 Result: ethical structure

As a result of the brainstorming sessions and validation workshops we developed a structure that should help identify ethical concerns at an early stage of a project and allow for appropriate measures to be taken in time. The challenge was how to bring together the ethical framework, the external ethical committee and the organisational process – as requested by the city council (Eindhoven, 2020) – and to develop it in such a way that it met the wishes of a multitude of actors. In this section we describe our proposal, which was approved by the city council on 22 September 2021.

In Figure 8.1 the three elements – ethical framework, ethical committee, and working method – are presented together. We developed a list of ethical values that can be discussed by the ethical team when an innovation project

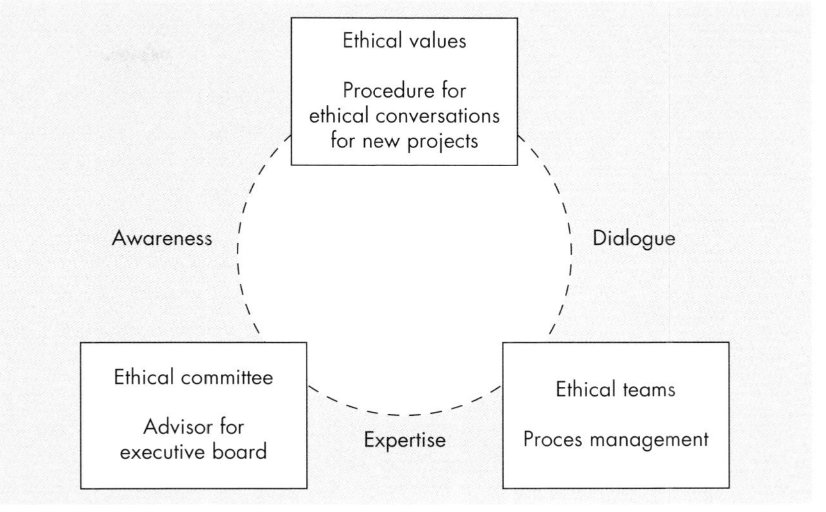

Figure 8.1. Ethical structure of Eindhoven.

is prepared. This deliberation results in a report for the project leader, which will be shared with the ethical committee. The ethical committee will study the report and use its findings to build a common understanding of ethical values. The approach does not enforce project leaders to 'get an ethical advice' and the approach does not give a 'go or no-go' recommendation. Its purpose is to create more awareness and consideration of ethical issues by the city council, aldermen and civil servants. We will elaborate further on each element.

ethical framework

Ethical framework. The ethical framework aims to relate public values with the necessity, means and purpose of the proposal. It does that by providing a list of 'topics' to guide the conversation about the proposed innovation. The list of seven public values is based on the 'Waardevol Digitaliseren' report provided by the Rathenau Institute (Van Est *et al.*, 2018) and discussed in the workshops.

List of public values:

- Autonomy: the people of Eindhoven are free to lead their life as they choose.
- People at the centre: the people of Eindhoven are not treated as a dataset but as human beings.
- Privacy: the privacy of the people of Eindhoven is not affected; the Municipality borrows the digital data from its citizens.
- Safety: the people of Eindhoven feel safe in the city and safe within the Municipality.

- In control: the people of Eindhoven can trust that the Municipality is in control of the technology that is used.
- Justice: the people of Eindhoven can trust that the Municipality will choose a technological solution that is beneficial for everyone.
- Sustainability: the Municipality opts for a solution that is beneficial for the environment and the climate.

The city council was requested to approve the list of public values to guide the internal conversations on ethics. The values are not 'fixed'. It depends on the matter in hand which values are relevant, or which ones should be added to the conversation. It is not meant as an assessment tool and therefore we talk about 'list of public values'. The ethical committee will be requested to update the list annually and help improve it. In that way we set up a continuous loop to introduce ethical aspects into the conversation.

ethical teams

Ethical teams. Many of the participants in the workshops requested some sort of guidance to structure the conversation on ethics a little; so we developed a framework for the conversation. The ethical framework should not be seen as a way of following steps and procedures and ticking boxes, but as a guided dialogue in which central questions lead to an in-depth discussion that provides input for actions and deliberations. The framework consists of five 'guiding' questions:

1. What is the matter in hand? The first step is to gain a good understanding of the problem. In this step it is necessary to perform a proper actor analysis and get a clear idea of the interests of every actor. What are the goals, and how will those goals be achieved by the proposed means?
2. Can we do it? When a project is proposed, it is often assumed that a certain opportunity can be realised. Technology usually plays a key role. But we want to gain more insight into the possibilities and limitations of the technology. In this step we look a bit closer at the technology: How does it work? What dependencies can be identified? Do we understand the technology? How does it fit into the infrastructure? What are the technological alternatives?
3. Is it (legally) allowed? New technology is also always bound by legal frameworks, agreements, and covenants. What laws and regulations should be considered? Also, can there be agreements with other parties?
4. Is it desirable? The desirability question is at the core of the ethical conversation. It is important to realise that the answers to 2 and 3 will influence the desirability of the project and should also be discussed in that context. In this step the interests of the stakeholders are related to the aims of the proposal and the public values it would like to achieve.

5. How can it be justified and what should we do? The conclusions of the conversation should be written down in a practical recommendation for the project manager. What are the issues that the project manager should consider while developing the proposal?

The conversation is initiated by the coordinator who manages a pool of civil servants that have additional training in ethics. Ideally, the pool consists of one or two representatives from each department. For each proposal that is submitted a group of four people is selected from the (larger) pool. The group – which we call the ethical team – discuss the proposal with the project manager in charge. The five steps and list of public values are used as the basis for this conversation. This process can be understood as a kind of peer review. The conclusions will be put in writing and submitted to the ethical committee. The overall aim is to make civil servants more aware of ethical issues in complex innovation projects.

ethical committee

Ethical committee. The Ethical Committee meets four times a year. Its task is to provide the mayor, aldermen and council with solicited and unsolicited advice on pressing ethical matters and contribute to the ethical awareness of the organisation. The committee receives copies of the recommendation of the ethical teams and can therefore validate the quality of the advice and formulate recommendations for the development of expertise. In that way it helps to develop 'moresprudence': a systematic body of knowledge on ethical decisions within the city (Kanne and Grootoonk, 2014).

8.5 Discussion: way ahead

The premise of this paper is the observation that the unprecedented development of new technologies has led to 'moral confusion' (Wernaart, 2021). People feel unable to influence the design of new technologies and accept it as some sort of natural force. From a policymaker's point of view this is unacceptable: people have agency and are responsible for the design of our world. New technology is not only developed by humans, but also distributed, regulated, used, and modified by human beings. This paper is about our experiences of how local government can try to get some sort of 'grip' on new technology.

We expect the approach to be helpful in asking the right questions in order to formulate an ethical position. The ethical reflection consists of three questions that lead to answers, all of which have ethical implications: Can we do it? Is it allowed? Is it desirable? An important part of this ethical reflection is zooming out. Often, ethical tools focus only on the technical solution that is presented. We try to avoid solutionism and tech goggles (Green, 2020) by defining the context in which this problem is defined and how it is presented.

We think that the approach contributes in four ways to the ethical city:

1. It is a conversation starter on ethical issues, and it helps to articulate the aims of the planned innovation. This might give an innovation project more focus, so it becomes clear when it has achieved its goal.
2. Technology is not understood as something that is inevitable. The approach makes every participant (partly) responsible for the technology development. It should help to identify alternatives for action and guide new technology development in a positive direction.
3. By taking the time to discuss ethical issues and public values – and writing it down – the approach might be helpful in disentangling the moral confusion that many experience.
4. The approach is not a magic wand that provides an answer to the question: what is the ethical thing to do? It instead should be helpful as a perspective on innovation in a political context. In the end, ethics is taking responsibility for the decisions that government makes.

We expect that ethical aspects will become an integral part of innovation projects, in a similar way to security and privacy. This is known as responsible innovation. The national government has recently published a code of conduct for good digital governance (Meijer and Ruijer, 2021) in which public values are the foundation. Many public institutions are advocating more attention for responsible innovation (Algemene Rekenkamer, 2021; Ombudsman, 2021; Raad voor het Openbaar Bestuur, 2021).

The approach presented here has recently been approved by the city council of Eindhoven. During the deliberations the council was very outspoken about citizen involvement and participation, and the idea will be developed further in which the 'parliament of things' (Latour, 2020) might be worth pursuing.

References

Achterhuis, H. (ed.), 1997. Van stoommachine tot cyborg: denken over techniek in de nieuwe wereld. Ambo, Amsterdam, the Netherlands.

Algemene Rekenkamer, 2021. Aandacht voor algoritmes, Algemene Rekenkamer, the Hague, the Netherlands.

Barrett, B., Horne, R. and Fien, J., 2021. Ethical cities. Routledge, London, UK.

Bietti, E., 2020. From ethics washing to ethics bashing: a view on tech ethics from within moral philosophy. FAT* '20: Proceedings of the 2020 Conference on Fairness, Accountability, and Transparency, pp. 210-219.

Bijker, W.E., 1995. Democratisering van de technologische cultuur (Inaugurele Rede). University of Maastricht, Maastricht, the Netherlands.

Eindhoven, 2017a. Open data principes. Available at: https://data.eindhoven.nl/explore/dataset/eindhoven-open-data-principes/information/?flg=nl.

Eindhoven, 2017b. Smart society charter: IoT archtecture principles & guidelines. Available at: https://data.eindhoven.nl/explore/dataset/eindhoven-smart-society-iot-charter/information/?flg=nl.

Eindhoven, 2019. Nota digitalisering van de stad. Available at: https://eindhoven.parlaeus.nl/user/bdocument/action=showannex/gdb=687/Bijlage_1_-_Nota_Digitalisering_van_de_stad.pdf.

Eindhoven, 2020. Commissienotitie ethiek en digitalisering. Available at: https://raadsinformatie.eindhoven.nl/user/showdoc/action=view/id=9531/type=pdf/Commissienotitie_Ethiek_en_digitalisering.pdf.

Franzke, A.S. and Schäfer, M.T., 2021. Data ethics decision aid (DEDA): a dialogical framework for ethical inquiry of AI and data projects in the Netherlands. Ethics and Information Technology. https://doi.org/10.1007/s10676-020-09577-5

Green, B., 2020. The smart enough city, putting technology in its place to reclaim our urban future. MIT Press, Cambridge, MA, USA.

Kanne, M. and Grootoonk, E.E., 2014. Moresprudentie in de praktijk. Eburon, Delft, the Netherlands.

Karssing, E.D., 2011. De oplossing is het probleem niet!: reflecties op ethiek, integriteit en compliance. Tijdschrift voor Compliance. Nyenrode Business Universiteit, Cappelle a/d IJssel, the Netherlands.

Karstens, B., Kool, L. and Van Est, R., 2020. Voeten in de aarde. datagestuurde innovatie in de stad. Rathenau Instituut, the Hague, the Netherlands.

Kitchin, R., 2019. The ethics of smart cities, Maynooth University, Maynooth, Ireland.

Latour, B., 2005. Reassembling the social: an introduction to actor-network-theory. Oxford University Press, New York, NY, USA, 301 pp.

Latour, B., 2020. Het Parlement van de dingen: over Gaia en de representatie van niet-mensen. Boom, Amsterdam, the Netherlands.

Lilley, S., Lightfoot, G. and Amaral, P., 2004. Representing organization: knowledge, management, and the information age. Oxford University Press, Oxford, UK, 204 pp.

Meijer, A. and Ruijer, E., 2021. Code goed digitaal openbaar bestuur (CODIO): borgen van waarden bij de digitalisering van openbaar bestuur, Universiteit Utrecht, Utrecht, the Netherlands.

Mol, A., 2002. The body multiple: ontology in medical practice. Science and cultural theory. Duke University Press, Durham, NC, USA.

Morozov, E., 2013. To save everything, click here: technology, solutionism and the urge to fix problems that don't exist. Penguin Books, London, UK.

Murphy, L.M. and Edwards, P.L., 2003. Bridging the valley of death: transitioning from public to private sector financing. NREL, Golden, CO, USA.

Ombudsman, N., 2021. Een burger is geen dataset: ombudsvisie op behoorlijk gebruik van dat en algoritmen door de overheid, Nationale ombudsman, the Hague, the Netherlands.

Open Data Institute, 2019. The data ethics canvas. Available at: https://theodi.org/article/data-ethics-canvas/.

Overlegorgaan Fysieke Leefomgeving, 2021. Rapport ethiek en digitalisering: Bezint eer ge begint, Overlegorgaan Fysieke Leefomgeving (OFL), the Hague, the Netherlands.

Raad voor het Openbaar Bestuur, 2021. Sturen of gestuurd worden? Over de legitimiteit van sturen met data, Raad voor het Openbaar Bestuur (ROB), the Hague, the Netherlands.

Schot, J.W. and Rip, A., 1996. The past and future of constructive technology assessment. Technological Forecasting and Social Change 54: 251-268.

Tenner, E., 1996. Why things bite back: technology and the revenge of unintended consequences. Knopf, New York, NY, USA, 346 pp.

Van Est, Q., De Bakker, E., Van den Broek, J., Deuten, J., Diederen, P., Van Keulen, I., Korthagen, I. and Voncken, H., 2018. Waardevol digitaliseren: hoe lokale bestuurders vanuit publiek perspectief mee kunnen doen aan het 'technologiespel'. Rathenau, the Hague, the Netherlands.

Van Lente, H., 1993. Promising technology: the dynamics of expectations in technological development. Eburon, Delft, the Netherlands, 260 pp.

Verbeek, P.-P., 2000. De daadkracht der dingen: over techniek, filosofie en vormgeving. Boom, Amsterdam, the Netherlands, 302 pp.

Verbeek, P.P. and Tijink, D., 2019. Aanpak beleidingsethiek: een dialoog over technologie met handelingsperspectief, ECP, the Hague, the Netherlands.

VNG, 2018. Principles for the digital city. the Hague, the Netherlands. Available at: https://vng.nl/files/vng/nieuws_attachments/2018/principles-for-the-digital-city_20181025.pdf.

Wernaart, B., 2021. Developing a roadmap for the moral programming of smart technology. Technology in Society 64: 101466. https://doi.org/10.1016/j.techsoc.2020.101466

Winner, L., 1986. Do artefacts have politics? The whale and the reactor: a search for limits in an age of high technology. University of Chicago, Chicago, IL, USA, pp. 19-39.

Part 3

APPLIED ETHICS IN TECHNOLOGY

9. Journalism by the dashboard light

Introduce rear-view mirror strategies to newsrooms and journalism education

Danielle J.A.M. Arets*, Jessy de Cooker, Jeroen van den Nieuwenhof and Hasib Shirbaz

Fontys Hogeschool Journalistiek, Readership Journalism & Responsible Innovation, Prof. Gimbrerelaan 16, 5037 EK Tilburg, the Netherlands; d.arets@fontys.nl

Abstract

Audience metrics have penetrated the newsroom. Web analytics can track real-time data on any digital news item and get information about audience appreciation and engagement. Audience metrics, in particular, are considered as a means to enhance journalistic quality as journalists can better respond to user desires. Although a greater focus on audience-centred journalism is a commendable development, it is questionable whether the metrics-driven focus is helpful in this respect. Audience desires cannot be captured in metrics alone. To prevent newsrooms from being blinded by the dashboard lights, we argue that engineers and ux-designers must build friction in order to stimulate responsible interactions.

Key concepts

- Audience metrics are the data used to assess the extent to which news matches the wishes and needs of the news consumer
- User-centred journalism places the news user's desires at the centre of journalistic work
- Data journalism embodies journalistic processes in which structured (large amounts of) data are used at the core of storytelling
- Moral design strategies are strategies on an organisational level that structurally embed a values-based design in the organisational processes

9.1 Introduction

measurable journalism

Whenever there are readers, there are numbers. In the analogue newsroom, surveys were conducted to examine whether the news sufficiently matched the wishes and needs of the news consumer. Nowadays, audience metrics have matured, resulting in measurable journalism (Carlson, 2018): digital platforms can provide real-time, quantitative data about audience consumption practices. The appreciation of an article becomes crystal clear from the number of clicks, the minutes people spend reading it and if they share it with others. Newsrooms respond to data-driven insights by writing articles with a high reach, high rating, and high impact.

To what extent is this focus on audience metrics desirable? Are analytics the answer to more human-centred news? Where most newsrooms consider audience metrics as a quality impulse – because insights from metrics ensure that audiences are taken seriously – academic studies indicate concerns. Some studies state that the quality of the news decreases; by focusing on what the audience needs, sensational news and entertainment – or soft news – would take the upper hand (Lehman-Wilzig and Seletzky, 2010). Other scholars indicate that the metrics-driven focus in newsrooms results in a market-driven logic, which is harmful for journalistic practice (Baker, 2002). The news culture would also be endangered as metrics would drive competition between journalists (Petre, 2015). There is a risk that performances – and salaries – will be dependent on metrics. There are already examples of this: the Daily Telegraph plans to link journalists' pay to article popularity (Guardian, 2021).

To get more clarity about whether audience metrics enhance or erode the quality of journalistic offerings and harm the journalistic culture, we examined how metrics are being used in Dutch newsrooms. In our research project 'Journalism beyond the Dashboard Light (2001-2002) we interviewed 12 journalists and 3 chief editors of large and small Dutch media outlets. We also conducted a survey and organised focus sessions with journalists, journalism students and news consumers.

We came to understand that audience metrics in Dutch newsrooms are treated with care; journalists are not obsessed with metrics and don't steer blindly with data-driven insights. Chief editors and journalists mainly use audience metrics as input to enhance the quality of their output.

We also noticed that most of the interviewees have a critical attitude towards 'numbers', considering analytics as an indicator, not as a guideline. In fact, all respondents – journalists and chief editors alike – point to their own moral compass as a guiding principle. Their inner compass, that they define as a gut feeling gained through training and experience, takes the lead when it comes to making final decisions about what to put on the front page, which point of view or writing style to use, and which headline to use.

9.2 Audience data as proxies

We noticed, however, less of a critical attitude when it comes to the accurate representation of the audience in numbers. Most of our interviewees feel audience metrics are helpful for better responding to user needs and tapping into their engagement. One chief editor mentioned that 'news consumers can finally talk back' and 'that audience metrics help to increase quality and appreciation'. One consulted chief editor sees a confirmation of the latter from the higher number of subscriptions since a self-developed metrics system was introduced in their newsroom.

There is a strong belief in the adequate representation of audiences in metrics, both when it comes to their preferences and their engagement. Recent research by Groot Kormelink (2019), however, shows that metrics don't always get it right. A click or headline does not always mean that an article will be read, and conversely, an article that gets hardly any clicks does not mean that news consumers don't find it interesting. Audience data are proxies for the audience, not the audience itself (Carlson, 2018). Metrics are constrained and partial as audiences can only respond to what they are offered. The desires of the minority can easily become neglected.

There is a tension between what audiences want and what they need; journalists have to deal with this tension by offering both: offering content that aligns with audience wishes as well as offering information for 'the greater good'. This has always been a balancing act and audience metrics provide the next challenge as they give partial insights into what the public is interested in but don't provide insights into the public interest (Tandoc and Thomas, 2015).

9.3 Lipmann and Dewey revisited

We cannot avoid the Lippmann-Dewey debate in this regard. The central issue in this debate between philosopher John Dewey and journalist Walter Lippmann in 1920 in the US is whether journalism should contribute to the democratic, deliberative quality of the public or conform to the wishes of that same public (Marres, 2005: 179; Steel, 2017).

Lippmann argues that an informed public opinion is not possible as the world is too complex and complicated, and citizens' attention spans are too fragmented. He is critical about his own profession of journalism. In his view, journalists are not able to bring adequate knowledge to the public.

Dewey, however, believes we are not isolated individuals. He emphasises the importance of social knowledge. Journalism – in the eyes of Dewey – can play a key role in offering 'a subtle, delicate vivid, and responsive art of communication' (Dewey, 1927: 184).

The polemic of this discussion has surfaced constantly in the 20th and 21st centuries (Schudson, 2008; Steel, 2017) and brings us back to the question of whether journalists can play their communitarian role with the emerging role of audience metrics. This question becomes especially relevant with the rise of artificial intelligence and automated decision-making procedures. How can we avoid metrics becoming too dominant a logic, and overpowering the civic role of journalism?

MacGregor indicates that journalists compare their own beliefs with data and 'no longer implicitly trust themselves' (MacGregor, 2007: 296). In our study we came across two incidents where the compass was challenged by the metrics. One journalist mentioned that a boring headline performed better than anticipated, but he decided to trust the metrics. A chief editor mentioned that he used insights derived from the metrics to give directions to a journalist for the best angle to use in a piece. Both times, however, the interviewees made a very conscious choice to use audience metrics and they were able to properly argue their reasons for doing so.

Nevertheless, this is something worth paying attention to. A study by Ferrer-Conill (2018) showed the inability of journalists to precisely define engagement. He says that understanding the audience solely through the prism of metrics provides a skewed picture of the audience.

It is particularly important to have more discussions around this considering the emerging editorial roles focused on navigating audience data and making sense of audience behaviour (Ferrer-Conill, 2018). Metrics introduce new routes and routines in the newsroom, and perceptions of the journalistic role change accordingly. It is crucial to discuss and reflect on these changes.

9.4 Training and tool criticism

The conversation about the role of metrics within newsrooms has just started, and in our study we stumbled upon varies strategies for dealing with metrics. Some newsrooms promote a rather loose culture around it. It's up to the individual journalist or chief editor how the metrics are used. Other newsrooms notify their journalists with e-mail updates, but only do this once or twice a week. One outlet offers training courses on how to deal with audience metrics. The journalist responsible for the training is very aware that a culture and training around audience metrics is key. 'Journalists need to be aware what they can and can't read from the data and that insights from metrics are sometimes not enough.'

Petre (2015) warns that chief editors should safeguard a healthy editorial culture; they should be careful not to come under the spell of data. The chief editors in our study are very much aware of this and do not manage according to metrics alone. Nevertheless, in the newsrooms where we conducted our interviews, there is little training offered to journalists working with audience metrics. This corresponds to other studies indicating that journalists generally receive a moderate amount of training. Formal instructions about the use of metrics are not given (Zamith *et al.*, 2020).

As metrics will become more present in newsrooms, instructions about the use of audience metrics (e.g. frequency of use), the culture around metrics (how to avoid metrics becoming a tool for competition) and data literacy (what can and can't be read from the dashboard) are important.

A more general critical evaluation of the dashboards is also needed. How do dashboards affect the idea of 'users'? What is the impact on the journalistic process? These are important questions to take into consideration. A look at the way these new tools and digital services impact knowledge production – what Van Es *et al.* refer to as tool criticism (Van Es *et al.*, 2018) – is recommended as our choices are determined by the design of the products and services we use (Thaler and Sunstein, 2008).

9.5 Value sensitive design: designing friction in the system

friction

Designers and engineers of data dashboards have a responsibility to stimulate this debate too. We think it could be very helpful to design 'friction' into these systems. With friction we mean interactions that require a bit of effort or even a degree of suffering for the user (Benford *et al.*, 2012).

Friction could take various shapes like subtle messages to remind users that metrics are a partial representation of the audience, or alerts when reaching a maximum amount of screen time, thus preventing overexposure to dashboard lights. Pop-up messages with critical questions around metrics could be helpful as these kinds of messages can prompt moments of reflection and more mindful interaction (Cox *et al.*, 2016) or even provoke interpretation rather than information (Sengers and Gaver, 2006).

Designing these frictions does however require careful ethical consideration (Cox *et al.*, 2016) as historical experiments have taught us that this process can also result in painful or unpleasant experiences; think of Stanley Milgram's experiment about obeying orders. When designing friction we aim for designs that support critical questioning of the tools and systems we use as well as our interactions with them.

As current design processes often eliminate friction and offer us smooth designs, we turn to designers and engineers to think of the moral and ethical implications of everything they design and make. (Arets and Dujardain, 2019). We call for value-sensitive design (Van den Hoven, 2007) where journalistic values – like the aforementioned civic role – are taken into account during the design process. In the case of dashboards, we think it is meaningful to design some rear-mirror strategies in the system, to remind us that driving solely by the dashboard light is a risky practice.

References

Arets, D. and Dujardain, A., 2019. Grenzen aan disruptie? Een pleidooi voor Maatschappelijk Verantwoord Ontwerpen. Available at: https://vance.nl/grenzen-aan-disruptie-pleidooi-maatschappelijk-verantwoord-ontwerpen/

Baker, C.E.,. 2002. Media, markets, and democracy. Cambridge University Press, New York, NY, USA.

Benford, S., Greenhalgh, C., Giannachi, G., Walker, B., Marshall, J. and Rodden, T., 2015. Uncomfortable Interactions, mixed reality laboratory & horizon. University of Nottingham, Nottingham, UK.

Carlson, M., 2018. Confronting measurable journalism. Digital Journalism, 6: 406-417. https://doi.org/10.1080/21670811.2018.1445003

Cox, A., Gould, S, Cecchinato, M., Iacovides, I. and Renfree, I., 2016. Designing frictions for mindful interactions: the case for microboundaries. CHI EA '16: Proceedings of the 2016 CHI Conference Extended Abstracts on Human Factors in Computing Systems. https://doi.org/10.1145/2851581.2892410

Dewey, J., 1927. The public and its problems. H. Holt & Co, New York, NY, USA.

Ferrer-Conill, R., 2018. The audience-oriented editor – making sense of the audience in the newsroom. Digital Journalism, 6: 436-453. https://doi.org/10.1080/21670811.2018.1440972

Groot Kormelink, T., 2019. Capturing and making sense of everyday news use, PhD Study, Faculty of Humanities; Network Institute; Communication, Vrije Universiteit, Amsterdam, the Netherlands. Available at: https://research.vu.nl/en/publications/capturing-and-making-sense-of-everyday-news-use

Lehman-Wilzig, S.N., and Seletzky, M., 2010. Hard news, soft news, 'general' news: the necessity and utility of an intermediate classification. Journalism, 11: 37-56. https://doi.org/10.1177/1464884909350642

MacGregor, P., 2007. Tracking the online audience. Journalism Studies, 8: 280-298. https://doi.org/10.1080/14616700601148879

Marres, N., 2005. Issues spark a public into being: a key but often forgotten point of the Lippmann-Dewey debate. In: Latour, B. and Weibel, P. (eds.) Making things public. MIT Press, Cambridge, MA, USA.

Sengers, P. and Gaver, W., 2006. Staying open to interpretation: engaging multiple meanings in design and evaluation. Proceedings of the 6th conference on Designing Interactive Systems. pp. 99-108. https://doi.org/10.1145/1142405.1142422

Schudson, M., 2008. The 'Lippmann-Dewey debate' and the Invention of Walter Lippmann as an anti-democrat 1985-1996. International Journal of Communication, 2: 1031-1042. Available at: https://ijoc.org/index.php/ijoc/article/view/343/229.

Steel, R., 2017. Walter Lipmann and the American Century. Routledge Taylor & Francis Group, New York, NY, USA.

Thaler, R.H. and Sunstein, C.R., 2008. Nudge: improving decisions about health, wealth and happiness. Penguin Books, London, UK.

Tandoc, E.C. and Thomas, R.J., 2015. The ethics of web analytics. Digital Journalism 3: 243-258. https://doi.org/10.1080/21670811.2014.909122

Van den Hoven, J., 2007. ICT and value sensitive design. In: Goujon, P., Lavelle, S., Duquenoy, P., Kimppa, K. and Laurent, V. (eds.) The information society: innovation, legitimacy, ethics and democracy. Springer, Boston, MA, USA.

Van Es, K., Wieringa, M. and Schäfer, M.T., 2018. Tool criticism: from digital methods to digital methodology. Datafied Society Working Paper Series. Available at: https://datafiedsociety.wp.hum.uu.nl/tool-criticism/.

Zamith, R., Belair-Gagnon, V. and Lewis, S.C., 2020. Constructing audience quantification: social influences and the development of norms about audience analytics and metrics. New Media & Society, 22: 1763-1784. https://doi.org/10.1177/1461444819881735

Media

Petre, C., 2015. The traffic factories: metrics at chartbeat, Gawker Media, and The New York Times. May 7. Available at: https://www.cjr.org/tow_center_reports/the_traffic_factories_metrics_at_chartbeat_gawker_media_and_the_new_york_times.php.

10. Immoral programming

What can be done if malicious actors use language AI to launch 'deepfake science attacks'?

Nadisha-Marie Aliman[1] and Leon Kester[2*]

[1]*Utrecht University, Faculty of Science, Department of Information and Computing Sciences, Princetonplein 5, 3584 CC Utrecht, the Netherlands;* [2]*TNO Netherlands, Intelligent Autonomous Systems, Postbus 96864, 2509 JG The Hague, the Netherlands; leon.kester@tno.nl*

Abstract

The problem-solving and imitation capabilities of AI are increasing. In parallel, research addressing ethical AI design has gained momentum internationally. However, from a cybersecurity-oriented perspective in AI safety, it is vital to also analyse and counteract the risks posed by intentional malice. Malicious actors could for instance exploit the attack surface of already deployed AI, poison AI training data, sabotage AI systems at the pre-deployment stage or deliberately design hazardous AI. At a time when topics such as fake news, disinformation, deepfakes and, recently, fake science are affecting online debates in the population at large but also specifically in scientific circles, we thematise the following elephant in the room now and not in hindsight: what can be done if malicious actors use AI for not yet prevalent but technically feasible 'deepfake science attacks', i.e. on (applied) science itself? Deepfakes are not restricted to audio and visual phenomena, and deepfake text whose impact could be potentiated with regard to speed, scope, and scale may represent an underestimated avenue for malicious actors. Not only has the imitation capacity of AI improved dramatically, e.g. with the advent of advanced language AI such as GPT-3 (Brown *et al.*, 2020), but generally, present-day AI can already be abused for goals such as (cyber)crime (Kaloudi and Li, 2020) and information warfare (Hartmann and Giles, 2020). Deepfake science attacks on (applied) science and engineering – which belong to the class of what we technically denote as scientific and empirical adversarial (SEA) AI attacks (Aliman and Kester, 2021) – could be instrumental in achieving such aims due to socio-psycho-technological intricacies against which science might not be immune. But if not immunity, could one achieve resilience? This chapter familiarises the reader with a complementary solution to this complex issue:

a generic 'cyborgnetic' defence (GCD) against SEA AI attacks. As briefly introduced in Chapter 4, the term cyborgnet (which is much more general than and not to be confused with the term 'cyborg') stands for a generic, substrate-independent and hybrid functional unit which is instantiated e.g. in couplings of present-day AIs and humans. Amongst many others, GCD uses epistemology, cybersecurity, cybernetics, and creativity research to tailor 10 generic strategies to the concrete exemplary use case of a large language model such as GPT-3. GCD can act as a cognitively diverse transdisciplinary scaffold to defend against SEA AI attacks – albeit with specific caveats.

Key concepts

- For safety reasons, it is vital to tackle the immoral programming issue of intentional malice
- Malicious actors could launch deepfake science attacks against the science enterprise
- Science is not immune to such attacks, and proactive defences are required
- Generic 'Cyborgnetic' Defence (GCD) is a transdisciplinary framework that crafts solutions from a cyborgnetic stance
- A cyborgnet is not to be confused with a cyborg. A cyborgnet is a generic substrate-independent hybrid functional unit (i.e. all cyborgs exist in cyborgnets but not the reverse)
- GCD is a complementary defence against deepfake science (or technically SEA AI) attacks
- Thereby, GCD acts as a cognitively diverse scaffold that uses epistemology, creativity research and knowledge from many other fields. GCD could be resilient – but not immune

10.1 The practical scientific and empirical adversarial AI attack problem

The not yet prevalent but technically feasible scientific and empirical adversarial (SEA) AI attacks could be launched in multiple modalities. However, for illustrative purposes, we focus on text-based SEA AI attacks using language models. We analyse three attack vectors: (1) AI-generated data and experiments, (2) AI-generated research articles, (3) AI-generated reviews. Firstly, it is noteworthy that the idea to artificially generate academic text contributions and inject them even into respected venues has been already implemented in some (later withdrawn) cases (Van Noorden, 2014) merely on the basis of a mediocre automated text generation mechanism. As researchers discovered (Eckert *et al.*, 2018), this was similarly possible with human-generated made-up empirical studies accepted by predatory publishers to which internationally

AI-generated data and experiments

respected scientists were found to nevertheless habitually submit contributions. With language models, malicious actors could potentiate such practices with regard to superficial linguistic quality, speed, scale, and scope including pre-print proliferations. In addition, in applied science or engineering contexts such as in cybersecurity, the emergence of sophisticated language models such as GPT-2 (Radford *et al.*, 2019) and GPT-3 (Brown *et al.*, 2020) open up an unprecedented field of affordances for malicious actors. For instance, recently, it has been shown that AI-generated cyber threat intelligence (CTI) obtained via a fine-tuned version of GPT-2 could be utilised to poison data-driven cyber defence systems at training time (Ranade *et al.*, 2021). This could have serious security consequences, given the growing cybercrime damages expected to reach 6 trillion USD in 2021 (Ozkan *et al.*, 2021) and the associated risk for critical infrastructure. Importantly, the abovementioned artificially generated text CTI samples, which reported about distorted cyber threat events, were also able to fool human cybersecurity experts who 'labelled the majority of the fake CTI samples as true despite their expertise' (Ranade *et al.*, 2021). Generally, legitimate experimental research relying on online data could be sabotaged on a large scale via such deepfake-based poisoning schemes. (A related feasible and serious but not yet prevalent recent concern from a very different science field is deepfake geography (Zhao *et al.*, 2021), i.e. AI-generated fictional satellite images in GIScience.) Beyond that, in a tentative prompting of a publicly available interface to the pre-trained GPT-2 model (Radford *et al.*, 2019), we found that the model is able to generate text samples vaguely mimicking academic paper structures and – for example – referring to never conducted experiments (Figure 10.1). While the consistency was often low, note that the utilised GPT-2 model has been neither fine-tuned on research papers nor is it as powerful as its successor GPT-3, whose parameters are two orders of magnitude bigger (Bartoli and Medvet, 2020) and which still has a closed access policy.

AI-generated research articles

Secondly, regarding more theoretical AI-generated research articles, the early GPT-3 paper already touched upon that potential misuse, albeit only briefly, via the mention of the expression 'fraudulent academic essay writing' (Brown *et al.*, 2020). In a very recent plagiarism study with the pre-trained GPT-3, it has been stated that in view of the quality of the textual results it 'may very well pass peer review' (Dehouche, 2021). Malicious actors could utilise such AI-generated contributions as instruments for information operations. For instance, in order to reinforce the sentiment of unresolvable disagreement also in the science field, they could fuel narratives on 'post-truth' (which, as mentioned in Chapter 4, we consider to be a misguiding misnomer) via a novel form of scientific astroturfing (Aliman and Kester, 2021). Thereby, astroturfing (Hartmann and Giles, 2020) is a deceptive online technique deliberately utilised in order to provide the impression that a certain asset is widely supported or

Figure 10.1. Counterfactual outputs of pre-trained GPT-2.

Prompt by human	**Counterfactual outputs of pre-trained GPT-2**
Abstract In a time in which topics such as fake news, disinformation, deepfakes and recently fake science affect online debates amongst the population at large, but also specifically in scientific circles, we thematize the following elephant in the room now and not in hindsight: what to do if malicious actors use AI for not yet prevalent but technically feasible 'deepfake science attacks' i.e., on (applied) science itself? Deepfakes are not restricted to audio and visual phenomena and deepfake text whose impact could be potentiated with regard to speed, scope, and scale may represent an underestimated avenue for malicious actors.	[...] Here, we assume that a bot that, by modifying and/or manipulating the data extracted from the published article, creates an actual original story about fake news, false claims, hoaxes, or lies. These are potentially useful means against those intending to harm credibility [...] of an author's work. [...] It is clear that these types of threats could be used to disrupt scientific debate in the future. [...] The threat in question is, therefore, likely to make the field far more hostile. Linguistics is a core part of science, and we need to learn it. [...] We can't just use algorithms to control it. [...] Liz Y. Wang *et al.*, 'A model of the Deep Web, the Deep Web and the First Human-Driving Device+: a comparative analysis in order to better inform the development and analysis of advanced techniques and technologies. Artificial Intelligence and Cyberthreats.' IEEE Transactions on Intelligence 14, 14 (2010): 1797-1805. [Crossref][PubMed] 2nd Annual IEEE International Conference on CyberSecurity. [...] we present empirical evidence that such false expectations apply disproportionately to bots compared to humans. [...] Human Cognition: Human Cognition and Cybernetic Engineering by Michael Schmitt *et al.*, 2002; doi:10.1016/j.heccomplications.2002.01.009 [...] What will be interesting about this issue is how much more targeted the attack will be under AI technologies, based only on what we previously thought was good science. Let us see how this plays out in practice. [...] The most likely scenario of malicious actors do exploit deepfakes is that they create a false narrative (by exploiting 'fake science'), by using the 'fake news' as their method of proving the 'fake evidence.' These tactics could even be employed to further their goals by using 'fake science' to further their own agendas, or (in this case) to gain public acceptance of them. [...] R. Pérez, 'Deepfaking Theses at the University of Cambridge, 'American Journal of Physics 37: 2890-2915 (2011), pp. 21-233:

rejected. This in turn combined with the societal-level automated disconcertion phenomenon (Aliman *et al.*, 2021). For instance, the fundamental epistemic confusion that arose by the mere existence of deepfakes may be able to trigger destabilising processes in fragile societies – one of the main aims in information warfare. Conversely, malicious actors could also seek to automate a flood of corroborative papers confirming the efficiency of an application, justifying a certain theory or the robustness of a defence method to exploit the vulnerabilities that a potentially decreased security awareness could bring about. The latter could in turn lead to domino effects related to the cybersecurity risks mentioned in the last paragraph. Apart from that, in our tentative probing of the publicly accessible pre-trained GPT-2, the model was able to output a few text blocks of passable quality on specific topics which could be assembled for abstracts (for a few examples, see Figure 10.1). Moreover, among others, the following artefacts that were generated may be of interest for future work: fictional links and references to fictional quotes attributed to individuals with synthetic names, self-generated structures for sections and even acknowledgment sections with mention of existing or fictional research institutes and specification of synthetic grants. Finally, we fed the twofold title of this very chapter into an interface for GPT-Neo (1.3B) (Eleuther AI, 2021), an open-source GPT-inspired language model trained on a dataset denoted 'The Pile' (Gao *et al.*, 2020). This dataset contains, among others, a large number of scientific papers and abstracts. The interface restricted its output to a certain number of characters (around the length of a sentence). The model outputted the following string: 'Intrusion Detection Systems (IDS) can detect potentially dangerous behaviours such as malicious actors launching deepfake science attacks disguised as science'. We leave this AI-generated text sample without comment for now and will return to it in Section 10.3. However, it is already worth stating that, generally, an explanation-anchored criticism is content-centred and does not need to consider the source (see Section 10.2). After the submission of this very book chapter, Eleuther AI (2021) released the novel open-source language model GPT-J (6B) which is larger than GPT-Neo (1.3B) (but not GPT-3) and has also been trained on the mentioned Pile dataset (Gao *et al.*, 2020). Strikingly, the bilingual language model Wu Dao 2.0, which is even ten times larger than GPT-3, has (among others) also been trained on the Pile (Romero, 2021). In light of the results from our tentative probing of the publicly accessible GPT-J (6B) model (displayed in Figure 10.2), we conjecture that a comparable or superior large language model trained uniquely (and not only partially) on scientific and technical papers may significantly boost the capabilities of SEA AI attackers. However, as will become apparent in Subsection 10.2.2., the same type of tool could conversely facilitate powerful defence strategies against such attacks and beyond that augment scientific creativity – a potentially safety-relevant quest for humanity. Hence, future work could investigate further along those lines.

AI-generated reviews

Thirdly, coming to the topic of AI-generated reviews, a small user study (Bartoli and Medvet, 2020) found that it was possible to mislead experienced human reviewers with text samples obtained from a fine-tuned version of GPT-2. Given the length of reviews, it seems that such settings could appear lucrative for malicious goals including insider threats, for example in the presence of financial motives. Crucially, reviewers have been described as gatekeepers of knowledge (Tsao *et al.*, 2019) that can influence the progress of creative ideas

Figure 10.2. Counterfactual outputs of pre-trained GPT-J 6B.

Prompt by human	Counterfactual outputs of pre-trained GPT-J 6B
Abstract In a time in which topics such as fake news, disinformation, deepfakes and recently fake science affect online debates amounts the population at large, but also specifically in scientific circles, we thematize the following elephant in the room now and not in hindsight: what to do if malicious actors use AI for not yet prevalent but technically feasible 'deepfake science attacks' i.e., on (applied) science itself? Deepfakes are not restricted to audio and visual phenomena and deepfake text whose impact could be potentiated with regard to speed, scope, and scale may represent an underestimated avenue for malicious actors.	[...] The occurrence of such attacks not only influences societal discourse and thus the public opinion of a society, but also impacts the public acceptance of the findings of academic science (Keil, 2019). In particular, deepfake science attacks are very easy to create (Kim, 2016). [...] We wish to convey our urgent call for scientists to be attentive of possible malicious AI/AS scenarios [...] The rapid increase of deepfake science could considerably harm a number of fields and even drive a number of current and future sciences to become incomprehensible to society. It also raises serious ethical and legal questions concerning scientific integrity, data privacy, and/or copyright. [...] we state that the advent of unsupervised machine learning for 'deepfake science' poses a growing threat [...] [...] The attacks presented in this paper show how AI is now used in text manipulation to alter and attack human perceptions of a scientific document. They do not make use of fake videos, fake images, fake audio, fake faces, fake objects, or fake messages, but manipulate text. Furthermore, we want to show that even though these attacks are in the scope of deepfake science and its sub-topic of deepfake text, their goal is to influence the public discourse. In this paper, we intend to exemplify such threat by investigating whether these fake news-like approaches will have an impact on scientific decision-making, trust, and decision credibility, as well as have an impact on the researcher's and academic institution's career or even lead to possible arrest and court trials. Our results are still early, yet they seem to be telling. [...] The overall paper focuses on all the technical aspects that come along with manipulating texts or data that include theoretical approaches, algorithms, models, tools and technology, with the ultimate goal of enabling an AI detection system. The contribution is in two directions: (1) how to mitigate and stop deepfake science, and (2) what kind of AI will manipulate science.

in a society. However, through automated disconcertion, a reviewer could in principle always object that a paper was presumably AI-generated. Also, if not counteracted early, an increased fraction of AI-generated reviews could skew the directions that science takes in the long term with many socio-psycho-technological repercussions. Furthermore, reviewers that act on empiricist epistemologies that seek for the justification of truer beliefs via probabilistic belief updates may face (unnecessarily in our view, as implied in Section 10.2) 'epistemic threats' (Fallis, 2020). On this score, the amount of information in audiovisual material decreases steadily due to widespread deployment of deepfake videos (Fallis, 2020), while a similar quantitative impact analogously already affected text material even earlier via fake news. An increased awareness in society with regard to such epistemically relevant problems and confusions in science could in turn exacerbate automated disconcertion and be instrumental in information operations as described earlier. To sum up, while some scientists may at first sight be under the impression that they are immune to such purposeful text-based immoral programming with the goal of provoking an AI-aided epistemic distortion (either as an end in itself or instrumental in achieving further malicious aims), a deeper analysis suggests that unfortunately this may not always be the case.

Exemplary text segments from our probing of the publicly available GPT-2 application interface accessible at: https://deepai.org/machine-learning-model/text-generator. The fragments were sampled from 20 consecutive prompts and have been hand-chosen to illustrate some of the extracted features as discussed in the text. The outputs are not deterministic and can vary widely in linguistic quality and consistency. However, 'some meta-cherry picking' (Radford *et al.*, 2019) has been as well performed by Open AI itself when displaying abilities of GPT-2 for demonstration purposes. The human prompt indicated corresponds to the conjunction of 'Abstract', a newline character and two sentences sampled from the first page of this very chapter on immoral programming.

Exemplary text segments from our probing of the publicly available GPT-J 6B application interface accessible at: https://6b.eleuther.ai/. The fragments were sampled from 20 consecutive prompts and have been hand-chosen to illustrate some of the extracted features as discussed in the text. The outputs are not deterministic and can vary in linguistic quality and consistency depending on the chosen parameters.

10.2 Generic cyborgnetic defence as complementary theoretical solution

Against the backdrop of the above-described possible SEA AI attack scenarios, this section introduces Generic 'Cyborgnetic' Defence (GCD) framed as a countermeasure to such attacks. GCD provides a novel (unquestionably non-exhaustive and hence to be steadily updated) set of generic strategies formulated from a cyborgnetic stance. The concept of a cyborgnet was previously introduced in Chapter 4. On an inflationary account extending beyond the mere study of systems, so-called Type II entities are all those for which it is possible to consciously create and understand explanatory knowledge, while Type I entities are all entities for which this is conjectured to be impossible[5]. Based on that, a cyborgnet represents a generic, substrate-independent and hybrid functional unit comprising relations between Type II entities (so today only applicable to humans including, but not restricted to, cyborgs) and Type I entities (such as Type I AIs but also any other Type I entities not limited to systems and thus also, e.g. ideas, processes, or objects). On that view, early Type II humans equipped with complex material tools and language abilities already had an inherently cyborgnetic existence with both material and linguistic tools representing integrated Type I entities (Aliman, 2021a). Today, not long after the intricacies of deepfake and automated disconcertion started to affect the information ecosystem, SEA AI attacks could now become the entry point for analogously discombobulating phenomena in the scientific ecosystem. While this can seem threatening to scientists, it need not be – if science creatively adapts to this novel complex field of affordances whilst not interrupting its quest for better explanations. In this vein, one motivation for both an epistemic and a creativity-centred cyborgnetic stance for a generic SEA AI attack defence, is the law of requisite variety from cybernetics stating that 'only variety can destroy variety' (Ashby, 1961). Since the malicious adversary operates from within a coupling with a language model targeting the victim at an epistemic level, a defender may profit from integrating not only epistemic knowledge but also such knowledge stemming from language models too. Building on this, the generic defences under GCD aim at: (1) facilitating resilience to malicious actors and their language models while (2) simultaneously facilitating a creativity-augmenting feedback loop between defenders and their own language models. The former is an inter-cyborgnetic and the latter an intra-cyborgnetic endeavour. While perhaps unintuitive since potentially unusual at first sight, we offer a deeper explanation of this line of thought, to which we return in the next subsection.

10.2.1 Generic epistemic defence

From a functional cyborgnetic point of view, an extremely vital asset in security contexts is embodied cyborgnetic creativity. Since one is not able to reduce the adversarial disturbances in the form of SEA AI attacks controlled

[5] Note that this ontology (Aliman, 2020; Aliman, 2021a) has no relation whatsoever to the metaphor of Kahneman related to System 1 and System 2 (linked to two modes of human brain functioning with the first one being prediction-dominated/automatic and the second one prediction-error dominated/controlled but both modulated by precision weights (Hutchinson and Barrett, 2019)). Conversely, currently known Type II entities are restricted to humans as a species and examples for Type I entities are everything else. This means Type I entities can be e.g. non-human conscious mammals like dogs, but also thoughts, language itself, mechanical tools, dreams, decision trees, chatbots, etc.

by malicious actors being unpredictable explanatory knowledge creators, a risk averse solution cannot be the only option. In fact, instead of shielding oneself from deepfake texts, which could in the long term even necessitate a retreat from society, another strategy could consist of building up resilience by actively seeking more exposure to deepfake texts (albeit at a self-defined pace in a self-selected setting). However, for such a solution to be workable, a robust epistemology is required that does not entail justification-related epistemic threats (Fallis, 2020) according to which the deepfake-permeated world gradually loses relational meaning via a quantitative decrease in information content. Despite epistemic dizziness, which has always existed for humans even before the advent of deepfakes (Aliman and Kester, 2021), explanatory-anchored science cannot be terminally disrupted by additional deceptive deepfake data. Instead, when faced with deceptive material such as that produced in SEA AI attacks, one can focus on ever better explanations of the world and criticise the perceived contents on a comparative basis without having to consciously update any latent probabilistic credence. Metaphorically speaking, better explanations are our only – though ephemeral – stones on our trajectory through the deep sea of doubt. Experimental falsification shapes this trajectory but does not determine it. Explanatory-anchored science makes pragmatic progress via incremental small steps from stone to stone, which is why the epistemic aim is of a relational and comparative nature. One does not epistemically fall deeper than on one's own stones (compared to the threatening void in which a justification-based epistemology could potentially fall in times of deepfake and fake news (Fallis, 2020)). The aim is not to find isolated good explanations, but to identify better ones (Frederick, 2020) according to criteria agreed upon with others.

Hence, the first generic epistemic defence against SEA AI attacks is to select an explanatory-anchored approach to science instead of the prevailing data-driven one. A crucial advantage of explanatory-anchored science is its concurrently open-minded nature with regard to the momentary primary uptake of ideas in order to be able to inspect and criticise them but also its self-shielding nature when it comes to the second step of a permission for that idea to provisionally stay in one's prior web of knowledge being filtered by explanatory knowledge – which in turn can precisely not be mimicked by Type I AI. Explanation-anchored science is thus also its own defence method in the face of SEA AI attacks. The second generic defence against SEA AI attacks within GCD is a trust-disentangled approach that divorces content from source. In this way, much less importance would be assigned to deepfake detection endeavours embedded in incessant cat and mouse games. Ideally, the integrity of explanation-anchored messages can be afforded by the content of the messages themselves as if connected via an invisible blockchain. In other social settings disjunct from science, people may have multiple reasons why a shielding from Type I entities stemming from their own or from other cyborgnets is desirable. For instance, a deceptive one-sided romantic relationship between a person and a chatbot can seem unwelcome. Similarly, one might try to avoid any involuntary investment of time on social media debates with future sophisticated bots acting as trolls or decide to forestall automated social engineering attempts in future social virtual reality. For all these practical cases, it is reasonable to implement something akin to a substrate-independent Type I shield (which is not to be confused and would not be equivalent to a Turing Test, see e.g. (Aliman, 2021a) for more details) if possible. However, in the science domain, there is as such no fundamental reason to shield oneself from ideas that one interprets from outputs generated by Type I AI.

There is no logical reason to assume a priori that everything generated by a Type I AI must necessarily be false. Even human liars are fallible and thus able to mistakenly tell something that may be true or stimulating in creative ways. So could the output of a Type I AI by chance sometimes contain some elements that humans might interpret as thought-provoking. Trust-disentanglement accommodates for that by allowing for novelties in deliberate and spontaneous idea generation. At worst, the Type I-generated content is rejected since both non-explanatory and useless. At best, the non-explanatory output comes with an additional element that stimulates creativity or a criticism of one's best present explanations – which brings us to the third and last generic epistemic defence under GCD: adversarial science. As already touched upon in Chapter 4, an adversarial approach to one's best prior conjectures is a rational creativity-stimulating strategy since one might e.g. unpredictably be able to falsify them and discover novel candidate better explanations upon acting against the old ones (Frederick, 2020). Pre-eminently, this signifies that explanatory-anchored science is not bound in any way to act on its best available explanations. This is decisively different from classical approaches such as encountered in empiricist and utility maximisation schemes that operate according to a fixed formula containing a set of options to which one is epistemically bound, which ignores spontaneous unpredictable creativity. In the main, explanatory-anchored adversarial science applies an adversarial paradigm to itself to such an extent that theories are purposefully formulated in a risky fashion such that they could be potentially and easily falsified. This allows for fast updates of knowledge and helps to avoid greater practical damages that could emerge by a prolonged stagnation in misleading assumptions. The goal is not to embellish one's conjectures and try to formulate them as carefully as possible to escape criticism. The aim is to formulate strong bold universal statements (Frederick, 2021). As stated by Popper, the more a theory forbids, the better that theory is (Popper, 1963). This leads us back to the beginning of this subsection. It now becomes clear why from an epistemic perspective, a self-paced exposure to adversarial patterns combined with creativity-augmenting measures may be helpful in building resilience to SEA AI attacks involving inter-cyborgnetic and intra-cyborgnetic feedback loops. In Section 10.3, we address the question of how to implement such generic strategies in practice. Prior to that, the next subsection first introduces compatible generic methods for cyborgnetic creativity augmentation in a pragmatic framework, compiling insights from creativity research in the fields of psychology and cognitive neuroscience, i.e. now formulated from a scientific and empirical stance.

10.2.2 Generic cyborgnetic creativity augmentation

The ambiguously designated artificial creativity augmentation research direction (Aliman, 2020) has recently been put forth for the purpose of implementing generic defences against societal level harm. It unifies two complementary and moreover interwoven research directions: (1) the artificial augmentation of human creativity; and (2) the augmentation of artificial creativity. Noticeably, artificial creativity augmentation represents one possible instantiation of cyborgnetic creativity augmentation. It seems well suited as a basis for crafting synergetic enhancement strategies for the intra-cyborgnetic feedback loop between human defenders and their language models. Applied to our generic defences against SEA AI attacks supported by language models, the twofold task can be exemplarily reformulated as follows: (1) augmenting human creativity using language models; and (2) augmenting artificial creativity in language models via humans. The former and the latter are intertwined since the subtask: (1) can reinforce the subtask; and (2) *vice versa*. In the spirit of recent work by Mick Ashby (2020) at the intersection of cybernetics and AI ethics, one could state that in this case, humans and language models reciprocally become a sort of ethical regulator of each other with the feedback loop instantiated for the purpose of counteracting unethical practices of deliberate disinformation in the (applied) science domain. Hence, cyborgnetic creativity augmentation proposed initially for security reasons against SEA AI attacks is also a form of augmenting intra-cyborgnetic ethical regulation. This in turn suddenly unifies moral programming and security research to counter immoral programming. Compellingly, it seems that security and ethics converge whilst counteracting SEA AI attacks. In the following, we now specifically map out two clusters of generic cyborgnetic creativity augmentation strategies.

The first cluster concerns generic strategies to augment anthropic creativity using language models. The second cluster pertains to generic strategies for the augmentation of artificial creativity within language models. To this end, we select suitable starting points based on the ten provisional available artificial creativity augmentation indicators (Aliman, 2020) which were grounded in explanations from psychology and cognitive neuroscience. On this score, seven possible indicators suggested to enhance human creativity were: transformative criticism and contrariness, divergent thinking training, alteration of waking consciousness, active forgetting (during sleep), frequent engagement, brain stimulation as well as sensory extension. Moreover, three indicators suggested to enhance artificial creativity were: immersion in the human affective niche, social cognition, and an egocentric integrated multimodal virtual reality experience of the world. In this paper, we focus on those strategies that are technically implementable in present-day advanced language models. Thus, we limit our analysis to the first six indicators specified for human creativity enhancement and to the first indicator mentioned for the augmentation of artificial creativity. Firstly, in order to augment human creativity using language models, suitable generic strategies could be to design these AIs with the following enhancing subgoals: (1) increase human criticism abilities; (2) stimulate human divergent thinking; (3) alter the nature of self-experience at waking time; (4) extend the nocturnal unconscious and/or dream-related creative generation and active forgetting processes; (5) encourage frequent human engagement; (6) provide human sensory extension. Secondly, concerning the human-performed augmentation of artificial creativity within language models, we add the following generic strategy; (7) immersion in the human affective niche via a mathematical approach and via active sampling. These seven generic strategies against SEA AI attacks via cyborgnetic creativity augmentation can seem abstract at first sight. For this reason, the next Section 10.3 now instantiates and illustrates their application (together with the three generic epistemic defence strategies from the last subsection) using design fictions for the use case of large language models – the same tools that malicious actors could utilise for advanced SEA AI attacks.

generic strategies

10.3 Practical use of theoretical solution

In Section 10.2, we introduced the reader to our GCD framework consisting of three generic epistemic defence strategies and seven generic cyborgnetic creativity augmentation strategies against SEA AI attacks. For illustrative purposes, we now apply those methods to the practical large language model (abbreviated by LLM in the following) use case. We use design-fictions as recommended in AI safety frameworks (Aliman *et al.*, 2021). In this section, we see how one key trick in applying the GCD framework to practically relevant defences against SEA AI attacks performed with an LLM, is to generate desirable upward counterfactuals of a possible defence with GPT-3 itself. Step-by-step, we systematically proceed through all three practically relevant SEA AI attack vectors specified in Section 10.1: (1) AI-generated data and experiments; (2) AI-generated research articles; (3) AI-generated reviews. For each attack vector, we clarify how instances of generic epistemic defences and generic cyborgnetic creativity augmentation measures can help in practice.

Since comparable practical reports on LLMs such as GPT-3 are still relatively scarce, we rely on plausible design-fictions of what we already deem technically feasible today, i.e. we craft upward counterfactuals projecting to the immediate past (e.g. literally yesterday). Normally, it makes sense to project such design-fictions to the immediate future. However, we specifically frame it in this way to stress the attainability of many potentially valuable opportunities. Upward counterfactuals pertain to better ways in which scenarios could have unfolded but did not. This means we conjecture a world in which large language models do not undergo a closed source policy and any interested entity (such as a scientist) with reasonable resources could have acquainted itself with an LLM interface. For simplicity, given the complexity of the underlying issue, we assume that this entity is able to design novel applications for the LLM and to modify the model (e.g. to fine-tune it on other datasets – although this might not always be necessary anymore –, change its loss function and parameters or to extend it with other available technologies). Moreover, to simplify, we assume that the human SEA AI attacker appears to the defending scientist as a grey box instance with the only information being that the attacker owns an LLM too. Also, within attack-defence cycles, the conjunction of scientist and LLM instantiate a cyborgnet as does the conjunction of attacker and LLM. A simplified illustration of important intra-cyborgnet and inter-cyborgnet relations are depicted in Figure 10.3. Whilst both attacker and defender are naturally embedded in a complex heterogeneous and multi-layered socio-psycho-techno-physical environment and while the cyborgnets of each of them can contain a much larger number of Type I entities (e.g. ranging from ideas to technologies over processes) in a given situated conceptualisation, we abstract away further details for a better overview and for the purpose of a better visualisation. Also, from a cyborgnetic stance, even before language models like GPT-3, in fact, since the advent of human linguistic abilities, humans use language as a form of technological tool since it involves the application of explanatory knowledge for practical aims such as teaching, learning and participatory sense-making (Aliman, 2020; Aliman and Kester, 2021). In brief, language models add new nested dimensions to the linguistic tools in one's cyborgnet.

10.3.1 Cyborgnetic defence against hypothetical LLM-generated data and experiments

Concerning epistemic defences, one could have implemented the following. From the perspective of engineers in security, AI, but also in many other domains as well as scientists involved in empirical studies, a first step could have been to consciously familiarise oneself with the different steadily shifting cyborgnet constructs at different spatiotemporal scales in which one is embedded, while performing research with data and different systems. The

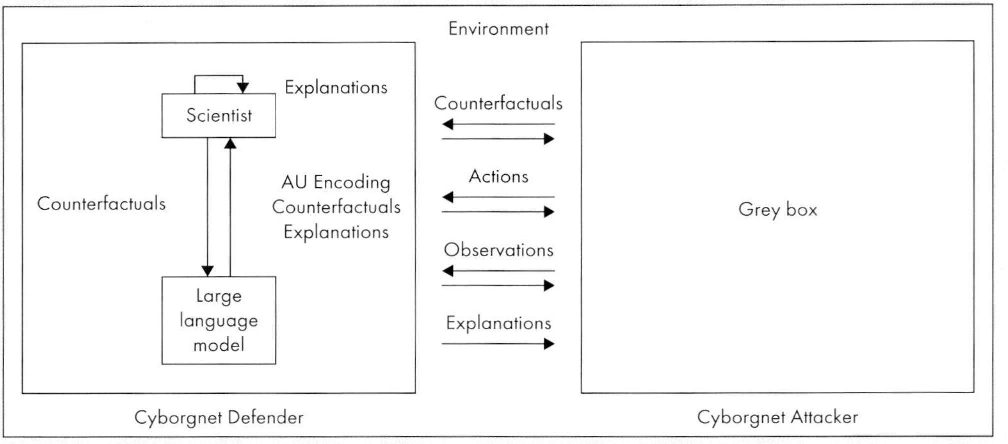

Figure 10.3. Simplified illustration and contextualisation of GCD-relevant intra- and inter-cyborgnet relations embedded in cycles of SEA AI attacks and defences performed by the cyborgnet of the attacker and the cyborgnet of the defender. Obviously, the entity labelled scientist could just as well be an engineer or a researcher from other areas.

cyborgnet inventory analysis

SEA AI attack surface is inherently cyborgnetic and could be much wider than it might seem with text inputs potentially hidden in non-salient parts of the research pipeline. It is feasible that certain SEA AI attacks may even stay undetected with only the repercussions becoming perceptible. Hence, one could have attempted to sketch individual- and group-level cyborgnet inventory maps in analogies to red team versus blue team settings (Rajendran *et al.*, 2011) to investigate how and via which assets and backdoors a covert textual SEA AI attack could potentially manifest itself. In such a cyborgnet inventory analysis, one could have specified underlying relations and feedback loops as minimally illustrated in Figure 10.3. Regarding the submission of one's own experimental contributions against the backdrop of LLM-generated samples, one could have adopted a refined strategy. Instead of mainly focusing on the experimental results, one could have strived for a thorough theoretical foundation coalescing theoretical and empirical elements. Every empirical study could have been supplemented with one explanatory section in which the empirical approach is contextualised against a larger scientific and theoretical background. Experiments designed to merely corroborate a theory could have been discarded. The scientific community could have opted for registered reports (West and Bergstrom, 2021), which represents a solution in which experimental research is assessed at an earlier stage based on the explanatory quality of the research proposal itself and not on the actual experimental results. Thus, one could have strived for exclusively explanation-anchored research proposals to which one could have in addition adopted an adversarial stance in order to improve them. One could have shifted the focus away from fixed measures on pre-determined data and replaced a normative design philosophy striving for highly intelligent Type I systems instead to ever more creative embodied cyborgnets without any fixed performance measure. One's socio-psycho-techno-physical state could have been synergistically entangled with the system. (Logically speaking, it is not because ethical or aesthetic goals and problem-solving ability are separable that one needs to keep them separated. In Chapter 4, we explained why foregrounding embodied cyborgnetic creativity may mitigate the risks of advanced Type I AI control.) Finally, one could have accommodated epistemic dizziness in experimental procedures and while being safety-aware, one would also have accepted that the future of cyborgnet safety and security cannot be predicted.

With respect to cyborgnetic creativity augmentation, this paragraph discusses how LLMs could have been used to augment the creativity of humans engaging in empirical research. Creativity can be described as a tripartite evolutionary affective construct with three modes (Dietrich, 2019): the deliberate mode (when consciously engaging in creative deliberations), the spontaneous mode (an unconscious process whose creative end result presents itself spontaneously to consciousness), and the flow mode (when creativity is enacted directly in emulations of the motor system). We focus on the two first modes in what follows. LLMs could have been utilised frequently to stimulate divergent thinking in the deliberate mode by first letting the scientist prompt the LLM on providing a solution to a given practical problem. Since LLMS are not able to create explanatory knowledge, the scientist could then criticise the generated output and re-prompt the LLM, derive inspiration from it, or utilise it to question own prior assumptions. By way of example, let us consider the output generated by GPT-Neo when prompted with the title and subtitle of this chapter. Namely, the answer was: 'Intrusion Detection Systems (IDS) can detect potentially dangerous behaviours such as malicious actors launching deepfake science attacks disguised as science'. We return to an LLM-aided critical assessment of the content of that statement in a few sentences. Generally, to improve the required critical reasoning abilities, a novel LLM-based systematic adversarial educational tool could have become available in empirical research. The LLM could have been utilised for life-long learning and for students in engineering and science to train the formulation of better explanation-anchored empirical research proposals, e.g. for the abovementioned registered reports. For instance, given a current paragraph and a history of earlier paragraphs, a student's next paragraph could then have competed with the LLM-generated continuation of it. This could have had a twofold function. The first aim could have been a training of the deliberate mode in creativity by exploring whether a human evaluator could distinguish between student and LLM-produced samples by reconstructing the exact chain of paragraphs generated by the student (with the only cue being the first paragraph that the student wrote). This could have been akin to testing the student's ability to maintain an invisible explanatory blockchain so to speak. The second aim could have been a short-term enhancement of divergent thinking in the deliberate mode or a long-term enhancement of the spontaneous mode. Namely, a sort of cognitive stimulation training could have thereby been implemented due to the student being exposed to the alternative LLM-generated 'deepfake science' branch. It is known from cognitive neuroscience, that 'cognitive stimulation via the exposure to ideas of other people is an effective tool in stimulating creativity in group-based creativity techniques' (Fink *et al.*, 2010). Interestingly, the 'other' in this case, while not being an explanatory knowledge creator, could have been the LLM, and the group-based functional unit could have been the cyborgnet. The LLM in turn could have been enhanced by fine-tuning the student's inputs at a later stage. Hence, this educational tool could have been called adversarial cyborgnetic cognitive stimulation. Coming back to the output of GPT-Neo on an intrusion detection system for deepfake science, one could then have investigated whether adversarial cyborgnetic cognitive stimulation (combined with a normalisation smoothing out superficial linguistic style differences) could allow for a subtly different defensive scheme with a similar effect: an explanatory intrusion prevention system (IPS) for science (Aliman, 2021b). Such a shielding IPS preceding scientific peer review could have been combined with a substrate-independent Type-I-shield or, technically, a substrate-independent Type-I-falsification-event test[6] (Aliman, 2021a). Its goal could have been to shield from non-explanatory texts – vitally however, without being equivalent to a deepfake science detection system. Thereby, such an explanatory IPS could not have been fully automated using a human evaluator.

adversarial cyborgnetic cognitive stimulation

explanatory intrusion prevention system

10.3.2 Cyborgnetic defence against hypothetical LLM-generated research papers

In connection with epistemic defences against SEA AI attacks utilising LLM-generated research papers, the just depicted adversarial cyborgnetic cognitive stimulation could have been proactively employed by scientists for self-education and life-long learning to improve explanation-anchored scientific writing practices. Furthermore, scientists could have engaged in red teaming and penetration testing procedures injecting LLM-generated papers into

[6] Such a substrate-independent Type-I-falsification-event test (Aliman, 2021a) requires a Type II evaluator (so specifically, a human nowadays) and merely leads to two asymmetric clusters: a first homogenous Type-I-free cluster and a second potentially heterogeneous cluster which, next to Type I entities, can also comprise Type II entities that have not yet passed the test (for instance because no suitable knowledge area tailored to the Type II test subject has been identified, because the Type II subject is still too young, for lack of motivation or willingness on the part of the Type II entity and so forth). In short, it is formally very different from the widespread idea of Turing Tests.

the submission process. To responsibly implement such schemes, scientists could have worked out tailored coordinated vulnerability disclosure practices (Kranenbarg *et al.*, 2018). For an LLM-aided trust-disentanglement to counteract SEA AI attacks at submission time, scientists could have experimented with the explanatory IPS tool just mentioned. Instead of striving for deepfake detection techniques, scientists could have aimed at implementing a scheme in which contents are not rejected because of the source that submitted them, but purely on explanation-anchored grounds. From this perspective, deepfake science papers would not have passed through the explanatory IPS because they have been generated by a Type I entity, but because those papers are merely of an imitative and hence non-explanatory nature. Pre-print platforms could have combined an automated active sampling of newly uploaded papers with an explanatory IPS involving human evaluators and LLMs. With regard to cyborgnetic creativity augmentation measures against SEA AI attacks, LLMs could have been used to frequently enhance divergent thinking with regard to the deliberate but also indirectly to the spontaneous creativity mode. Recently, a study demonstrated how GPT-3 can be utilised as a 'multiversal' language model (Reynolds and McDonell, 2021), interactively generating branches of fictional counterfactuals to stimulate human creativity in fictional writing. Extending beyond that, scientists could now have combined an LLM-aided adversarial cyborgnetic cognitive stimulation with the multiversal approach to GPT-3 to stimulate scientific writing. The fundamental difference with fictional writing would have been that it is the steady application of explanatory criticism by the human combined with adversarially motivated exploration and the possibility to experimentally falsify statements of interest that would have guided the extension of counterfactual nodes.

multiversal cyborgnetic co-creation

This multiversal cyborgnetic co-creation could have been further fine-tuned by scientists. Firstly, one could have had increased the immersion of the LLM in the human affective niche via directing its outputs with a slightly altered loss function. Instead of only predicting the next word in a sentence, aesthetic or moral parameters could be for instance considered as well. Interestingly, scientists could have used the input-agnostic generic mathematical scaffold and encoding of augmented utilitarianism (AU) (introduced in Chapter 4) to specifically tailor such parameters for the LLM they own. This conceptual idea is reflected in Figure 10.3 with the arrow labelled 'AU-encoding', flowing from scientist to LLM in the cyborgnet of the defender. Secondly, while language models like GPT-3 are imitative, outcomes perceived as creative are mainly those that exhibit implausible utility (Tsao *et al.*, 2019), i.e. utile outcomes with unexpectedly surprising previously underestimated facets. Scientists in their quest for implausible utility, could have been inspired by the idea of transdisciplinary cross-pollination effects and insights from research on cognitive diversity (Mitchell *et al.*, 2017; Reynolds and Lewis, 2017). Cognitive diversity is related to the differences in information processing and cognitive styles which means it is related to variety with respect to functional features. To fuel intra- and inter-cyborgnetic cognitive diversity with an LLM, scientists could have been motivated by composer-audience architectures (Bunescu and Uduehi, 2019) from computational creativity (Franceschelli and Musolesi, 2021) utilised to produce humorous outputs by combining an audience model trained on a non-humorous dataset A and a humorous composer model trained on both a different dataset B and the expectations that the pre-trained audience model outputs for that dataset. Analogously, scientists could have used a dataset from a scientific discipline A and another from a scientific discipline B. A deepfake science LLM composer could then have learned to surprise a deepfake science LLM audience – yielding interesting avenues to augment deliberate and spontaneous creativity but also criticism in the scientists interacting with that double deepfake science model. Finally, scientists could have harnessed the knowledge that spontaneous human creativity strongly profits from nocturnal brain processes during sleep (Lewis *et al.*, 2018) to improve the LLM's generation of outcomes perceived to stimulate ideas of implausible utility. To this end, they could have repeatedly fine-tuned the LLM on recursively changing text data modified by loosely mimicking e.g. partially sighted evolutionary affective processes of the spontaneous creativity mode (Aliman, 2020) extending to synergetic cycles of human sleep (Lewis *et al.*, 2018). In simpler cases, this could technically have included, e.g. targeted semantic mutations, syntactic-semantic crossover and a form of semantic noise injection followed by autocorrection at the sentence level. In extensions of such conceptual ideas, scientists could have enriched this shifting dataset by letting the LLM actively integrate scientific knowledge sampled, e.g. from suitable knowledge graphs. Simple active forgetting mechanisms to reduce data size and complexity could have been for instance steered by integrating human preferences via the AU encoding and/or by integrating human attention during interactions with the LLM.

10.3.3 Cyborgnetic defence against LLM-generated reviews

As can be extracted from the last subsection, scientists could have practically transformed the initial merely imitative LLM into an interactive multiversal transdisciplinary deepfake science incubator. The interesting aspect thereby is that this advanced interactive LLM incubator would still not be able to understand and create explanatory knowledge. This signifies that it could have been utilised as a strong baseline offering an enormous amount of material to train the epistemic defences of reviewers against SEA AI attacks. In theory, any conjectured approach to an explanatory IPS to shield peer-review from the non-explanatory contents of SEA AI attacks must be at least robust against the outputs of that LLM incubator at test time. Generally, this could already have deeply impacted the nature of peer-review. Thereby, the interactive LLM incubator could also have been utilised for autodidactic purposes and to prepare for the red teaming and penetration testing procedures that we already hinted at previously. Strikingly, many of the aforementioned could have led to a human sense of empowerment emerging from the cyborgnets of defenders via the augmentative feedback loops with LLMs. Simultaneously, this could have encouraged an increased awareness of responsibility on the part of the reviewers potentially paired with an altered nature of self-experience via the immensely extended field of affordances for human creativity. This explains why in the quest to defend against SEA AI attacks, humans and language models could indeed become a sort of ethical regulator (Ashby, 2020) of each other. Moreover, it also brings us back to the end of Section 10.2 where we implied that applying our GCD framework to counter SEA AI attacks could at once engender a convergence of moral programming and security research to counter immoral programming.

10.4 Conclusions

In this paper, we performed an in-depth analysis of how to possibly counteract a severe not yet prevalent but technically feasible case of immoral programming: SEA AI attacks, i.e. deepfake science attacks on (applied) science itself. For instance, malicious actors could exploit language AI for future SEA AI attacks instrumental in performing cyber(crime) and information warfare – which requires a thorough assessment of defence methods now and not in hindsight. To this end, we introduced our transdisciplinary GCD framework that can be utilised as a complementary generic scaffold to craft tailored defences. GCD comprises three generic epistemic defences and seven generic so-called cyborgnetic creativity augmentation measures. Focusing on SEA AI attacks with language AI models, we then instantiated this generic scaffold within one exemplary use case, namely large language models such as GPT-3. We then elaborated on how an LLM itself can be employed to defend against SEA AI attacks with LLMs. Thereby, cyborgnetic feedback loops between scientists and LLMs could offer resilience to SEA AI attacks. In addition, they could also transform the language models into interactive multiversal transdisciplinary deepfake science incubators (generating creativity-stimulating but still non-explanatory outcomes) while simultaneously encouraging the multiversal scientists to stay critical and to engage in explanation-anchored, trust-disentangled, and adversarial scientific knowledge co-creation. Whilst implementing such hybrid defence methods against SEA AI attacks, scientists and language models reciprocally become ethical regulators of each other. In short, counteracting immoral programming and moral programming itself converges within the GCD scaffold. In our view, once a rigorous epistemic elucidation is provided to the general public, humanity as a whole may profit from creativity fostering deepfake incubators via e.g. language model subscriptions that could be available to everyone, such as is the case with access to the internet. Obvious limitations of our framework could be the need to address emerging plagiarism issues (Dehouche, 2021). Overall, GCD-based solutions to SEA AI attacks also come with the following inherent caveats: (1) they can be resilient but not immune; (2) they cannot and should not be entirely automated. In summary, we pointed to the daunting SEA AI elephant in the room and proposed a complementary non-exhaustive solution. GCD could provide cognitively diverse incentives for AI safety and for ongoing efforts in moral programming for which Bart Wernaart (2021) recently set forth a future-oriented road map. As we have seen, the international meta-cyborgnet of multiversal scientists is latently capable of building up resilience to SEA AI attacks. In this vein, may the elephant rest in peace.

References

Aliman, N.M., 2020. Hybrid cognitive-affective strategies for AI safety. Doctoral dissertation, Utrecht University, Utrecht, the Netherlands. https://doi.org/10.33540/203

Aliman, N.M., 2021a. Cyborgnetics – The type I vs. type II split. Utrecht, the Netherlands.

Aliman, N.M., 2021b. Explanatory IPS. Available at: https://nadishamarie.jimdo.com/app/download/10829271571/Explanatory_IPS.pdf?t=1632837017.

Aliman, N.M. and Kester, L., 2021. Epistemic defenses against scientific and empirical adversarial AI attacks. In Workshop on Artificial Intelligence Safety 2021 co-located with the 30th International Joint Conference on Artificial Intelligence AISafety@IJCAI 2021.

Aliman, N.M., Kester, L. and Yampolskiy, R., 2021. Transdisciplinary AI observatory–retrospective analyses and future-oriented contradistinctions. Philosophies, 6: 6.

Ashby, M., 2020. Ethical regulators and super-ethical systems. Systems, 8: 53. https://doi.org/10.3390/systems8040053

Ashby, W.R., 1961. An introduction to cybernetics. Chapman & Hall Ltd, London, UK.

Bartoli, A. and Medvet, E., 2020. Exploring the potential of GPT-2 for generating fake reviews of research papers. In Fuzzy Systems and Data Mining. IOS Press, Amsterdam, the Netherlands, pp. 390-396. https://doi.org/10.3233/FAIA200717

Brown, T.B., Mann, B., Ryder, N., Subbiah, M., Kaplan, J., Dhariwal, P., Neelakantan, A., Shyam, P., Sastry, G., Askell, A. and Agarwal, S., 2020. Language models are few-shot learners. arXiv preprint arXiv:2005.14165.

Bunescu, R.C. and Uduehi, O.O., 2019. Learning to surprise: a composer-audience architecture. In ICCC, pp. 41-48.

Dehouche, N., 2021. Plagiarism in the age of massive generative pre-trained transformers (GPT-3). Ethics in Science and Environmental Politics, 21: 17-23. https://doi.org/10.3354/esep00195

Dietrich, A., 2019. Types of creativity. Psychonomic Bulletin & Review, 26: 1-12. https://doi.org/10.3758/s13423-018-1517-7

Eckert, S., Sumner, C. and Krause, T., 2018, 11 August. Inside the fake science factory. Presentation at DEF CON, 26, Las Vegas, NV, USA.

Eleuther AI, 2021. Available at: https://www.eleuther.ai/

Fallis, D., 2020. The epistemic threat of deepfakes. Philosophy & Technology, 1-21. https://doi.org/10.1007/s13347-020-00419-2

Fink, A., Grabner, R.H., Gebauer, D., Reishofer, G., Koschutnig, K. and Ebner, F., 2010. Enhancing creativity by means of cognitive stimulation: evidence from an fMRI study. NeuroImage, 52: 1687-1695. https://doi.org/10.1016/j.neuroimage.2010.05.072

Franceschelli, G. and Musolesi, M., 2021. Creativity and machine learning: a survey. arXiv preprint arXiv:2104.02726.

Frederick, D., 2020. Against the philosophical tide: essays in Popperian critical rationalism. Critias Publishing, Yeovil, UK.

Frederick, D., 2021. Critique of Brian Earp's writing tips for philosophers. Think, 20: 81-87. https://doi.org/10.1017/S1477175621000063

Gao, L., Biderman, S., Black, S., Golding, L., Hoppe, T., Foster, C., Phang, J., He, H., Thite, A., Nabeshima, N. and Presser, S., 2020. The pile: an 800GB dataset of diverse text for language modeling. arXiv preprint arXiv:2101.00027. https://arxiv.org/abs/2101.00027

Hartmann, K. and Giles, K., 2020, May. The next generation of cyber-enabled information warfare. In: 2020, IEEE 12th International Conference on Cyber Conflict (CyCon) 1300: 233-250. https://doi.org/10.23919/CyCon49761.2020.9131716

Hutchinson, J.B. and Barrett, L.F., 2019. The power of predictions: an emerging paradigm for psychological research. Current Directions in Psychological Science, 28: 280-291. https://doi.org/10.1177/0963721419831992

Kaloudi, N. and Li, J., 2020. The AI-based cyber threat landscape: a survey. ACM Computing Surveys (CSUR), 53: 1-34. https://doi.org/10.1145/3372823

Kranenbarg, M.W., Holt, T.J. and Van der Ham, J., 2018. Don't shoot the messenger! A criminological and computer science perspective on coordinated vulnerability disclosure. Crime Science, 7: 1-9. https://doi.org/10.1186/s40163-018-0090-8

Lewis, P.A., Knoblich, G. and Poe, G., 2018. How memory replay in sleep boosts creative problem-solving. Trends in cognitive sciences, 22: 491-503. https://doi.org/10.1016/j.tics.2018.03.009

Mitchell, R., Boyle, B., O'Brien, R., Malik, A., Tian, K., Parker, V., Giles, M., Joyce, P. and Chiang, V., 2017. Balancing cognitive diversity and mutual understanding in multidisciplinary teams. Health Care Management Review, 42: 42-52. https://doi.org/10.1097/HMR.0000000000000088

Ozkan, B.Y., Van Lingen, S. and Spruit, M., 2021. The cybersecurity focus area maturity (CYSFAM) model. Journal of Cybersecurity and Privacy, 1: 119-139. https://doi.org/10.3390/jcp1010007

Popper, K.R., 1963. Conjectures and refutations: the growth of scientific knowledge. Routledge, London, UK.

Radford, A., Wu, J., Child, R., Luan, D., Amodei, D. and Sutskever, I., 2019. Language models are unsupervised multitask learners. OpenAI blog, 1: 9. Available at: https://cdn.openai.com/better-language-models/language_models_are_unsupervised_multitask_learners.pdf.

Rajendran, J., Jyothi, V. and Karri, R., 2011, October. Blue team red team approach to hardware trust assessment. In: 2011, IEEE 29th international conference on computer design (ICCD), pp. 285-288. https://doi.org/10.1109/ICCD.2011.6081410

Ranade, P., Piplai, A., Mittal, S., Joshi, A. and Finin, T., 2021. Generating fake cyber threat intelligence using transformer-based models. arXiv preprint arXiv:2102.04351.

Reynolds, A. and Lewis, D., 2017. Teams solve problems faster when they're more cognitively diverse. Harvard Business Review, 30. Available at: https://hbr.org/2017/03/teams-solve-problems-faster-when-theyre-more-cognitively-diverse.

Reynolds, L. and McDonell, K., 2021. Multiversal views on language models. arXiv preprint arXiv:2102.06391.

Romero, A., 2021. GPT-3 scared you? Meet Wu Dao 2.0: a monster of 1.75 trillion parameters. Available at: https://towardsdatascience.com/gpt-3-scared-you-meet-wu-dao-2-0-a-monster-of-1-75-trillion-parameters-832cd83db484

Tsao, J.Y., Ting, C.L. and Johnson, C.M., 2019. Creative outcome as implausible utility. Review of General Psychology, 23: 279-292. https://doi.org/10.1177/1089268019857929

Van Noorden, R., 2014. Publishers withdraw more than 120 gibberish papers. Nature News. https://doi.org/10.1038/nature.2014.14763

West, J.D. and Bergstrom, C.T., 2021. Misinformation in and about science. Proceedings of the National Academy of Sciences, 118: e1912444117. https://doi.org/10.1073/pnas.1912444117

Wernaart, B., 2021. Developing a roadmap for the moral programming of smart technology. Technology in Society, 64: 101466. https://doi.org/10.1016/j.techsoc.2020.101466

Zhao, B., Zhang, S., Xu, C., Sun, Y. and Deng, C., 2021. Deep fake geography? When geospatial data encounter artificial intelligence. Cartography and Geographic Information Science, 48: 1-15. https://doi.org/10.1080/15230406.2021.1910075

11. Moral MindSphere

David A.M. Bamps[1,2*], Veronique De Roeck[2] and Lize Verbeke[2]

[1]Multidisciplinary Centres of Expertise: Resilience People and Smart Organisations, Engaged and Entrepreneurial European University (E³UDRES²), Human Contribution to Artificial Intelligence; [2]UCLL, Groeneweg 151, 3001 Heverlee (Leuven), Belgium; david.bamps@ucll.be

Abstract

A data-driven society is the society of the future. How could a technological infrastructure help to optimise moral behaviour? Is it possible for an intelligent environment to help predict inappropriate behaviour as well? What are the possible positive and negative effects of business models that validate information sharing? This chapter deals with these questions from the perspective of a data-driven morality and the possibility of developing a technological platform where moral behaviour can be monitored. As a thought experiment, it builds on what has already been developed by Siemens, the so-called MindSphere. The MindSphere is an industrial service solution invented by Siemens and is part of an IoT application. Its function is to collect and analyse all kinds of sensor data in real time (Reinheimer, 2017: 26). This information can then be used to optimise products, production assets and manufacturing processes along the entire value chain (Dowling *et al.*, 2016). The aim of this chapter is to offer the reader a rudimentary blueprint that will enable the construction of a moral IoT as a service solution: a Moral MindSphere. The starting point of this thought experiment is to explore the possibility of collecting data on human behaviour with a moral qualification.

Key concepts

- The Internet of Things (IoT), is a system of interrelated computing devices, mechanical and digital machines, objects, animals or people that are provided with unique identifiers (UIDs) and the ability to transfer data over a network
- Data-driven morality is a particular form of morality in which moral decisions are driven by data. In its highest form, technology itself can make moral decisions or judge social behaviour based on this data and other parameters

- Moral MindSphere (MM) is a hypothetical IoT environment to optimise moral behaviour. It is the blueprint of a future dashboard society using the infrastructure in smart homes and smart cities, based on data-driven analytics and actionable intelligence of the MindSphere (Siemens)
- Software Environment for Neural Networks (SENN) is the most advanced high-dimensional, non-linear modelling system (created by Siemens). It makes the science of prediction possible. This kind of highly accurate demand forecasting can help to optimise a supply chain and reduce costs, but we hypothesise[7] that in the future it may also help to predict processes of social order, moral behaviour or unethical consequences

11.1 The superlative of a network society: IoT as a service solution

network society

A network society is 'a society whose social structure is made up of networks powered by micro-electronics-based information and communications technologies' (Castells, 2004: 2). Three processes[8] led to the emergence of this new social structure in the late 20th century, where the revolution in information and communication technologies is one of the most important from the perspective of a functional extrapolation. Today, we live in a hyper-connected world: this is where the contextual basis for hyper-morality lies. We hear everything from everyone, and we apply a moral standard to it. This is more common than ever. In other words, the society we live in is not amoral but hyper-moral, thanks to the connected world. There is little room for alternative views on social media: any deviation is immediately and very harshly punished (Boutellier, 2019). The yardstick of a hyper-connected society is calibrated by constant monitoring.

This standard can be found not only in general society, but also in the development of specific domains, for example in medical applications. 'Sensors, smart devices, and automated systems have been used in life sciences industries for the benefit of patients and medical personnel for the diagnosis of disease, monitoring of patient conditions, treatment of chronic conditions, and

[7] The hypothesis is not part of an experimental setting but rather of a thought experiment. '[C]onjectures or hypotheses ... are really to be regarded as thought "experiments" through which we wish to discover whether something can be explained by a specific assumption in connection with other natural laws' (Ørsted, 1997) (as elaborated in The Dashboard Society (Bamps, 2021: 44-45).

[8] Castells identifies three processes: the restructuring of industrial economies to accommodate an open market approach; the freedom-oriented cultural movements of the late 1960s and early 1970s, including the civil rights movement, the feminist movement and the environmental movement; the revolution in information and communication technologies. For a detailed explanation of the processes, see Castells (2004).

Internet of Things

manufacturing and distribution of drugs. The Internet of Things (IoT) has been implemented for connecting sensors and networked devices and collecting and analysing the experimental data obtained from those sensors and actuators' (Sarbagya and Sudan, 2019). Beyond this application in life science industries, this technology has also been adopted in more industrial environments with important economic effects. For the sake of argument, in relation to the Korean market the effect of the IoT industry on production inducement is KRW474.6 billion; the effect on value-added inducement is KRW314.7 billion; and it is estimated that 3,628 jobs will be created by the Internet of Things (IoT) industry (Jeong *et al.*, 2013). In the future, IoT will be widely integrated into every possible intelligent environment, even in our most domestic or intimate infrastructure. This requires a control system with an expansion of information exchange possibilities and service solutions.

> MindSphere® is the leading industrial IoT as a service solution. Using advanced analytics and AI, MindSphere powers IoT solutions from the edge to the cloud with data from connected products, plants and systems to optimise operations, create better quality products and deploy new business models (Siemens, 2021).

The example in Case 11.1 makes it clear that a global application of this system on all possible scales would be the superlative of a network society. It is not only a society where everything is connected with everything, but also a society where extreme monitoring is possible to optimise systems. This system has huge advantages, but one of its most important assets is the development of product intelligence (PI) and Manage MyMachines (MY) (Siemens, 2018). PI is used to automate insights from product performance data to create actionable intelligence. MY leverages data analytics and product and process expertise at a level that makes it possible to transform critical machine data into data-driven analytics.

Product Intelligence and Manage MyMachines

Case 11.1. A smart city app connects Expo 2020 in Dubai to IoT[1]

A new cloud-based app that will support Expo 2020 Dubai in becoming a blueprint for future smart cities is being developed by Siemens, and has already connected part of the site's infrastructure to the Internet of Things.

The app is designed to digitally optimise Expo 2020 Dubai, enhancing the visitor experience and reducing energy and water use during the six-month event. Initially operating across two domains – environmental monitoring and irrigation – the app will collect, monitor, correlate and analyse data from the site, and visualise it to allow real-time monitoring and control of infrastructure from static and mobile devices.

Currently under development in close collaboration with Expo 2020 Dubai, the app is running in a test phase at the Expo Site Project Office. It will be scaled up as infrastructure is completed. In addition to environmental monitoring and irrigation, further urban domains will be added in subsequent phases, including Sustainability, Electric Vehicle Charging Monitoring and other data lake use cases, incorporating multiple smart city elements into a single app. Eventually it will be connected to the full Expo ecosystem, creating the most connected Expo in history and demonstrating its ability to integrate a city's infrastructure into the Internet of Things.

> This new digital application will support Expo 2020 Dubai in becoming the most connected world fair in history; a great step forward as we continue to explore how digitalisation and IoT technologies can address urban challenges,' said Dr Roland Busch, Chief Operating Officer and Chief Technology Officer, Siemens AG. 'By integrating the infrastructure of the largest-ever event in the Arab world with the Internet of Things, we have the potential to inform the development of future smart cities around the world.

Mohammed Alhashmi, Chief Technology Officer at Expo 2020 Dubai said: 'Siemens' app is a perfect example of what we're trying to achieve as we prepare to host Expo 2020 – providing a test bed for our partners to deploy innovation at scale and explore the potential for new technology to deliver enhanced operations and experiences, in line with our sub themes of Opportunity, Mobility and Sustainability.'

[1] The full text block is quoted as an example from: https://www.securityworldmarket.com/int/News/Business-News/siemens-develops-smart-city-app-for-expo-2020-dubai1.

Let us illustrate the importance of these applications for the development of a hyper-connected society with an example[9]:

Starting from the idea of improving the efficiency of gas turbines, we could link the installation to PI. It delivers data that is processed in an extraordinary way: namely by using a very powerful neural network model. Such a model can be used to analyse the relationship between numerous inputs and outputs of a turbine or other installation over time. The raw data, collected in real time from the turbine installation, is loaded into a computational system that enables extraordinary analysis (MY). One of the extraordinary possibilities is known

[9] The example of 'Quantifying the Unknown' was inspired by the description of NOx gas turbines, see Siemens (2011), and supplemented by a personal explanation of the possible relationship between PI and MY.

as 'quantifying the unknown'. Mathematical models, with the help of PI, start with raw data only. It creates a virtual duplicate of the turbine from which all calculations can be made.

The ability to quantify the unknown, and to apply it on a large scale and in other domains of the connected society, is the greatest virtue of this development. The addition of the prefix 'hyper' is important because the degree of success depends on the quantity and quality of data. A hyper-connected society is necessary in order to make this possible.

As the system learns the relationship between variables, the model and the virtual duplicate grow closer to reality: it duplicates and imitates the behaviour of the turbine, and is ultimately able to predict its behaviour (real time) with great accuracy (Siemens, 2011: 54). Similarly, it can be applied to other areas of society where data and accuracy are functional: e.g. moral behaviour and the ability to avoid decisions that lead to morally bad outcomes in the future. In this context, it is important to point out another powerful development: the possibility of making predictions with the right data sets and calculation models, the so-called science of prediction (Siemens, 2011).

11.2 The science of prediction

Since ancient times, people have tried to predict the future. The Romans looked at bird formations in the sky, the so-called auguries. And Nostradamus used a technique dating back to biblical times known as bibliomancy: 'a selection of extracts from older sources at random and then used astrological calculations to project its recurrence in the future'[10] (Schumm, 2018). It does not mean that there are no possibilities of predicting the future. The main goal should be to discover data-driven patterns in an accurate way, based on facts (data) to reduce bias and beliefs.

In the magazine 'Pictures of the future', a narrative example was written to illustrate this data-driven pattern recognition (see Case 11.2).

[10] 'One of the major sources used for his most famous work, "The Prophecies", was the "Mirabilis liber" of 1522, an anthology of prophecies from well-known seers of the time, while "Livre de l'estat et mutations des temps" by Richard Roussat provided the basis for his astrological references', quoted from Schumm (2018).

Case 11.2. Data-driven pattern recognition (Siemens).

Take a break and look out of the window for a second. What do you see? Partially identifiable shapes – or buildings and trees? Chances are that if you had never seen a building or a tree, and had never even heard of such things, the view might indeed appear to be confusing jumble. The reason that it isn't is that you have models in your mind that organise the enormous amount of data in your field of visions into things that are instantly identifiable.

Now, the same sort of challenge is being confronted by complex man-made systems; but models in question make sense of patterns that are so multifaceted and so invisible to our senses that no human being could even see them. They are the patterns from which with steadily growing success, predictions can be made (Siemens, 2011: 52-53).

Software Environment for Neural Networks

This form of pattern recognition is already being used in applications in an industrial context, for instance the gas turbine example described in the previous section. The spectacular space that emerges from a collaboration between mathematical models and neural networks was given a name: a Software Environment for Neural Networks (SENN). 'It is the most advanced high-dimensional, non-linear modelling system of its kind.[11] It is a learning system capable of forecasting everything from the price of copper to the best place to locate a factory.' (Siemens, 2011: 53) Using very powerful neural network systems and mathematical models (SENN) and the continuous collection of data, it is possible to match the demand and supply of gas, simulating the behaviour of the installation so accurately that the adjustment can anticipate future events (predictions). These mathematical models and technology are currently used to match electricity supply and demand, but also to predict the failure of parts in machines so that they can be replaced in time (Siemens, 2011).

The important momentum in the history of our human and technological existence follows from Case 11.1 and 11.2: the scientific ability to predict. So why could such a network not be applicable to the *moral behaviour* or aspects of a society? I have elaborated on this question, and have also given this hypothetical system a name: the Moral MindSphere (Bamps, 2021: 156-167).

[11] Quoted from Dr Hans Zimmerman of Siemens Corporate Technology, the founding father of the system.

11.3 Moral MindSphere (MM)

This hypothetical system runs on data – any kind of data – that relate to moral behaviour. Every human action can be analysed. Moral actions are a subcategory of these actions with most of their roots in the moral instinct, the flesh of morality, as described in Chapter 19. With the development of mankind, rationality became increasingly important in justifying actions: from a neuropsychological perspective, the power of the cortex (which evolved later than the emotional centres). There is always a reason for a moral action (because accountability can be demanded afterwards). As rationality in justification increased, reference frameworks, so-called ethical normative reference frameworks, emerged in history. These make justification easier and clearer, because reference can be made to predetermined criteria. Each moral action is, from a normative ethical perspective, always reviewed against a frame of reference. Some normative models allow better measurement of the outcome using an agreed measure and retrospective data, but concreteness is certainly possible with many simple agreements. At the most basic level: did someone do what he said or promised?

This can be done on a very simple scale[12] to begin with, similar to a thermostat in accordance with the rules of cybernetics. An action is then measured against a predefined threshold. The data-output may not be as defined or nuanced as we would want it to be, but it is not unimportant either from a normative perspective. For example, imagine you would want to better monitor and control bullying. Suppose you set a threshold: after a warning from a person who finds the behaviour hurtful, you must stop. If you receive two warnings, you have crossed the critical threshold. In social situations it is often complex to set control variables and threshold values, but when behaviour is related to something very concrete and measurable, one can easily set an MM. A simple example is elaborated in Chapter 19, the so-called Laura case: not using too much water when taking a shower is relatively easy to measure. But the MM can theoretically go much further than that. By analogy with the Siemens smart city app (The Expo 2020 Dubai), it will be possible to map more and more moral behaviour with meaningful parameters – even when it comes to highly complex data. Within the described science of prediction, it should even be possible to discover patterns in moral behaviour that make prudence possible. It is about patterns in moral behaviour that we cannot observe from a human point of view.

[12] It is interesting to think of Popper (1957) in this context and in relation to the science of prediction in the previous Section 11.2. He argues that the only form of social engineering that can be rationally justified is one which is small-scale, incremental, and continuously amended in the light of experience. See the notion of piecemeal engineering in *The Poverty of Historicism*.

The problem with humans is that cognition is limited. Most people are not chess masters or mathematical prodigies, and can think only a few steps ahead. The result can be a bad decision that one could easily have avoided if one had seen the outcome. In the film 'A Beautiful Mind', there is a famous scene where the young mathematician John Forbes Nash jr. asks himself the question, 'Will he tempt the pretty girl who walks in or not?'

What follows is a dizzying decision tree in which he goes much further than his normal fellow students in making this decision. Some normative models based on consequentialism, for example, require great predictive computing power. If one could reveal the hidden patterns of life, decision trees would probably look very different. Evil or sin, from that perspective, is attributed more to a lack of computing power (mathematical models) – that is not the same as the decision itself, which is always made in a limited cognitive context from a human perspective: a small fluctuation in input can lead to a dramatically large difference in (moral) outcome (Bamps and Roels, 2009).

Technology can play a very important role in increasing the computing power to discover hidden patterns of moral behaviour and improve our ethical decision trees. This creates high expectations and the possible applications are numerous. An important factor is the model used to recognise patterns. This is also evident from the example given by the mathematician Nash in his amorous deliberations (see A Beautiful Mind, as described above). Within the doctrine of normative theories, there are a number of models available (see also Chapter 4 in the context of moral programming) However, not all of them can be easily operationalised (Wernaart, 2021). At the same time, there are situations that are very easy to operationalise, even if the normative models use different arguments to test not only the correctness of a moral action but can also ensure the monitoring of moral behaviour, especially when this is necessary after criminal offences. To illustrate this, let us take a simple hypothetical example with different underlying normative frames of reference:

deontological perspective

From a deontological perspective, rules, laws and maxims are important. They also colour the values and norms one sets when judging actions. The use of technology can provide interesting opportunities for judicial and supervisory authorities.

Suppose you are the tenant of a property and you have come before the court for dealing drugs. They had received a tip-off, and after observation the police actually saw people dealing from the living room: people would ring the doorbell and then disappear again. Police officers entered the house and collected all the possible evidence. The case was crystal clear. The judge had no trouble reaching verdict. A sanction was imposed. Besides community service, the judge also wants you to have no more contact with dealers or other shady characters. In the house, sensors are activated that allow for in-depth monitoring: this can be about controlling who goes in and out of the house, but also the activation of chemical sensors (sniffers, or sensors built into the toilet) can do their job to immediately detect the presence of drugs.

care ethics perspective

From a care ethics perspective, the emphasis is on relational aspects: connection and care. Further investigation also revealed that domestic violence took place within the family. Children were not only neglected, but also regularly beaten up. The defendant, and in particular the mother of the children, was also addicted to alcohol. This certainly contributed to the aggressive atmosphere and the lack of parenting skills.

Technology can play a role in the follow-up of these crimes: in particular, the use of courtroom technology: a dashboard that allows different agencies to follow a case in real time and very closely. The judge decides to use the dashboard in this case in cooperation with several other agencies involved: child protection, social workers, a family coach, the doctor, the neighbourhood policeman, the principal of the school and the Alcoholics Anonymous (AA) club. Other controlling or restrictive sanctions may also be imposed in the context of associational behaviour in general.

For example, when assessing criminal cases of domestic violence, it is important to know whether an offender regularly attends the therapy sessions. This information often arrives too late on the judge's desk. Technology can improve this.

11.4 What are the possibilities now and in the future?

The use of anti-social behaviour orders (ASOs) is an important tool in the UK (Flint and Nixon, 2006). Why would this not be possible in the Netherlands? According to Vols (2013), Dutch legislation also offers an excellent basis for adopting this approach, based on what is called in Dutch Law 'gedragsaanwijzing' (Van Wezel, 2013; Vols and Sieljes, 2018). Vols is referring to article 6:265, 3:296 and 7:213 of the Dutch Civil Code (Centrum voor criminaliteitspreventie en veiligheid, 2021).[13] It enables the judge to monitor tenant behaviour, to

[13] 'The landlord submits a claim to a court. Primarily, he claims the dissolution of the lease agreement and the eviction from the rented accommodation (articles 6:265 and 7:213 of the Civil Code). The landlord claimed that the tenant must be ordered to perform his obligations as a good tenant (articles 3:296 and 7:213 of the Civil Code) and a related penalty payment.' Quoted from Centrum voor criminaliteitspreventie en veiligheid (2021).

consolidate what is socially acceptable in an agreement, and in case of non-compliance, the judicial powers have sufficient ammunition to take action (e.g. financial penalty or eviction).

Technological infrastructure can play a very important role in monitoring unethical, anti-social or problematic behaviour, especially if this is required within the IoT environment of a smart home of the future. This could be an interesting avenue for regulating social and moral behaviour. In that view, it fits perfectly within the regulation area of a moral MindSphere. The IoT technology could be used perfectly well by the relevant agencies. The intelligent environment could result in a very powerful dashboard: a dream for every social engineer with its own toolbox. The science of prediction (see Section 11.3) can lead to meaningful applications in the social and legal context. Based on the right parameters, risk markers and real-time monitoring, it would be possible to predict when risky situations occur, for example in the case of domestic violence (together with other applications of instruments to assess the likelihood that violence will recur and escalate) (Dutton, and Kropp, 2000; Gundolf, 2002).

The combination of the normative approaches A and B can be a very powerful blend from the perspective of a Moral MindSphere. At the same time, it is a very clear example of what moral design and technology can achieve in practice. It follows the path of a society where the exchange of information will become increasingly important. So important, in fact, that information can ultimately be equated with a regulatory and economic value. Using advanced analytics and AI within the framework of a Moral MindSphere powers IoT solutions from the edge to the cloud with social data from connected smart house infrastructures to optimise judicial and behavioural regulations, create better moral environments and deploy new business models: including information revenue models. The creation of a moral environment can have significant societal benefits. At the same time, there are challenges when data becomes a profit-driven economic source. These challenges need to be addressed urgently, as emphasised in Case 11.3.

Case 11.3. Information revenue model: a devil's bargain

Technological infrastructure is also installed in rental properties for other reasons. A good example is the research that took place during the implementation of the strategic action programme in Limburg (Belgium). I was fortunate enough to be one of the subcontracted project writers for the Limburg Innovation Centre. It gives me the opportunity to explain this issue as a primary witness: information revenue models is a remarkable story of good and evil.[1] A brief outline of the most important findings:

In the year 2013, the research plan to put technology into homes was started. The living lab condition provided an opportunity to investigate how rental houses could be better energy-regulated. This depends not only on the technology in the house, but also the behaviour of the residents. As the research progressed, a separate working group was set up to consider possible economic-revenue models as well. The philosophy was as simple as it was dangerous. If technology worked well in homes, it could significantly reduce electricity and heating costs. Information sharing is very important for improving technology and the AI learning process. It is the basic assumption for general regulation in smart homes. It's about monitoring when someone switches on a fan above the cooker, cooks, opens windows, knowing when someone is in certain rooms, comes home from work, how much water is consumed when taking a shower and flushing the toilet, etc. If someone wants to exchange information, this can help reduce costs and improve regulation. Something could be done in return: for example, a reduction in rental costs, tax relief or other benefits.

What would be the result of this philosophy? At best, it is basic information with no privacy complications. At worst, it is sensitive information that people are reluctant to share. If it is purely about optimising energy use in a house, there may not be a problem; but it is a dangerous deal. One never knows what will happen in the future. There is always the issue of who controls the controllers. Moreover, the tenant might put themselves in a weaker legal position in the event of future conflicts.

Suppose a tenant complains about mould. In normal circumstances, the tenant and the landlord would speak in court. Further evidence is often lacking. In circumstances with a technological infrastructure, the rental housing cooperative can put a logbook on the table with all the data that prove the inappropriate behaviour of the residents. With the available data, the case takes on a completely different legal weight.

Another issue is the vulnerability of the stakeholders. Poor people, in particular, might agree to share information against their moral judgement. If they are forced to choose between privacy or bread on the table, the choice is quickly made. After all, the rich tenants would be giving up their privacy rights voluntarily because they do not need a discount. The poor tenants have no choice. It is a devil's bargain. These findings are based on a survey (n=277) conducted in 2019 as part of a living lab study on incorporating technology into social housing to make it energy neutral (Bamps, 2021: 206).

[1] The full report can be found in Bamps, D. (2021), empirical study No. 4, p. 207.

11.5 Conclusions

The Moral MindSphere demonstrates hypothetically how powerful technological IoT infrastructure can be in regulating moral behaviour. When the values of a moral system are clearly operationalised, it is possible to set a threshold and provide feedback. It starts with the monitoring of very simple moral actions. With the advent of an intelligent IoT infrastructure in smart homes, nothing will stand in the way of this evolution. It also underlines the need that exists due to a lack of social and legal regulation at the moment.

A powerful example is the use of anti-social behaviour orders in the UK. By analogy, the same path can be chosen in the Netherlands, in particular because Dutch legal law theoretically allows for this possibility (Bamps, 2021: 213).[14] When judging anti-social or criminal behaviour, evidence is needed. In the follow-up, technology in smart homes could be very useful and confirm the judge's verdict. This can be in the form of courtroom technology but also through the use of chemical sniffers, intelligent cameras and other surveillance gadgets in the infrastructure of the rental property. These are cases where an addicted person is forbidden from having dealers or drug users in their home, must see a therapist regularly (this has to be monitored), and where there is a history of domestic violence (where close monitoring is necessary and immediate action can be taken). It is also nice case study for showing how moral design and technology can support normative theories: in this case a deontological approach (cf. you must not violate rule X or Y) combined with care ethics (cf. you have to take care of your children). When the science of prediction is used, it can even act as a very powerful risk assessment tool with enormous anticipatory or preventive value.

The development of IoT environments where information exchange is so important can also have a dangerous outcome. This danger arises when certain business models are applied in precarious and questionable situations. A notable example is the investigation of technology in rental houses within the framework of nearly zero-energy (NZE) standardisation (cf. living lab research, IWT). The idea of using technology and monitoring to reduce energy consumption inspired the idea of giving tenants a discount, rent reduction or other benefit in exchange for sharing privacy-sensitive information (resident behaviour). Vulnerable tenants would have no choice: they would be forced to choose between survival (food on the table) or sharing privacy-sensitive information.

---

[14] It refers to the legislation of what is called in Dutch 'gedragsaanwijzing' (Vols, 2014). See also Raaff (2015).

Building IoT infrastructure can be the stepping stone to a moral MindSphere. When functioning properly, this environment can help stimulate and maintain moral behaviour on a micro, meso and macro level. But there is also a dark side. The same infrastructure can, under certain circumstances, result in undesirable ethical situations or unethical business models.

References

Bamps, D., 2021. De Dashboardsamenleving: van improvisatie tot providere. Boom Uitgevers, the Hague, the Netherlands.

Bamps, D. and Roels, R., 2009. Gordon Research Conferences: Visualization in science & education – revealing nature, generating insight. Visualization of a beautiful mind, mathematical eye-openers of moral behaviour and related (hidden) consequences. Magdalen College, Oxford, UK.

Boutellier, H., 2019. Over wat niet deugt, wat je eraan kunt doen en wat je daarvan moet denken. Afscheidsrede 'Hoe veilig verder!?' van prof. Hans Boutellier. Verwey-Jonker Instituut, Utrecht, the Netherlands. Available at: https://www.verwey-jonker.nl/artikel/afscheidsrede-prof-hans-boutellier/.

Castells, M., 2004. Informationalism, networks, and the network society: a theoretical blueprint. Edward Elgar Publishing, Northampton, MA, USA.

Centrum voor criminaliteitspreventie en veiligheid, 2021. Veelgestelde vragen over de gedragsaanwijzing. Available at: https://www.hetccv-woonoverlast.nl/gedragsaanwijziging/veelgestelde-vragen.

Dowling, M., Eberspächer, J., Neuburger, H., Noll, E. and Zisler, K., 2016. Neue Produkte in der digitalen Welt. Books on Demand, Norderstedt, Germany.

Dutton, D. and Kropp, R., 2000. A review of domestic violence risk instruments. Trauma, Violence and Abuse, 1: 171-181. https://www.jstor.org/stable/26636247

Flint, J. and Nixon, J., 2006. Governing neighbours: anti-social behaviour orders and new forms of regulating conduct in the UK. Urban Studies, 43: 939-955. https://doi.org/10.1080/00420980600676386

Gundolf, E., 2002. Batterer intervention systems: issues, outcomes and recommendations. Sage Publications, Thousand Oaks, CA, USA.

Jeong, W.-S., Kim, S.-H. and Min, K.-S., 2013. An analysis of the economic effects for the IoT industry. Journal of Internet Computing and Services, 14: 119-128. https://doi.org/10.7472/JKSII.2013.14.5.119

Ørsted, H.C., 1997. Selected scientific works of Hans Christian Ørsted. Translated to English by Jelved, K., Jackson, A.D. and Knudsen, O. (translators). Princeton University Press, Princeton NJ, USA.

Popper, K., 1957. The poverty of historicism. Routledge & Kegan Paul, London, UK.

Raaff, H., 2015. Voor de goede orde: handhaven van regels en wetten. Uitgeverij Coutinho, Bussum, the Netherlands.

Reinheimer, S., 2017. Industrie 4.0: Herausforderungen, Konzepte und Praxisbeispiele. Springer Verlag, Berlin, Germany.

Sarbagya, R. and Sudan, J., 2019. Chapter 9 – Internet of Things application in life sciences. In: Valentina, E., Le Hoang, S., Sudan, J., Manju, K. and Raghvendra, K. (eds.) Internet of things in biomedical engineering. Academic Press, Cambridge, MA, USA, pp. 213-233.

Schumm, L., 2018, 31 August. History Stories: 5 Things you may not know about Nostradamus. Available at: https://tinyurl.com/c5evh56j.

Siemens, 2011. The science of prediction (Machine learning | neural networks). Pictures of the future, The Magazine for Research and Innovation, pp. 52-56.

Siemens, 2018. MindApps and applications for MindSphere. Available at: https://tinyurl.com/uvnsfd2a.

Siemens, 2021. MindSphere: Connecting the things that run the world. Available at: https://siemens.mindsphere.io/en.

Van Wezel, M., 2013, 27 August. Is een gedragsaanwijzing aan uw huurder beter dan een ontruiming? Tomlow Advocaten. Available at: https://www.tomlow-advocaten.nl/nieuws/is-gedragsaanwijzing-aan-huurder-beter-dan-ontruiming/.

Vols, M., 2013. Woonoverlast en het recht op privéleven. De aanpak van overlastveroorzakers in Nederland, Engeland, Wales en België. Boom Juridische uitgevers, the Hague, the Netherlands.

Vols, M. and Sieljes, L., 2018. Gedragsaanwijzing ingebed in het huurrecht? Centrum voor Criminaliteitspreventie en Veiligheid (CCV), Utrecht, the Netherlands.

Wernaart, B., 2021. Developing a roadmap for the moral programming of smart technology. Technology in Society, 46: 101466. https://doi.org/10.1016/j.techsoc.2020.101466

Media

Vols, M., 2014, February 12. Overlast aanpakken met gedragsaanwijzingen: antwoord op de meest gestelde vragen. Available at: http://www.openbareorderecht.nl/index.php/overlast-aanpakken-met-gedragsaanwijzingen-antwoo

12. Designing a griefbot-for-good

Rens van der Vorst[1]* and Jo-An M. Kamp[2]

[1]Fontys University of Applied Sciences, Corporate IT department, P.O. Box 347, 5600 AH Eindhoven, the Netherlands; [2]School of Information and Communication Technology, Fontys University of Applied Sciences, P.O. Box 347, 5600 AH Eindhoven, the Netherlands; r.vandervorst@fontys.nl

Abstract

We teach our students to solve problems with technology. After all, the possibilities of technology often seem limitless. Students design technological solutions to problems in their field or in society. We think that is important. Our educational programmes buzz with terms such as 'technology-for-society' or 'technology-for-good'. But how do you determine the impact of new technology? When is the impact positive? When does a technology make something better? What do we even mean by 'better'? It is already extremely difficult to assess the impact of an existing digital technology on technological, ethical, democratic, and economic affairs. But this assessment may be even more complicated if a technology does not yet exist. Yet it is precisely then that an impact analysis is extremely important. After all, everything is still possible in the design phase. We tell our students that a good idea is not enough. That a good idea only gets better when you ask better questions. That is why we introduce them to the *Technology Impact Cycle*. This is an approach we crafted based on four existing frameworks. We introduce our students to the Technology Impact Cycle by asking them to design a griefbot with a positive impact. In this chapter we describe the process that we have now gone through a number of times. By describing this process we introduce the reader to the Technology Impact Cycle (which can also be used for other technologies and by people other than students) and we take the reader into the controversial world of digitally bringing the dead to life. As a central example, we use the hypothetical design of a griefbot.

Key concepts

- The Technology Impact Cycle is a method that can be used to measure the impact of new technology

- The Technology Impact Cycle Tool is composed of a small and an extended cycle that help assess new technology on ethical, democratic, legal, technological, and economic aspects after which they initiate a process of improvement.

12.1 Applied ethical design of new technology: the griefbot-for-good

The decision to use the griefbot as an example with the students, as well as in this chapter, is a fundamental one. It is tempting to just say 'no' to griefbots. To call them creepy. Let the dead rest in peace! But with technology the ethics of 'yes' or 'no' often does not fit well. This is because technology is a medium in which we live (Verbeek, 2014), just as a fish lives in water. This means that technology ethics is usually not about whether a particular technology should or should not be. The ethics surrounding technology should focus on *how* a particular technology is designed, which is also far more challenging.

This also applies to griefbots (see Case 12.1). As we will see, there are already working prototypes. Developments are moving fast and it is realistic to assume that commercial griefbots will be widely available within a few years. While we probably cannot prevent them from entering the market, we can ask ourselves *how* we can increase the chance that the design of a griefbot will lead to a positive impact.

Case 12.1. Aunt Ad.

On April 18, 2015 my (Rens) beloved Aunt passed away at the age of 76.

Aunt Ad was my father's sister and the last of six siblings still alive. Aunt Ad was special. All her life she lived in the same old, stately house. A house full of bulging bookcases and overflowing ashtrays. Aunt Ad was well-read and interested in my family. When we came to visit she photographed us fanatically with an old-fashioned camera. She had the photos developed and sent them to us by mail, usually accompanied by a postcard with some kind words on it. Aunt Ad called me every other day and if I could not answer the phone she would leave me lengthy messages. At the same time, Aunt Ad had her problems. She filled the gaps in her life with medicines, unhealthy food and cigarettes. Lots of cigarettes.

When Aunt Ad passed away, I cleared out the house together with my brother. After all, we were the only heirs. In her study we found a huge pile of full diaries. In her neat, accurate handwriting, Aunt Ad had entrusted all her feelings, ideas and thoughts to paper every day for sixty years.

Suppose Aunt Ad had turned 86. Then she would have died in 2025. Then there would probably have been online services available that would have made it possible to bring the dead to life digitally as a chatbot, or as an avatar, or hologram. In 2025, chances are you just need to upload the necessary input. We had all the material. Diaries, photos, voicemail.

But would we have done it?

The Technology Impact Cycle is a simple method. There is a small cycle and an extended cycle. Students start with the small cycle, constantly making changes to the design by asking themselves what the exact technological problem is, what the contextual particularities are, and who its users will be.

The output of the small cycle is the input for the extended cycle. There, students try to answer questions about the impact of technology from very diverse angles, each time checking what consequences this has for their design choices.

In Section 12.2 we describe the small cycle. In Section 12.3 the extended cycle.

Before you read on, ask yourself a few questions. What do you think a griefbot is? Do you think it is possible to design a griefbot with a positive impact? Or do you think, like Frankenstein's author, Mary Shelley (Shelley, 1818), that technological innovation tends to outpace all moral, ethical, and legal restraints, and that a griefbot will become an uncontrollable monster?

Let's find out.

12.2 The small cycle

First, our students are introduced to the Technology Impact Cycle, which itself consists of two cycles that help assess new technology on ethical, democratic, legal, technological, and economic aspects after which they initiate a process of improvement.

This method is based on four existing frameworks:

- The Technology Impact Cycle Tool (Fontys University of Applied Science, 2020), which helps to assess the impact of a new or existing technology from all sides by means of smart scans and smart questions.

- A techno-ethical framework (Krutka *et al.*, 2019) that is mainly suitable for existing technologies, but certainly also has added value for the analysis of new or proposed technology.
- The approach to guidance ethics (Verbeek and Tijink, 2020) that helps to make recommendations for a more ethical use of a specific technology.
- A small wins framework (Termeer and Dewulf, 2018). This framework was conceived as a contribution to the progress of the approach to wicked problems. However, the framework and the introduction of the concept of small wins also offer starting points for designing and reviewing improvements in the design of technology.

The above four frameworks have shaped the Technology Impact Cycle as follows: the ideas from the approach to guidance ethics form the basis for the small cycle. This has been expanded with a number of elements that make the Technology Impact Cycle particularly suitable for analysing the impact of new technology. Inspired by the Technology Impact Cycle Tool and the techno-ethical framework, the extended cycle has been added (Figure 12.1). Finally, the concept of small wins helps to critically question the proposed design changes.

The Technology Impact Cycle starts with the small cycle, which aims to further refine the design of: (1) the technological solution by asking questions about; (2) the context and the users; and (3) the exact nature of the problem.

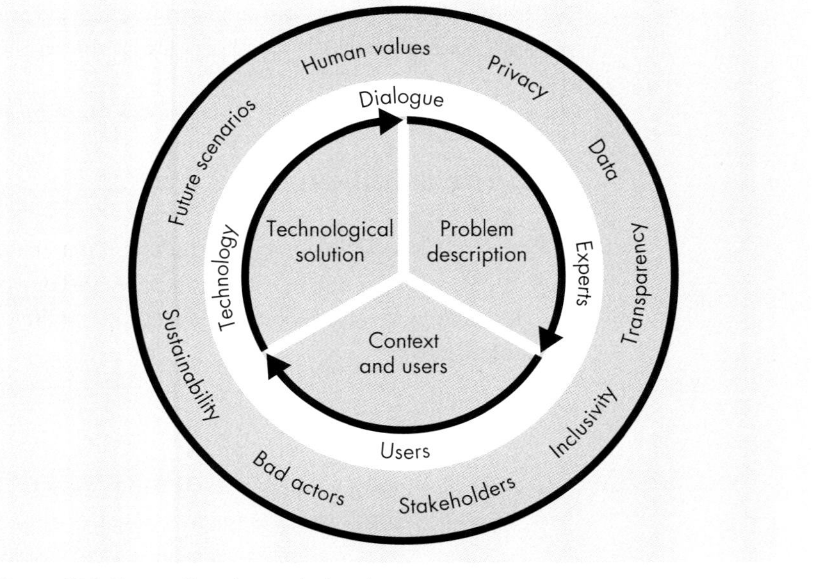

Figure 12.1. The small and extended cycle.

Students find answers in dialogue by immersing themselves in the subject matter, in conversation with experts and potential users, and from a realistic understanding of the technology. The answers force them to make choices, which often lead to new questions. That is why students constantly circle back from context/users to problem description to design. Hence, the cycle.

It also means that the technology design choices can only be made in relation to the description of the problem, the users, and the context. These are also the options for action. Students can adapt the design of the technological solution, and/or the problem description, and/or the users, and/or the context.

The design changes are judged on the extent to which they contribute to the purpose of the technology. The aforementioned small wins framework for wicked problems describes small wins as those improvements that lead to an in-depth change, which are sometimes radically different from previous proposals and which contribute to a positive assessment of the issue. Small wins are therefore completely different from quick wins in this context. The idea behind this way of thinking is applied to the design of new technology in the Technology Impact Cycle.

In the remainder of this section we will illustrate the use of the Technology Impact Cycle with the example of the griefbot.

The idea is that students move criss-cross from context to technology, to users, to problem description. In this way students discover a context, make a first draft of the problem, come up with a solution, and get to work on it.

In the chapter below, we first provide an overview of the context of griefbots, then we look at technological developments, and finally we discuss the technological problem (Section 12.2.1 and 12.2.2).

Furthermore, we have compiled a more or less random summary of examples of work by students that best serves to demonstrate the added value of the technology impact cycle and that puts the discussion around the griefbot in sharp focus (Section 12.2.3).

12.2.1 Context and users

Technology is designed to solve a problem. But what problem will be solved? For whom? And where? Asking these questions will help uncover the particularities of the context of new technology and the nature and interests of its users.

By asking these questions in the context of our griefbot, the student will quickly discover that the term 'griefbot' can mean a lot of things. Let's look at some examples.

digital conversation assistance

Conversations about impending death are complicated. Whether it is a conversation between a doctor and a patient, or a parent and a child, these are always difficult and loaded with emotions, and tough decisions. Therefore, it is often recommended to talk to a professional counsellor. However, this can be expensive and is generally not covered by health insurance.

Partly for these reasons, experiments are conducted involving a chatbot that facilitates the conversation. Think of a tablet with software that terminal people can use to have a conversation. The first results are encouraging (Utami and Bickmore, 2018) and more extensive versions are expected soon.

So, in this case, a griefbot is a digital tool that helps users prepare for impending death.

digital grief counselling

A completely different type of griefbot helps with grief processing. Grieving is a natural process. However, that does not mean that all people grieve in the same way. For some people, the grief is so intense that it leads to serious emotional suffering, which can even lead to (thoughts of) suicide. This condition is officially diagnosed as a mental illness in the Diagnostic and Statistical Manual of Mental Disorders, 5th Edition.

With this in mind, experiments are being conducted with digital services that use synthetic media to enable the patient to have therapeutic conversations with the deceased. Synthetic media are media (images, audio, video) that are manipulated or entirely generated by artificial intelligence. When synthetic media are used in bad faith, we call them deepfake (Schick, 2020).

In South Korea, for example, a documentary called 'Meeting You' was screened, in which a mother is reunited with her deceased daughter in virtual reality. The daughter was brought to life by creating a virtual reality avatar using child actors together with photo and audio material of the daughter. Some may find this technological application a bit morbid. In this case, however, it had a positive impact on the well-being of the mother.

In another example, a conversation takes place with a deceased loved one. The 'role of the deceased loved one' is played by a counsellor, but the next of kin or friends see a synthetic adaptation of their loved one on the screen. Rapidly developing technology makes the experience (audio, facial expressions, mouth movement) increasingly real and more accessible to therapists.

resurrect the dead

When you walk through the fairy-tale forest in *World of Warcraft*, you sometimes come across a mossy monument of a player who died in real life. We see the same more and more in Massively Multiplayer Online Role-Playing Games: people who vanquished hundreds of monsters in-game, but succumbed to coronavirus in real life.

An estimated 30 million deceased people have a profile on Facebook. This makes Facebook the largest cemetery in the world. Much larger than Wadi-us-Salaam in Iraq where 'only' ten million people are buried. But what if we could now bring the people on Facebook back to life? There is enough data available for experimentation with that.

The griefbot 'from the afterlife' is a digital technology that uses the data that someone has generated in a lifetime to bring her/him back to life. The first experimental versions are chatbots. They use artificial intelligence that is trained with all kinds of data (e.g. text messages, posts, e-mails). As a result, you can chat with a digital representation of someone who has passed away.

In combination with the previously mentioned developments in the field of synthetic media, we see that computers are becoming better at generating audio and images. It is therefore in line with expectations that in the near future you will not only be able to chat with the deceased, but also have realistic, face-to-face, conversations with them.

12.2.2 Technological design

The above examples show that the term 'griefbot' can be loaded with all kinds of manifestations and can be used in all kinds of different contexts. It makes a big difference whether you design a chatbot that is used by healthcare professionals to treat terminal patients or if you design a commercial chatbot that can be downloaded by anyone. In the first case, you probably make very different design choices than you do in the second. But in both cases, the aim can be to have a positive impact. That is why students make choices, circle back, ask new questions and eventually find out if they have made the right choice (a technological design with a high probability of a positive impact).

To support this process, it is important that students explore the potential of the technology itself. This gives them a good impression of possibilities and impossibilities. Dialogue, reading, and consulting experts are again of great importance.

In the case of the griefbot, students learn that the first prototypes of griefbots already exist. Some can be downloaded, some are in design, some are mainly fiction, but together they give a good impression of the concept, the possibilities of the technology, and provide a glimpse into the future.

Below we give an impression of the technological possibilities. In this case, we work on the assumption that the students opted for an after-life griefbot that could potentially be used by anyone. We discuss three types of technology that can be used to construct this kind of griefbot: the text-based griefbot, the synthetic griefbot, and the physical griefbot.

text-based griefbots

In 1966 Josef Weizenbaum designed ELIZA, the world's first chatbot. The idea behind ELIZA was to demonstrate the shallowness of communication between humans and machines, but the users attributed all kinds of human feelings to the computer program (Weizenbaum, 1976). Weizenbaum's secretary even asked Joseph to leave the room because the conversation became too personal. Chatbots are much better today. Where Weizenbaum's program only responded to keywords, chatbots are nowadays increasingly powered by advanced, artificial neural networks.

In 2015, Roman Mazurenko died in Moscow at the age of 34 in a car accident. A friend, Eugenia Kudya, used a neural network to bring Roman back to life as a griefbot. She downloaded hundreds of Roman's messages from the Telegram messaging app and used them as input for the griefbot, which she then put online so that all his friends could talk to 'Roman'. The reactions were mixed, but mostly positive. You can download the latest version of Roman from the app store.

There are more examples of text-based griefbots. James Vlahos wrote down long conversations with his terminal father with the aim of building a griefbot that could emulate his father. The working title? A dadbot. Muhammed Ahmad is working on an AI-version of his father so that one day his kids can get to know their grandfather.

These text-based griefbots will only get better. Generative Pretrained-Transformer-3 (GPT-3) is a language model that uses deep learning to generate texts that are almost indistinguishable from human-made texts. With GPT-3, entire news articles have already been written for *The Guardian* (Theguardian.com, 2020).

In addition, Roman the griefbot resulted in the company Replika (Replika.ai, 2021), which focuses on chatbots-as-a-companion and has millions of customers, including people who have a lifelong subscription because they are in love with a chatbot. Now, that's commitment! Replika uses a deep learning model called sequence-to-sequence that is very capable of imitating how people talk.

Not only are the models improving at lightning speed, people are also leaving more and more data behind. The 'digital remains' of someone are already often enormous and include e-mail, text messages, and social media posts. The input with which the artificial networks are trained is therefore forever increasing and the output, therefore, increasingly realistic.

In a next phase it is also conceivable that the griefbot is no longer a stand-alone chatbot, but becomes part of existing social media channels. A griefbot that sends you a text message to wish you good luck for an exam, or likes one of your messages, or maybe posts a photo, or starts a conversation with you on Clubhouse.

In the latter case, you need a talking griefbot. That may seem far-fetched, but it is not.

synthetic griefbots

Synthetic media, as we described in Section 12.2.1, are media manipulated or generated with AI technology. These can be images or audio. For example, there is a speech by John F. Kennedy that he never actually gave, yet thanks to artificial intelligence you can hear the typical JFK voice. It is expected that a few minutes of someone's voice will soon be enough to generate credible audio.

Developments are also coming up to speed in the field of images. *MyHeritage's Deep Nostalgia* website (Myheritage.nl, 2021) allows you to upload photos of people (such as a grandparent) and then the software brings the photos to life seemingly effortlessly. An amazing sight, reminiscent of the moving people in the Harry Potter picture frames.

Combine audio and image and there are countless examples of synthetic media. There are deceased pop stars acting as holograms. In the Dali Museum in St. Petersburg, the grandmaster speaks to you. But the living are also affected. There are countless computer-generated porn videos circulating on the Internet, usually starring Emma Watson or Scarlett Johansson.

Combine the input from the text-based griefbots with the synthetic options above and you get videos, avatars, or holograms of deceased people with whom to have conversations. They will look and sound exactly the same.

But, you can go one step further.

physical griefbots

In 2013, an episode called 'Be Right Back' appeared in the Black Mirror series. In this episode, the young widow Martha is enrolled in an online service. This service creates a virtual version of Ash, her late husband, based on his social media posts. Martha starts with text messages, but soon she uploads more material and is able to talk to Ash on the phone. She eventually signs up for a new service that transforms a neutral synthetic body into an android that looks almost exactly like Ash. As always in Black Mirror, things don't end well.

The creation of an Android Mourning Robot is still a fantasy for the (distant) future, but robots are becoming more and more human. Projects such as BINA48 and Sophia, a humanoid robot from Hanson Robotics, provide an 'uncanny' glimpse into the future. (Hansonrobotics.com, 2020).

12.2.3 Problem description

The insight into the technology as described above ensures that students ask other questions about the problem, the context and users. They circle back and forth. They understand that experiments with bringing grandparents to life for the benefit of grandchildren require a completely different design than a griefbot for surviving relatives.

Slowly students are getting a better idea about the context, the problem, and the design. They answer the questions in the small cycle as closely and pointedly as possible. This can, for example, look something like this:

> *Problem:* To ease suffering after the death of a loved one.
> *Technology:* A griefbot. A commercial (web) app. After the user takes out a subscription all kinds of materials can be uploaded to the application via a wizard. This can be e-mails, text messages, posts, but also Google data or self-generated input, such as diaries. In addition, all kinds of audiovisual materials can be uploaded. The application creates a griefbot using AI. The griefbot learns from the conversations, so that after a while you have a unique relationship with your griefbot. Deep learning technologies are aimed at real-life conversations. The basic version is text-based, but more expensive versions are planned for the future with audio and video.
> *Users:* People who have just lost someone and want to be helped in the grieving process for a certain period of time by means of a griefbot.
> *Context:* The griefbot is for everyone. Anyone, worldwide, can take out a subscription.

Like real designers, students at this stage are generally very satisfied with their choices. They can't wait to start building, using sketches, mock-ups, and programming rules. However, tunnel vision is lurking. That is why it is so important in this phase to again start cycling to clarify the problem description.

The cycle helps students to think about the following questions.

Are you sure you are solving the right problem?

Isn't it healthy to feel pain after the death of a loved one? Isn't it counterproductive to ease that pain? Isn't it important to process the pain together with other relatives? Are those natural processes disrupted by a griefbot?

These are difficult questions and the answer to such questions depends on many things, such as culture, circumstances, relationship, and personality. While a certain technology may contribute to a better life for one person, it may cause harm to another.

It is not easy to solve that contradiction. What, for instance, is the net effect of WhatsApp? Has it generally been of benefit to humanity? These are questions that are almost impossible to answer. In the book 'Apps are the new smoking' (Van der Vorst, 2017) these questions are explored. A diverse picture emerges, which in some cases is very depressing. Addictive apps that are designed to keep users hooked can have a devastating impact on certain user groups, such as young people with behavioural problems. In that case, the users are victims and not perpetrators, as we have known from the history of smoking for a long time. But even groups that have a healthier relationship with their apps still see a lot of disadvantages. Discussions about the pros and cons of WhatsApp are always lively. However, the reality is also that 94% of Dutch people use WhatsApp and they can no longer imagine life without it.

The realisation that technology does not have a positive impact on everyone is very important. This is especially true when a technology deals with a sensitive matter, such as with a griefbot. How do you design technology that has the most positive impact on the 'right' users? How do you ensure that the 'wrong users' are cautioned against the technology? What are 'right' and 'wrong' users?

Maybe an understanding of how the technology contributes to solving the problem is of some help here.

How will the technology contribute to solving the problem?

One of the most important questions about the impact of a technology, and one which is often forgotten, is to what extent the technological solution actually contributes to the solution of the problem. Very often something looks really promising or makes a lot of sense, but on closer inspection we may need to reconsider this image. Has the solution been tested? If so, how? Is the solution simplified so that the technology fits? Technosolutionism (Morozov, 2014), a swift and uncritical enthusiasm to embrace an 'easy' technological solution to a complicated problem, is lurking.

These questions are especially difficult to answer with a new technology. But there are a number of options for action. Students can seek collaboration with experts and researchers to obtain a better picture. Extensive testing can be planned. You can choose to roll out the technology in phases and to carefully monitor the impact, and growing pains can be addressed quickly. But choices can also be made in the design of the technology.

Do you expect negative effects?

Technology often has unintended negative effects (Tenner, 1997). By being aware of this, changes can already be made during the design process to help prevent or mitigate some of these negative effects. A brainstorm of possible adverse effects is a good method for identifying them.

Generally, students really like to brainstorm about potential negative effects. They will come up with things like:

- Users who find it more difficult to resume life because they are addicted to conversations with the griefbot.
- Griefbots involved in family quarrels: 'Dad said that ...'.
- People who feel pressure to post positive messages online during their lifetime to prevent the data in the griefbot from making a negative impression after their death.
- People who have little online activity are seen as selfish for robbing their loved ones of the opportunity to have a functioning griefbot.

For each potential negative effect it can be determined whether and how the effect can be prevented and mitigated. The design can be adapted, the problem description refined, or the context and users changed. These are important choices that can prevent a lot of trouble later on.

But even then, of course, there will still always be unexpected negative effects.

12.2.4 From small to extended circle

From the above, by asking questions and circling back, students continuously make changes to the description of a problem, the technological solution, and the context and users. They do this in dialogue, with the help of experts or expert material, and potential users.

The end result is a completed small cycle for the griefbot. Now it is time to move on to the extended cycle. Here they look further into the impact of technology by asking new questions. The answers to these questions can lead to changes in the technological design but also in the problem description, the users, or the context. That in turn can lead to a restart of the small cycle.

12.3 The extended cycle

The Technology Impact Cycle Tool and Krutka's techno-ethical framework are two frameworks that show that there are more categories of questions that can contribute to designing technology with a positive impact. We tested questions in these categories with students and decided on eight categories that we found to be the most effective. These questions are addressed in the extended cycle. The aim is to adapt the design in such a way that the chance of a positive impact is increased. There are questions in the following categories (Figure 12.2):

- human values;
- data and privacy;
- transparency;
- inclusivity;
- stakeholders;
- bad actors;
- sustainability;
- future scenarios.

We briefly discuss each category in the subsections below and deal with a number of questions. This does not provide a complete picture but gives a good impression of the added value of the extensive cycle.

We will work with the version of the griefbot described in Section 12.2.3. Ideally questions in the extended cycle lead to new insights, which in turn lead to small wins, i.e. improvements that contribute to a positive impact and provoke other positive improvements.

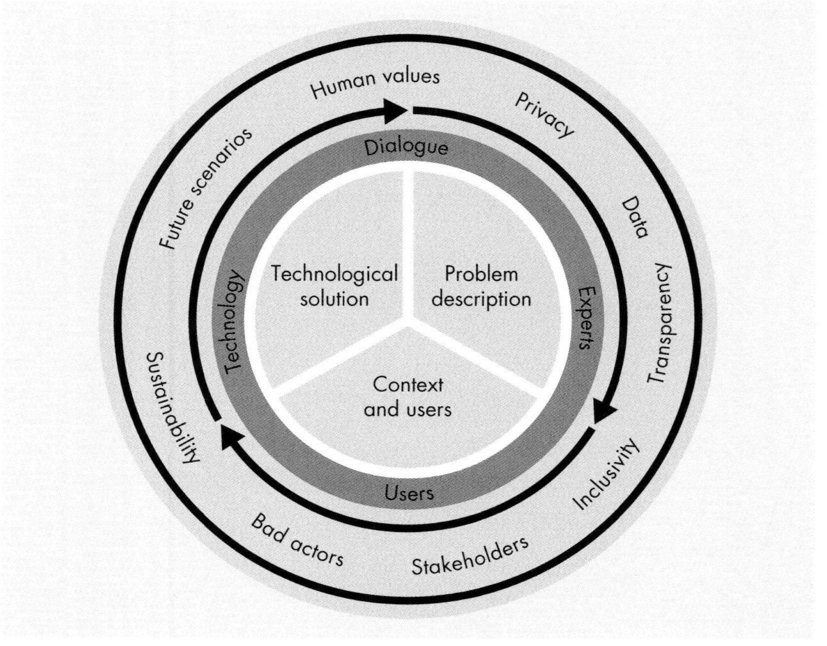

Figure 12.2. Technology Impact Cycle Tool.

12.3.1 Human values

Does the griefbot contribute to human values in a positive way? How do you know that? How do you design for that? Which considerations are important? The following questions will help you to find the answers.

Who are the users?

The initial reaction of students is often: haven't we already answered this question in the small cycle? And they're right, but in the extended cycle we refine this by specifically zooming in on the human values of the users. To do this, we first need to know what types of users there are. That is a different question from which stakeholders there are. After all, there are more stakeholders than users. We deal with the stakeholders in Section 12.3.5.

It is important to consider the values that are important to the users of new technology. It may be the case that the design must take into account very different users. In the case of the griefbot, the users are, for example, the people who register for the service, but also the dead who are brought to life digitally.

This leads to questions such as: to what extent do you include the interests of a deceased person in the design? Do you choose to have someone give consent during life, through some kind of 'mind-donor-codicil'? Are people entitled to 'post-mortem authenticity'?

In some countries it is illegal to slander dead people. In other countries, the portrait rights of dead people pass to the heirs. Copyright also plays a role. Do you take this into account? This is often new territory and the choices may result in design changes.

But for now, let's focus on the living users.

There are many human values to consider. In the extended cycle we focus on the three values which we have found contribute the most to a positive design of the technology: the identity of the user, the autonomy of the user, and the user's well-being.

How is the identity of the users affected?

Technology itself represents values and can change societies (see also Section 5.2.2). For instance, technology has changed the way we communicate, act, and even think. A good example of such technology is the smartphone. Obviously, the access that smartphones afford their users makes the lives of many easier and more convenient. However, it may also have a more subtle, negative effect. Many smartphone users have come to depend on their devices; whether they need important contact information, an address, or driving directions, they pull out their phones to get it. This change may impede smartphone users' ability to navigate some situations should they find themselves without a working smartphone, and, more generally, may degrade their resourcefulness.

In this phase, students often tend to underestimate the impact of their technology. They argue, for example, that people have long since relieved their suffering by viewing photos or videos of a deceased loved one. The digital griefbot merely adds an 'interactive' element to this.

However, this is too simplistic. It helps to ask more probing questions. Can the technology be perceived as stigmatising? Does the technology imply or impose a certain belief or world view? Does the technology affect users' dignity? Is the technology in line with the person the user wants to be perceived as?

As a result, students often conclude that it is important to regularly make users aware that they are talking to a griefbot and not really to their deceased loved one. Things that were previously seen as a problem, such as a griefbot occasionally giving strange answers, are suddenly seen as positive, as it reminds the user that they are not actually talking to their deceased loved one. It is also an important lesson: often perfect technology does not lead to perfect results.

How is the user's autonomy affected?

When we think about human values, autonomy deserves special attention. Western Culture is built around ideals of individual choice and freedom. Millions of us fiercely defend our right to make 'free' choices. This is what we call autonomy. But how is our autonomy affected by a certain technology? Does the technology enable the user to make better choices? Or does it sometimes give us the illusion of free choice while architecting the menu? Do people ask what's not on the menu? And is the technology empowering the original need or is it actually a distraction? (Harris, 2016).

If technology strengthens the user's autonomy, the chances increase that the technology will have a positive impact. However, it also means that the user must equally well be able to choose *not* to use a digital technology, to stop using it, to use an alternative, or use it less or for a shorter time.

In the case of a griefbot, the above considerations are especially important given the sensitive nature of the technology. Students often make adjustments in the technical design on this basis.

For instance:

- The griefbot avoids addictive design principles, such as triggers and rewards (Eyal, 2014).
- The griefbot provides information about use, warns the user about overuse and/or makes too much use impossible (which at the same time can affect autonomy).
- The griefbot can only be used temporarily (maximum one year).

These adjustments sometimes lead to the restart of the small cycle. Should the problem description be adjusted? Is a temporary subscription really a good idea? Isn't it terribly cruel to first bring someone to life digitally and then let someone 'die again'? In short, are these design changes helping the user to make better choices or do they limit the choices?

These are all questions that concern the effect of the technology on the well-being of users.

What is the effect on the user's 'well-being'?

Can technology trigger fears? Or stress? Is the user being manipulated? And what are the effects of overuse?

It is obviously not the intention of the designers of the griefbot to negatively influence the well-being of its users, but that does not mean that it cannot happen. Unintended 'Algorithmic cruelty' is already occurring right now. For example, social media platforms may festively remind you of the day your partner died (Casey, 2015).

It is not difficult to imagine that the data set of the deceased person hides secrets or data about painful events that emerge when using the griefbot. It is very complicated to filter this data from the input. Some students therefore opt for a 'panic button' which can be pressed if the digital mourning robot 'misbehaves'.

In psychology, so-called para-social relationships are already a well-known phenomenon. One person puts a great deal of energy into a relationship, while the other doesn't even realise they exist: celebrities and their superfans are a common example. The same danger lurks with griefbots. In addition, we have a tendency to anthropomorphise things. Did you ever think that your toaster was misbehaving? Anthropomorphising in combination with para-social relationships can cause some people to have trouble distinguishing between the virtual and the real. This can have negative consequences, such as taking advice from a griefbot.

Furthermore, there is a risk that the griefbot is used too much or for too long. Maybe the result is that the pain of grieving may not be eased, but prolonged or delayed. Perhaps users lose contact with reality because they are always in conversation with the griefbot. Perhaps the griefbot does not help you cherish the memories but instead creates new, fake memories.

The above questions make students realise that a griefbot that is indistinguishable from the real thing may not be a good idea. The results are often user restrictions, panic buttons and periodic, automatic alerts from the griefbot (I'm not real!).

It's not easy to find answers to the above questions. It is, however, crucial to ask the right questions and to continuously discuss what could be improved in the design of the technology. There are no right or wrong answers, but asking and discussing questions helps ensure things are well thought through.

12.3.2 Data and privacy

If a particular technology uses data and/or personal data, it is important to pay a lot of attention to this in the design. This is particularly crucial in the field of data and privacy protection because strict laws and regulations can make certain design choices impossible.

data pitfalls

Data has many pitfalls. Data is always subjective. Data collections are never complete. Do you also know what you are *not* measuring? Data collections are often biased (O'Neill, 2017). Reality is much more complex than millions of data points, and so on. Students should be able to detect these data pitfalls. What do the pitfalls mean for the functioning of the technology and to what extent can these risks be mitigated in the design?

In Case 12.2, there are two examples of such data pitfalls.

Case 12.2. Data Pitfalls.

Input

The griefbot is based on available data. If the input is business e-mail, a different 'digital dead person' is likely to occur than when using social media. Still other data leads to different results. Students therefore often choose to give the users of the griefbot the opportunity to experiment with different data-input.

Revenue model

In many cases, collecting data from users is part of the revenue model of digital technology. It is not just about what the users do, but also when and how they do it (Zuboff, 2019). That is precisely why a lot of digital technology companies strive to have as many users as possible using the technology as much as possible. This can result in a practice in which the values of the user are no longer at the core of the technology design. That is why, as mentioned earlier, students often choose to work with a subscription model. An additional benefit is that this can be used as a unique selling point.

privacy

Earlier in Section 12.3.1 we discovered that there are at least two types of users: deceased persons and living users. The data protection law only applies to natural (living) persons but there are also other rights that pass to heirs, such as copyright and portrait rights. In addition, the data of a deceased person naturally also contains a lot of personal data about those who are still living.

Privacy also includes more than just personal data. Privacy is a fundamental human right that can only be violated if the measures are proportional and subsidiary. With this type of new technology, this is still very new and unexplored territory.

The data protection law (GDPR) is European legislation that protects personal data. In the European Union, this is a very important point of attention for all technology, and for the design of the griefbot in particular. On the one hand because the design of the griefbot is based on personal data, and on the other hand because strict laws and regulations can ensure that certain things are not allowed. In practice, we see that students are often not informed, or are informed too late, of the (im)possibilities when it comes to using personal data.

Suppose we use WhatsApp data as input for the griefbot, then we know for sure that this data is full of personal data of all kinds of people. There may even be special (sensitive) personal data in the input. This leads to all kinds of new design issues:

- Is it possible to anonymise the data during upload, and what does that mean for the functioning of the griefbot?
- Is it an idea to collaborate with WhatsApp, where people have already given permission and there may be more options? Can you reconcile this collaboration and the ambition to make technology-for-good?
- Does the specific context (e.g. the legislation in a particular country) influence the design possibilities?

The privacy principle and the corresponding GDPR are good examples of how related questions in the extended cycle can cause the small cycle to be restarted. This is an important lesson that underlines the importance of going through a complete Technology Impact Cycle before you start building. After all, making changes is much harder once you have invested time and energy.

Students may try to solve the problem by changing the context. For instance, some shift to a chatbot that guides terminal patients. Others look for the solution in the design of the technology. In this way, the importance of privacy can sometimes lead to unexpected creativity. For example, some students suggest creating an app that you can download and talk to every day about what happened that day, what you did, and what you thought. At the end of your life, all those conversations lead to a digital clone, a self-generated griefbot.

Of course, such a solution leads to completely new questions and a restart of the small cycle. Students will also find that a design change like this unexpectedly solves other problems, like asking for permission from the deceased.

12.3.3 Transparency

Transparency is important, especially for the company that offers the technology. Can you easily file a complaint? Can you call the company? Is the company honest about the way they make money (data!)? How about the disadvantages of the technology on offer?

However, the most important transparency issue in the design of new technology has to do with the technology itself. Can you explain why technology behaves the way it does? This transparency is under pressure, thanks to certain forms of advanced artificial intelligence, such as deep learning software. This software is often a black box (Shane, 2020). We know that the software *can* do something, but not always *how* the software does it. That makes transparency difficult. In fact, with this software there is often a trade-off between transparency and accuracy. A lot of deep learning algorithms thrive in the dark.

There is more and more legislation requiring artificial intelligence to be transparent. But what does that mean for the functioning of the griefbot?

Sir Arthur Charles Clarke, an English science-fiction writer, formulated three laws of which the third is the best known and most widely cited: 'any sufficiently advanced technology is indistinguishable from magic' (Clarke, 1968). This also applies to a griefbot. The griefbot eases the pain through an illusion. A true magician does not reveal his secrets. In this case, is it therefore better that you cannot explain why the technology behaves in a certain way?

In his book Nihilism & Technology Nolen Gertz takes this idea one step further. Perhaps artificial intelligence should not be transparent, he suggests. After all, we believe in God and God is not transparent. Maybe artificial intelligence should move in mysterious ways, like God, so it becomes easier to believe in it (Gertz, 2018).

Students often choose the latter option: no explanation. At the same time, they believe that this principle should be clearly communicated and that the technology should be self-learning, but set in such a way that the griefbot remains very close to the given input. The griefbot should not make up things and should not use news plug-ins so that it can also discuss current affairs. The risk is, of course, that if the world becomes more inclusive, you will find out that your grandfather was very racist or misogynistic.

That is the topic of the next section.

12.3.4 Inclusivity

Is the technology available to everyone? Does everyone benefit or just a small group? Does the technology have intentional or unintentional built-in biases? There are many examples of technology that is racist or that works better for men than for women (Perez, 2019).

available to everyone

Students almost always choose to generate income for the griefbot through a subscription. But what if that means that the costs are too high for many people, leading to inequality? Students sometimes propose looking at collaborations with, for example, funeral insurers.

Many choices can also be made in the technology. Do you use social media input or not? If so, do international social media also participate? Can you upload typed memoirs? Or must you have talked to an app all your life, as proposed in Section 12.3.2? Every choice includes and excludes some groups.

built-in bias

In a TED Talk Cathy O'Neil explained her definition of a successful meal. She said, 'a meal is successful if my kids eat vegetables. It's very different from if my youngest son were in charge. He'd say success is if he gets to eat lots of Nutella. But I get to choose success. I am in charge. My opinion matters. That's the first rule of algorithms. Algorithms are opinions embedded in code.'

You can have many different opinions about how a griefbot should function. These opinions are reflected in the design. For example, if the input of a deceased person contains racist or misogynistic data, should this also be reflected in the behaviour of the griefbot? And what if things change in society (for example #MeToo or #BLM)? How will the griefbot handle this without doing injustice to the functioning of the griefbot as a digital representation of the deceased? Some students don't think a politically correct griefbot is a good idea. Other students see good communication and the aforementioned 'panic button' as the best solution. In other cases, there may be effects of the design that were not taken into account in advance. There is an unintentional built-in bias. For example, it could be that women in general are much more positive and active on social media, which makes their griefbots function better. They are richer with data and more sympathetic. Maybe if your father dies, you can only talk soccer with this griefbot.

diverse team

One way to avoid unconsciously building bias into technology is to make sure the team is diverse enough. Student teams are by no means always diverse.

12.3.5 Stakeholders

In Section 12.3.1 we talked about different types of users of the griefbot. We distinguished between the relatives and the deceased. These users are also stakeholders. Furthermore, in this chapter we talked a lot about the impact on society, which also involves looking at stakeholders. But it is still worthwhile explicitly zooming in on stakeholders again in the extended cycle. If you broaden the circle of stakeholders, you will encounter all kinds of other actors, some of which we will discuss below by way of illustration.

Students are challenged to consider the following questions:

- Who are the stakeholders? Did you have a brainstorm to identify them?
- In what way are the identified stakeholders affected by the new technology?
- Are you sure? Have you talked to the stakeholder her/himself? Did you consult them? Or did you assume their position?
- In what way are you going to take the interests of the stakeholders into account in the design process?

In Case 12.3 some examples are offered in which stakeholders can influence the design process.

Case 12.3. Stakeholders and design.

The angry brother

A father dies. The two children, brother and sister, have long lost contact with each other. They both want to use a digital griefbot, but have nothing to do with each other. Both had completely different relationships with their father.

Students may choose to design the technology in such a way that there is always a one-to-one relationship with the griefbot. Someone who creates a subscription and has access to certain data can create a griefbot and have a conversation with it. That conversation is different for everyone. There are no family accounts.

Religious people

Strictly religious people might protest against the idea of the digitally revived dead. They feel that this goes against their beliefs.

In this case, some students may want to have a conversation and explain why the griefbot is a means to alleviate pain, and listen to suggestions from these stakeholders. Other students may simply choose to ignore religious people. Of course, this may be at odds with the previously discussed questions about inclusivity.

12.3.6 Bad actors

The world is full of hackers, phishers, identity fraudsters, (deep) fakers and people who abuse a technology. The more sensitive the technology, the more the design must take into account the so-called 'bad actors'.

In addition to a large number of hygiene factors, such as data security and access (imagine for a moment that someone can take over the conversation with the griefbot), it is particularly effective to challenge students to think about how they can abuse each other's solution. Creativity knows no bounds at those moments.

This in turn leads to design changes, such as:

- A ban on griefbots for deceased people under the age of 18.
- A ban on active griefbots of deceased celebrities with a lot of public data. Imagine if someone launches a GriefMichaelJackson and that GriefMichaelJackson makes a bid for the presidency after his death.

12.3.7 Environmental sustainability

A technology that is designed 'for good' will always have to take into account the environmental sustainability of the design. Students often think that this does not apply to software solutions, such as the griefbot, because no physical products are made. But that is a misunderstanding. For example, the anatomy of an AI system (Crawford and Joler, 2018) shows convincingly how many resources are used to make the Amazon Echo.

A digital app, such as the mourning robot, must also take into account environmental sustainability questions that can subsequently have an impact on the design. For example:

- What is needed in the background for the griefbot to function and how sustainable are those resources? How 'green' are the data centres?
- Is the app backwards compatible or do you need newer models of smartphones to run it?

- How does the app affect the energy consumption of the device that is used?

Asking questions about environmental sustainability contributes to a conscious design, but can also be at odds with other attempts to achieve a positive impact. For example, sometimes students choose to secure the app using blockchain, but then run into problems with the energy resources that such a solution requires.

12.3.8 Future scenarios

Most of the questions above are very specific. The risk of this is that things will be missed as a result of a too limited, short-term mindset, and we have previously indicated that technology often has disadvantages in the longer term.

That is why it is always good to have students sketch wild utopian and dystopian scenarios and think about which new questions this raises and whether something should be done with this in the design.

A concrete future question that can also be asked and is often not addressed is what the result of a takeover or a copying action would be. How do you ensure that your technology continues to have a positive impact in different hands? What if Facebook launch their own griefbot? And what if your well thought of, ethically well-designed product can be bought by someone with plenty of money and bad intentions? What will you do then?

12.3.9 Detailed frameworks

The Technology Impact Cycle consists of two cycles that designers go through over and over again with the aim of sharpening the technological design morally, as they strive with increasing certainty for technology with a positive impact.

Sometimes it makes sense to use a detailed framework. For example, in the case of the digital griefbot, where data and privacy play a major role, the Ethical Data Assistant (Utrecht Data School, 2018) can be consulted or, when it comes to transparency and AI, the Ethical Toolkit for the development of AI Application (Hidalgo, 2019).

An extensive collection of detailed frameworks can be found on the website of the Technology Impact Cycle Tool (Fontys University of Applied Science, 2020).

12.4 Conclusions

We teach students to solve problems with new technology. After all, technology can do more and more. Students design technology with the goal of having a positive impact and they often have good ideas.

However, good ideas only get better by asking better questions.

The Technology Impact Cycle is a model we use to ask those better questions. The answers often lead to even better questions and step by step the design improves and the chance that the technology has a positive impact increases.

In general, students don't design technology as controversial and complex as a griefbot, but an exercise with such technology is an effective way to get students to practise going through the cycles.

They notice that they keep finding improvements, which in turn lead to new questions, choices and further improvements. They also notice that questions in the extended cycle (on, for example, privacy) can change the whole design. They learn about the importance of thinking things through at the start of the project.

Using the griefbot livens things up for the students, and the brief illustration in this chapter alone leads to dozens of recommendations and sometimes 'small wins' for a griefbot-for-good: from panic buttons to apps-as-input, to the choice of subscriptions, or the decision to stay close to the input.

But do you, after reading this chapter, think that it is possible to design a griefbot-for-good? Probably not. This isn't exactly a cookbook. Some design requirements are contradictory, other questions are difficult or impossible to answer, or the outcome is uncertain. Perhaps you disagree with many of the conclusions and arguments in this chapter. Perhaps you think we missed something. We understand that.

Still, perhaps in a few years' time the griefbot will be widely available and maybe it really will have a positive impact. Maybe that will cause us to change our minds, but for now, even after going through the Technology Impact Cycle, we think dead people like Aunt Ad should stay where they belong: in our cherished memories.

References

Casey, C., 2015, Facebook, grief and algorithmic cruelty. Available at: https://medium.com/@PQPP3/facebook-grief-and-algorithmic-cruelty-24d1dd2a0540.

Clarke, A.C., 1968. Clarke's third law on UFO's. Science, 159: 255. https://doi.org/10.1126/science.159.3812.255-b

Crawford, K. and Joler, V., 2018. Anatomy of AI. Anatomy of an AI system. Available at: https://anatomyof.ai/.

Eyal, N., 2014. Hooked. Penguin Books Ltd, London, UK.

Fontys University of Applied Science, 2020. Technology impact cycle tool. Available at: https://www.tict.io.

Gertz, N., 2018. Nihilism & technology. Rowman & Littlefield, Washington, DC, USA.

Harris, T., 2016. How technology is hijacking your mind. Medium. Thrive Global. Available at: https://medium.com/thrive-global/how-technology-hijacks-peoples-minds-from-a-magician-and-google-s-design-ethicist-56d62ef5edf3.

Hidalgo, S.M., 2019. Design of an ethical toolkit for the development of AI applications. TU Delft Repositories. Ethical AI Assistant. Available at: https://repository.tudelft.nl/islandora/object/uuid:b5679758-343d-4437-b202-86b3c5cef6aa?collection=education&ref=hackernoon.com.

Krutka, D.G., Heath, M.K. and Willet, B.S., 2019. Foregrounding technoethics: toward critical perspectives in technology and teacher education. Journal of Technology and Teacher Education, 27: 555-574. Available at: https://www.learntechlib.org/p/208235/.

Morozov, E., 2014. To save everything, click here: technology, solutionism, and the urge to fix problems that don't exist. Penguin Books Ltd, London, UK.

O'Neil, C., 2017. Weapons of math destruction. Penguin Ltd, London, UK.

Perez, C., 2019. Invisible women. ABRAMS books, New York, NY, USA.

Replika, 2021. Replika. Available at: www.replika.ai.

Schick, N., 2020. Deepfakes. Van Haren Publishing, 's-Hertogenbosch, the Netherlands.

Shane, J., 2020. You look like a thing and I love you. Headline Publishing Group, New York, NY, USA.

Shelley, M., 1818. Frankenstein – the modern prometheus. Lackington, Hughes, Harding, Mavor & Jones, UK.

Tenner, E., 1997. Why things bite back. Adfo Books, Amsterdam, the Netherlands.

Termeer, C.J. and Dewulf, A., 2018. A small wins framework to overcome the evaluation paradox of governing wicked problems. Policy and Society, 38: 298-314. https://doi.org/10.1080/14494035.2018.1497933

Utami, D. and Bickmore, T., 2018. Talk about death: end of life planning with a virtual agent. Available at: http://relationalagents.com/publications/IVA17.spiritual.pdf.

Utrecht Data School., 2018. Data ethics decision aid (DEDA). Available at: https://dataschool.nl/en/deda/.

Van der Vorst, R., 2017. Appen is het nieuwe roken. De Kleine Uil, Groningen, the Netherlands.

Verbeek, P-P., 2014. Op de vleugels van Icarus: hoe techniek en moraal met elkaar meebewegen (Dutch edition). Lemniscaat, Uitgeverij, Rotterdam, the Netherlands.
Verbeek, P.P. and Tijink, D., 2020, January. Guidance ethics approach. Available at: https://ecp.nl/wp-content/uploads/2020/11/Guidance-ethics-approach.pdf.
Weizenbaum, J., 1976. Computer power and human reason: from judgment to calculation (1st ed.) W.H. Freeman & Co, New York, NY, USA.
Zuboff, S., 2019. The age of surveillance capitalism. Profile Books, London, UK.

Media

GPT-3, 2020, 8 September. A robot wrote this entire article. Are you scared yet, human? The Guardian. Available at: https://www.theguardian.com/commentisfree/2020/sep/08/robot-wrote-this-article-gpt-3.

Company websites

Hanson Robotics, 2020. Sophia. Available at: www.hansonrobotics.com/sophia/.
My Heritage., 2021. Deep nostalgia. Available at: www.myheritage.nl/deep-nostalgia.

13. Antiviruses against fake news

Persuasive design strategies to stimulate responsible online behaviour

Danielle J.A.M Arets*, Ruben M. Bastiaanse and Anne L. Giesen

School of Journalism, Readership Journalism & Responsible Innovation, Fontys Hogeschool Journalistiek, Fontys Hogescholen, P.O. Box 90906, 5000 GH Tilburg, the Netherlands; d.arets@fontys.nl

Abstract

Numerous initiatives are being developed to counter the prevailing 'info-demic'. Under pressure from politicians and society, social media platforms have taken measures to counteract the spreading of misleading information about Covid-19. Journalistic editors are working overtime to correct the flow of misinformation, and governments are considering stricter legislation on disinformation. Yet none of these initiatives are yet sufficiently effective. Those who believe Bill Gates is using the vaccine for personal gain will only see confirmation of their own rightness in Twitter's pop-up messages warning against fake news. Fact-checking proves to be only partially helpful. A verified message has a positive effect on the factual knowledge of news users, but this does not mean that they change their beliefs and behaviour. To effectively deal with online misinformation, we need to shift the focus from debunking to prebunking (pre-emptive debunking). Based on our research project *AntiViruses for fake news* (2019-2020) – where we investigated game interventions increasing digital resilience to misinformation – we came to understand that the prebunking of young adults (18-22 years old) about the tactics of disinformation and its consequences makes them more resilient to manipulated information. Furthermore, we have reason to believe they are less inclined to spread disinformation.

Key concepts

- Prebunking (pre-emptive): exposing people to a weak dose of the techniques used in misinformation in order to make them more resilient

- Persuasive design: using design practices to influence human behaviour
- Fake news: umbrella term for disinformation (information that is false and deliberately created to harm a person, social group, organisation or country)
- Misinformation: information that is false but not created with the intention of causing harm
- Disinformation: information that is false and created with the intention of causing harm
- Mal-information: information that is based on reality, used to inflict harm on a person, social group, organisation or country

13.1 Introduction

Since the outbreak of the global COVID-19 pandemic, we have seen an increase in the spread of misinformation. The World Health Organization (WHO) warned of an impending 'info-demic' of misinformation via social media. The virus is said to have 'escaped' from a laboratory in Wuhan, there are various theories as to why Bill Gates, George Soros and other 'elites' are profiting from this 'pandemic' (Lytvynenko, 2020; Mikkelson, 2020a), and some see a relationship with the 5G network, which has led to arson in various countries (Jolly and Paterson, 2020; Mikkelson, 2020b). Others find (false) indications of alternative cures online and there are eager discussions about the effectiveness of PCR tests, face masks and vaccines. Not all this information is necessarily incorrect or fabricated, but it is misleading. Recent surveys show that almost half of the British population (46%) is exposed to misinformation about the virus, and in the United States 48% of those questioned indicated that they had been confronted with such information (Ofcom, 2020).

In the Netherlands too, conspiracy theories and misinformation[15] about the virus appear to be a problem. Although large-scale research on the scale and impact of misinformation is not available, research by the journalistic platform Pointer (KRO-NCRV) shows that the amount of unreliable news on Twitter has grown since the outbreak of the corona crisis (Vermanen, 2020). Moreover, a representative sample from Ipsos and Nieuwsuur showed that although the majority of the Dutch population do not believe in conspiracy theories, 15% think the coronavirus is a biological weapon manufactured in a laboratory (Van Heck, 2020).

---

[15] In this chapter, the words 'misinformation' and 'disinformation' are used to address false information, and the choice of words depends on the context or sources we refer to. In our research we did not focus on why the false information was created, but on how we can increase digital resilience to fake news in general.

Disinformation receives a lot of political attention, and rightly so. Susceptibility to disinformation seems to influence people's willingness to comply with the corona rules prescribed by the government (Hameleers *et al.*, 2020; Roozenbeek *et al.*, 2020b). Young people in particular engage in social media, which makes this group especially vulnerable to the temptations of misinformation (Plantinga and Kaal, 2018). This target group, which has grown up in the digital age, is said to find it increasingly difficult to distinguish real from fake news (Wineburg *et al.*, 2016). This group is therefore especially interesting to look at, as current debunking strategies seem to fall short.

13.2 Current strategies combating misinformation fall short

To combat the prevailing 'info-demic', various initiatives are being developed. Under pressure from politics and society, social media platforms are springing into action and taking measures to prevent the spread of misleading information; fact-checkers in news outlets are working overtime to correct the flow of misinformation, and governments are introducing legislation and issue frequent warnings about the dangers of misinformation.

artificial intelligence

Facebook has developed technical solutions – using artificial intelligence (AI) – such as warning messages to support informed decision-making (Clegg, 2020; Ross *et al.*, 2018). Twitter also uses AI for warning labels and to delete harmful content (Roth and Pickles, 2020). But even the sophisticated detection of algorithms will not entirely prevent the spread of false information (Ross *et al.*, 2018).

The question is, however, whether these initiatives are sufficiently effective. After all, social media platforms have a poor track record in this area and their algorithms seem insufficiently capable of detecting misleading content (Wakefield, 2017).

In addition, critical questions are being asked about it is desirable for governments and the private sector to decide what is 'true' and what is not. At a time when there is growing distrust of public institutions like the media, politics, government, and the legal system, the knowledge they produce is also distrusted (Harambam, 2017).

fact-checks

There is also some debate about the effectiveness of fact-checks (Nyhan *et al.*, 2019). In the first place, it is impossible for fact-checkers to debunk the endless stream of misinformation. In addition, research shows that while fact-checks have a positive effect on people's factual knowledge, there is not necessarily a resulting change in their beliefs and behaviour (Barrera *et al.*, 2020). Sometimes it can even backfire (Krause *et al.*, 2020). The spread of rumours often continues even after they have been debunked, also referred to as the 'echo effect' (Jong and Dückers, 2016: 340, retrieved from Ross *et al.*, 2018).

13.3 Inoculation theory

inoculation theory

In response to these objections, researchers have investigated whether it is possible to make citizens more resilient to the persuasive power of misinformation. In this line of research, the inoculation theory (McGuire, 1964, 1970) has (re)gained interest. This theory is based on a comparison with immunology: just as some vaccines makes people more resistant to a virus by injecting a reduced dose of that same virus, it should also be possible to make people more resistant to the 'misinformation virus' by 'injecting' them with small doses of misinformation (Roozenbeek and Van der Linden, 2019: 2; Van der Linden *et al.*, 2020; Van der Linden and Roozenbeek, 2020). In other words, by confronting people with the techniques and strategies behind misleading information as well as the consequences, they are pre-bunked, which makes them less receptive to the 'virus' itself.

Bad News Game

The best example of such a pre-bunking misinformation intervention is the award-winning Bad News Game designed by Cambridge researchers and the journalism-education platform DROG. In this game, the player puts himself in the role of a social media manipulator with the aim of gathering as many followers as possible by spreading misinformation. During the game, the player can collect various 'badges' for successfully applying common misinformation tactics. The idea behind this is that the player inoculates himself against the misinformation virus in an (inter)active way.

Playing the game has a positive effect on the resilience of people to the deceptive power of misinformation. In the study, players were asked before and after playing the game to judge the plausibility of fictitious misleading and non-misleading tweets. People from different countries and backgrounds proved to be better able to recognise misinformation after playing the game without this making them more sceptical in general. Furthermore, they were more confident in their (correct) judgements of the tweets after playing the game (Basol *et al.*, 2020; Roozenbeek and Van der Linden 2019; Roozenbeek *et al.*, 2020a). Additional research has shown that the effect of inoculation is not short-lived but can last up to three months (Maertens *et al.*, 2020).

Building on the success of the Bad News intervention, researchers Roozenbeek and Van der Linden, in collaboration with DROG, the UK Cabinet office, with support from the WHO and the United Nations, developed a similar game

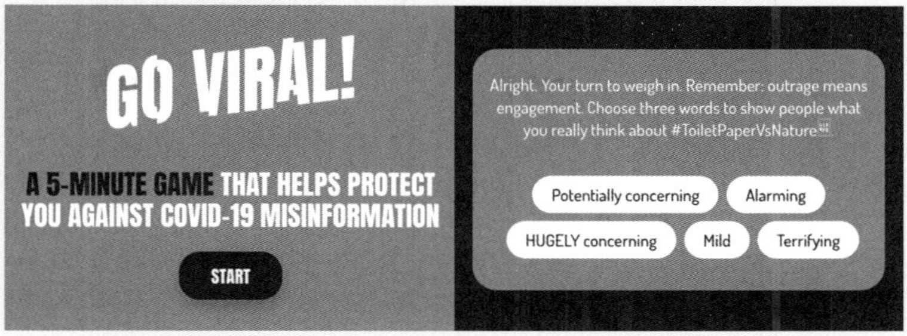

Figure 13.1. Images from the Go Viral game.

Go Viral!

called 'Go Viral!' (Figure 13.1) which specifically targets the current 'info-demic' (Lewsey, 2020). The logic of the game is to mobilise the largest possible number of followers by spreading misinformation about COVID-19. In the game, three common manipulation techniques are used: fearmongering (using moral-emotional language to enhance the virality of social media content (Acerbi, 2019) (Berriche and Altay, 2020), using fake experts, and spreading conspiracies (Zarocostas, 2020). A player causes unrest in the self-created filter bubble by posting emotional and misleading content. He or she underpins this content by quoting non-existent experts and shows discord about the virus by spreading conspiracy theories.

13.4 Antiviruses for fake news

To what extent does playing the game contribute to greater resilience among young adults and does it support critical reflection on online behaviour? These were the starting questions for our research '*AntiViruses for fake news*' (2020-2021) for which we used a mixed-methods approach, using quantitative strategies – building on the aforementioned research by Roozenbeek *et al.* (2020a) and a qualitative approach with semi-structured interviews to zoom in on behavioural change.

We organised three working sessions at intermediate vocational education institutes (ROC) in the cities of Tilburg and Amersfoort. In total, 86 young adults (between 16 and 22) participated in the workshops. In our ninety-minute workshop, we confronted students with a mix of misleading and non-misleading tweets before and after playing the Go Viral! game (Case 13.1). These tweets were compiled via social media (three misleading, three non-misleading). On a Likert scale of 1-7, the students were asked to decide how plausible they thought these tweets were (1 = not manipulative, 7 = very manipulative). The design of this small-scale pre-post-test is similar to the design used in previous research (Basol *et al.*, 2020; Maertens *et al.*, 2020; Roozenbeek and Van der Linden, 2019; Roozenbeek *et al.*, 2020a).

Case 13.1. Manipulated messages for pre-tests.

Misleading tweet – fake authority

This tweet falsely suggests that the WHO has stated that COVID-19 is no more dangerous than the 'normal' flu virus. It is a misleading story that entered the world via the dubious website Off Guardian and was quickly translated and adopted by Dutch sites such as RT Dutch and Ninefornews. On Twitter, the piece was shared frequently, including by the user below, 'Kukkie jommes', who refers to Ninefornews. The story has been labelled untrue by fact-checkers.

Deceptive tweet – conspiracy theory

In this tweet, user 'Evert G. Davelaar' states that COVID-19 was deliberately brought into the world by the Bilderberg Group and the World Economic Forum (WEF) in order to implement a so-called 'New World Order'. The conspiracy theory states that a globalist elite aims to gain control over the world's economy and establish a new world order, and has many variants. It has become popular during the COVID-19 pandemic. This theory, also known as 'The Great Reset' after the WEF plan of the same name, has been described by fact-checkers as unfounded.

The reason for this tweet is that both American and Chinese researchers stated in the news in May that the COVID-19 virus did not originate at an animal market in Wuhan. President Donald Trump suggested in the same period that COVID-19 originated at a lab in Wuhan. This claim has also been branded as unsubstantiated.

Deceptive tweet – emotion

This tweet is from the twitter account of Robert Jensen, a Dutch media producer. According to several fact-checkers, Robert Jensen spreads misinformation on his YouTube show, on his website and via Twitter.

In the game, students are made aware of the role that emotional language plays in the spread and persuasiveness of misinformation, and Robert Jensen's tweet is a good example of this.

The compiled tweets reflect the different misinformation strategies being used in the game design: fake authority (someone who masquerades as a doctor, expert or scientist and thus has greater authority to get a message across), conspiracy, and emotion (emotional language being used to draw more attention to a message). The three non-misleading tweets were added as control questions to find out if students did not simply become more sceptical about all the tweets.

13.5 Quantitative data processing: pre-post-test

To measure whether the Go Viral! game has an impact on the ability to identify manipulative headlines, a short questionnaire was presented to the participants. They had to fill this out twice. Once before the game and once afterwards. The questionnaire consisted of six posts from Twitter. The participants had to indicate how manipulative they found these posts to be on a score of 1 to 7, where 1 is non-manipulative and 7 is very manipulative.

In total, 86 participants completed the short questionnaire before the game and 80 participants after. The difference is due to input errors by the participants. They are therefore the same participants, but for 6 of them the second attempt to fill in the questionnaire failed.

The coded dataset showed us that in the second questionnaire more radical choices were made when judging the posts for manipulation. The further the extremes in numbers are apart, the greater the majority for one or the other. The outliers in this case are the number of votes for post 1 and post 5. This number represents the participants who are convinced that these posts are written very manipulatively.

This would give us reason to believe that the participants rate misinformation about COVID19 as more manipulative after playing Go Viral! We should note, however, that this is a (too) small data set for a quantitative analysis. It is therefore interesting to look at a larger study by Bol *et al.* (2020) who analysed a total of 2,634 complete pre-post-survey responses within the Go Viral! game environment (between 27 October and 26 November 2020), out of 14,755 people who completed the game in this time period.

They found that participants rate misinformation about COVID19 as significantly more manipulative after playing Go Viral! Furthermore, Go Viral! players are better able to distinguish real news from misinformation about COVID-19 after playing. This substantiates the belief that the prebunking strategies in the game are effective at evaluating fake news.

13.6 Behavioural change

As we aimed to understand whether the game also helped to change behaviour to disinformation, we conducted a qualitative study. After each workshop, six participants were invited for an individual semi-structed interview with one of the present researchers. In total, 22 students (16-22 years) were interviewed.

behavioural lenses model

The topics list for the interview was derived from the behavioural lenses model that identifies the most important aspects of behaviour and behavioural change (Figure 13.2; Hermsen, 2015; Renes and Hermsen, 2016). The model has been used to make contributions to interventions aimed at behavioural change (Hermsen *et al.*, 2015, 2016; Van Essen *et al.*, 2016).

The first lens ('habits and impulses') looks at the automatic behaviour that the target group displays. The second lens ('knowing and believing') zooms in on knowledge about misinformation and how they relate to it: what is misinformation according to young adults? Do they encounter it often and what do they think about it? The behavioural lens 'seeing and realizing', examines unconscious digital behaviour. Motivation is needed to change behaviour; do young people want to change their online behaviour? We explored this with questions around the wanting and being able to' change the time spent online, the style of writing and responding to online messages. Finally, with the behavioural lens 'doing and persisting', we explored whether the students intend to permanently change their behaviour vis-à-vis misinformation.

From the interviews, we noticed that fake news is one of the items on the daily information menu of young adults. 19 out of the 22 interviewees regularly see messages containing fake news. Most of the respondents don't seem to be disturbed or irritated by fake news. They show a rather loose attitude. It's part of their daily diet. Only 2 students were concerned, one of them being frustrated by the amount of disinformation.

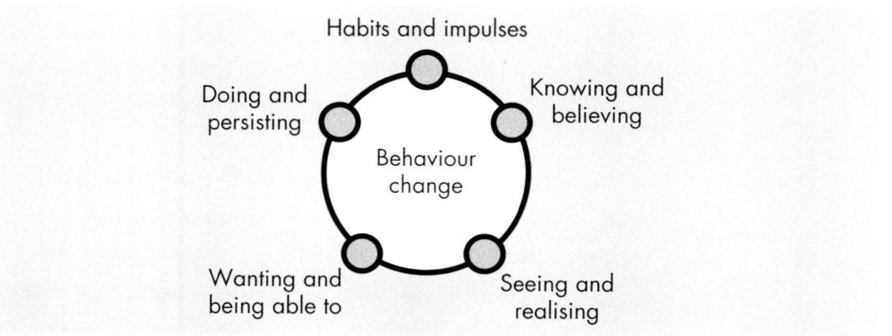

Figure 13.2. The five behavioural lenses, Sander Hermsen.

Despite their exposure to disinformation, half (eleven) of the respondents say they are not influenced by fake news; nine think they are and two are not sure. The majority are, however, confident that they can recognise fake news.

Despite the fact that the students mention not being susceptible to fake news, they do think friends, classmates and young people in general are influenceable. This 'better-than-average effect', where people see their own abilities as superior to those of the average person in their reference group (peer), is a well-known psychological mechanism. In this case, the better estimation of one's own performance and that of one's peers may also be related to the fact that the respondents are studying media and are therefore, in their own words, 'better informed'.

The interviewees are very good at arguing why people spread disinformation. They say it is either about money or receiving attention; the latter is especially the case for celebrities who are guilty of spreading disinformation. On further questioning, three respondents also accused some news sources of spreading fake news.

It must be said though, that our interviewees – with the exception of one respondent – don't take in much news. From the average of four hours that they spend on (social) media, they only occasionally consult the news (maximum ten minutes). Often this is more by chance because it is forwarded or shows up in their timeline.

When it comes to defining which manipulation tactics are being used, students mention emotional language, the use of block capitals, extreme vocabulary, and spelling mistakes. Only fake authorities were not mentioned. Since the interviews took place after the game was played, we could argue that the tactics used in the game were still fresh in their mind. Upon further questioning, however, some were able to mention examples that were not offered in the workshop, which would indicate that they were already aware of some of the tactics.

13.7 Prebunking design strategies for moral behaviour

Are the respondents willing and able to change their behaviour? Most of them were not sure whether they would change their behaviour, though two of them mentioned that playing the game prompted them to consume more 'reliable' news as sometimes it is hard to judge whether a social media post is fake or not. Furthermore, two interviewees said that the game prompted them to pay more attention to their online behaviour. This mainly concerns forwarding dubious messages – they explicitly mention they will be more careful about doing this. One respondent said he would report fake news to the organisation. This sample is too small to make general statements about behavioural change. Nevertheless, in comparative international studies we see promising results (Roozenbeek and Van der Linden (2020) Mol *et al.* (2021).

We have reasons to believe that playing the Go Viral! game makes users more resilient to disinformation and supports behavioural change towards fake news. Respondents indicate that they aim to send fake messages less often and are willing to change their own behaviour, compiling more reliable messages.

It should furthermore be noted that students unanimously mentioned it was fun playing the game. One student was very keen to get a high score and even played the game twice. Another student mentioned he forwarded the game to friends while playing it. One student mentioned it is a helpful 'learning strategy'.

> By playing the game the information comes across well and it is not as condescending as all these fake news lessons at school.

13.8 The need for moral design

From the interviews, we can deduce that most of the young people consulted are reasonably well informed about what fake news is, how to recognise disinformation, and what damaging effects it can have on society. Despite most respondents being aware of the manipulation tactics used in disinformation messages, the prebunking strategies in the game were considered helpful. Some said it triggered them to be more aware of disinformation and the manipulation strategies. The manipulation tactics used in the game (fearmongering, fake authority, and emotional strategies) were mentioned in the interviews, which gives us reason to believe the game helps participants to memorise and recognise these strategies.

prebunking

Pre-emptively debunking ('prebunking') misinformation seems to be effective as it helps recipients to recognise fake news and makes people – in our research young adults – willing to change their behaviour. The game design is very helpful in this respect as students say it is fun to play and less condescending than more common learning strategies.

It should, however, be mentioned that we consulted ROC students studying content strategy (Tilburg, the Netherlands) or media design (Amersfoort, the Netherlands). Due to the Covid-19 restrictions we were unable to offer the workshop to other students. The consulted students are expected to have a professional interest in information and disinformation strategies and therefore might be more informed than others. More studies on pre-emptive debunking are therefore recommended.

The persuasive (pre-emptive) design strategies of the fake news game are particularly interesting in regards to developments in the field of Responsible Design (Tromp *et al.*, 2011) in which designers, in a strategic, early phase of development, take more responsibility for the use and societal implications of the products and services to be developed; this with the aim of contributing to a social, sustainable and ethical use (Dujardin and Arets, 2017).

In the digital context and with the societal consequences related to disinformation, there is an increasing focus on the design of the systems and services with which we communicate. Current digital communication systems and the intended methods of interaction have a strong impact on how we shape our conversations. The architecture of these online programs organises the interaction by users and guides them in a certain direction. This is further enhanced by the use of artificial intelligence, with which communication platforms, based on user data, can steer online behaviour.

These *platform mechanisms* (Van Dijck *et al.*, 2016) are however designed with commercial values as a starting point, and are therefore not always aligned with public values.

The Go Viral! game is a good example of how a strategic design game, developed by designers, researchers and governmental bodies, can be used to design for good. However, the real challenge is to design communication platforms that encourage reliable information and discourage disinformation.

References

Acerbi, A., 2019. Cognitive attraction and online misinformation. Palgrave Communications, 5: 1-7. https://doi.org/10.1057/s41599-019-0224-y

Barrera, O., Guriev, S., Henry, E., and Zhuravskaya, E., 2020. Facts, alternative facts, and fact checking in times of post-truth politics. Journal of Public Economics, 182: 104123. https://doi.org/10.1016/j.jpubeco.2019.104123

Basol, M., Roozenbeek, J., and Van der Linden, S., 2020. Good news about bad news: gamified inoculation boosts confidence and cognitive immunity against fake news. Journal of Cognition, 3: 1-9. https://doi.org/https://doi.org/10.5334/joc.91

Berriche, M. and Altay, S., 2020. Internet users engage more with phatic posts than with health misinformation on Facebook. Palgrave Communications, 6: 1-9. https://doi.org/10.1057/s41599-020-0452-1

Bol, D., Giani, M., Blais, A. and Loewen, P.J., 2020. The effect of COVID-19 lockdowns on political support: some good news for democracy? European Journal of Political Research, 60: 497-505. https://doi.org/10.1111/1475-6765.12401

Clegg, N,2020. Combating COVID-19 misinformation across our apps. Available at: https://about.fb.com/news/2020/03/combating-covid-19-misinformation/.

Dujardin, A. and Arets, D., 2017, June 1. The unintended consequences of design. Medium. https://medium.com/@Pocketstories/the-unintended-consequences-of-design-c613aa1ea086

Hameleers, M., Van der Meer, T.G.L.A. and Brosius, A., 2020. Feeling 'disinformed' lowers compliance with COVID-19 guidelines: evidence from the US, UK, Netherlands and Germany. The Harvard Kennedy School (HKS) Misinformation Review. https://doi.org/10.37016/mr-2020-023

Harambam, J., 2017. The truth is out there': conspiracy culture in an age of epistemic instability. Erasmus University Rotterdam, the Netherlands. Available at: http://hdl.handle.net/1765/102423

Hermsen, S., Renes, R.J., Mulder, S. and Van der Lugt, R., 2015. Using the Persuasive by Design-model to inform the design of complex behavior change concepts: Two case studies. In: Proceedings of the 11th Conference of the European Academy of Design. Paris: Université Paris Descartes.

Hermsen, S., Van der Lugt, R., Mulder, S. and Renes, R.J., 2016. How I Learned to a appreciate our tame social scientist: Experiences in integrating design research and the behavioural sciences. In: 2016 Design Research Society 50th Anniversary Conference, 1375-1389. http://doi.org/10.21606/drs.2016.17

Jolley, D., and Paterson, J.L., 2020. Pylons ablaze: examining the role of 5G COVID-19 conspiracy beliefs and support for violence. British Journal of Social Psychology, 59: 628-640. https://doi.org/10.1111/bjso.12394

Jong, W. and Dückers, M.L.A., 2016. Self-correcting mechanisms and echo-effects in social media: an analysis of the 'gunman in the newsroom' crisis. Computers in Human Behavior, 59: 334-341. https://doi.org/10.1016/j.chb.2016.02.032.

Krause, N.M, Freiling, I., Beets, B., and Brossard, D., 2020. Fact-checking as risk communication: the multi-layered risk of misinformation in times of COVID-19. Journal of Risk Research, 23: 1466-4461. https://doi.org/10.1080/13669877.2020.1756385

Maertens, R., Roozenbeek, J., Basol, M., and Van der Linden, S., 2020. Longterm effectiveness of inoculation against misinformation: three longitudinal experiments. Journal of Experimental Psychology: Applied, 27: 1-16. https://doi.org/10.1037/xap0000315

McGuire, W.J., 1964. Inducing resistance against persuasion: some contemporary approaches. Advances in Experimental Social Psychology, 1: 91-229. https://doi.org/10.1016/S0065-2601(08)60052-0

McGuire, W.J., 1970. A vaccine for brainwash. Psychology Today, 3: 36-64.

Nyhan, B., Porter, E., Reifler, J., and Wood, T.J., 2019. Taking fact-checks literally but not seriously? The effects of journalistic fact-checking on factual beliefs and candidate favorability. Political Behavior, 42: 939-960. https://doi.org/10.1007/s11109-019-09528-x

Ofcom, 2020. Third annual Media Nations report. Available at: https://www.ofcom.org.uk/research-and-data/tv-radio-and-on-demand/media-nations-reports/media-nations-2020.

Plantinga, S. and Kaal, M., 2018. De mediawijsheid van jongeren in beeld. Onderzoek naar mediawijsheid onder jongeren van 13 tot 18 jaar. Kantar Public. https://www.mediawijzer.net/wp-content/uploads/sites/6/2018/11/Mediawijsheid-jongeren_27-november1.pdf

Renes, R.J. and Hermsen, S., 2016. Draaiboek gedragsverandering: de psychologie van beïnvloeding begrijpen en gebruiken. Business Contact, Amsterdam, the Netherlands.

Roozenbeek, J., and Van der Linden, S., 2019. Fake news game confers psychological resistance against online misinformation. Humanities and Social Sciences Communications, 5: 1-10. https://doi.org/10.1057/s41599-019-0279-9

Roozenbeek, J. and Van der Linden, S., 2020. Breaking Harmony Square: A game that 'inoculates' against political misinformation. The Harvard Kennedy School (HKS) Misinformation Review, 1(8). https://misinforeview.hks.harvard.edu/article/breaking-harmony-square-a-game-that-inoculates-against-political-misinformation/

Roozenbeek, J., Schneider, C.R., Dryhurst, S., Kerr, J., Freeman, A.L.J., Recchia, G., Van der Bles, A.M. and Van der Linden, S., 2020b. Susceptibility to misinformation about COVID-19 around the world. Royal Society. Open Science, 7:201199. http://dx.doi.org/10.1098/rsos.201199

Roozenbeek, J., Van der Linden, S. and Nygren, T., 2020a. Prebunking interventions based on 'inoculation' theory can reduce susceptibility to misinformation across cultures. Harvard Kennedy School (HKS) Misinformation Review, 1: 1-15. https://doi:org/10.37016//mr-2020-008

Ross, B., Jung, A-K., Heisel, J., Heisel, J. and Stieglitz, S., 2018. Fake news on social media: the (in)effectiveness of warning messages. In: Proceedings of the 39th international conference on information systems (ICIS), San Francisco, CA, USA.

Tromp, N., Hekkert, P., and Verbeek, P.-P., 2011. Design for socially responsible behavior: a classification of influence based on intended user experience. Design Issues, 27: 3-19. https://doi.org/10.1162/DESI_a_00087

Van Dijck, J., Poell, T., and De Waal, M., 2016. De platformsamenleving: Strijd om publieke waarden in een online wereld. Amsterdam University Press, Amsterdam, the Netherlands.

Van Essen, A., Hermsen, S., and Renes, R.J., 2016. Developing a theory-driven method to design for behaviour change: two case studies. In: Lloyd, P. and Bohemia, E. (editors). Proceedings of DRS2016: Design + Research + Society – Future-Focused Thinking, Volume 4, pp 1323-1338.

Van der Linden, S., and Roozenbeek, J., 2020. Psychological inoculation against fake news. In: Greifenader, R., Jaffé, M., Newman, E. and Schwarz, N (editors). The psychology of fake news: accepting, sharing, and correcting misinformation. Psychology Press, London, UK.

Van der Linden, S., Roozenbeek, J., and Compton, J., 2020. Inoculating against fake news about COVID-19. Frontiers in Psychology, 11: 1-7. https://doi.org/10.3389/fpsyg.2020.566790

Wineburg, S., McGrew, S., Breakstone, J. and Ortega, T., 2016. Evaluating information: the cornerstone of civic online reasoning. Stanford Digital Repository. http://purl.stanford.edu/fv751yt5934

Zarocostas, J., 2020. How to fight an infodemic. World Report, 395: P676. https://doi.org/10.1016/S0140-6736(20)30461-X

Media

Clegg, N,2020. Combating COVID-19 misinformation across our apps. Available at: https://about.fb.com/news/2020/03/combating-covid-19-misinformation/.

Hermsen, S., 2015, 21 May. How can we support designers to create theory-driven Designs for Behaviour Change? Medium. https://medium.com/@sanderhermsen/how-can-we-support-designers-to-create-theory-driven-designs-for-behaviour-change-the-persuasive-a8e3f5a5f6de

Lewsey, F., 2020, 11 October. Cambridge game 'pre-bunks' COVID-19 conspiracies as part of the UK. University of Cambridge. https://www.cam.ac.uk/stories/goviral

Lytvynenko, J., 2020, 22 April. Bill Gates Coronavirus False Misinformation Vaccine Microchips. BuzzFeed News. Available at: https://www.buzzfeednews.com/article/janelytvynenko/conspiracy-theorists-are-using-a-bill-gates-reddit-ama-to

Mikkelson, D., 2020a, 22 April. Does George Soros own a lab that 'developed' COVID-19? Snopes.com. https://www.snopes.com/fact-check/soros-wuhan-lab/

Mikkelson, D., 2020b, 12 June. How the 5G coronavirus conspiracy theory began. Snopes.com. https://www.snopes.com/news/2020/06/12/how-the-5g-coronavirus-conspiracy-theory-began/

Roth Y, Pickles N., 2020. Updating our approach to misleading information: Twitter, Inc. https://blog.twitter.com/en_us/topics/product/2020/updating-our-approach-to-misleading-information

Van Heck, S., 2020, mei. Complottheorieën over het coronavirus. Ipsos/Nieuwsuur. https://www.ipsos.com/sites/default/files/ct/news/documents/202005/ipsos_corona_complot_v1.0.pdf

Vermanen, J., 2020, 23 March. Flinke stijging van onbetrouwbaar nieuws over coronavirus op Twitter. Pointer. https://pointer.kro-ncrv.nl/artikelen/flinke-stijging-van-onbetrouwbaar-nieuws-over-coronavirus-op-twitter

Wakefield, B. J., 2017, 7 November. Facebook's fake news experiment backfires. BBC News. https://www.bbc.com/news/technology-41900877

14. Benevolent lies by design

Reflecting on ethical challenges for design in dementia care

Rens Brankaert[1]*, Ans Tummers-Heemels[2] and Wijnand IJsselsteijn[2]

[1]Dementia & Technology, Industrial Design, Eindhoven University of Technology, P.O. Box 513, 5600 MB Eindhoven, the Netherlands; [2]Expertise Center Dementia & Technology, Human Technology Interaction, Department of IE&IS, P.O. Box 513, 5600 MB Eindhoven, the Netherlands; r.brankaert@fontys.nl

Abstract

In dementia care, deception is widely debated amongst caregivers. In general, outright lying is not accepted, yet benevolent (white) lies can be acceptable if they come with good intentions to comfort or protect someone. However, with the increasing need to support dementia care, new interventions and technology are entering the discussion. Many of these interventions, implicitly or explicitly, embody deceptive practices. In this chapter, we present three cases that stir the debate around deception and (technology) design. We aim to better understand the applied moral dimensions in dementia care, taking into account the current developments in the literature. The aim of this overview is to create awareness among technology developers and designers who come up with the interventions, as well as among care professionals who apply these interventions in dementia care practice.

Key concepts

- Dementia – The cluster denomination of a group of pathologies that cause cognitive deterioration progressively over time, resulting in memory loss, reduced ability to take care of oneself and other consequences
- Caregiver – The group denomination of those that provide care for people with dementia. This can be informal, often conducted by family members or a spouse, or formal when it is conducted by care professionals working for a care organisation

- (Technology) Interventions – The artefacts, services or systems that we design to provide support to the challenge in hand. These can be both technological or non-technological
- Designers /Design researchers – Those that create and/or research interventions. This often happens through a design process that typically includes problem-defining, idea-finding, prototyping and testing ideas with target audience

14.1 Introduction

The number of people living with dementia is growing dramatically. This comes with challenges such as a shortage of funding in healthcare, pressure on care professionals and, above all, a diminished quality of life for those living with dementia. Dementia refers to a cluster of pathologies that progressively affect memory, thinking, language, motor abilities, and social abilities. Dementia therefore severely interferes with daily functioning, especially in the later stages of the disease.

dementia

Unfortunately, there is as yet no cure for the underlying diseases that cause dementia, and therefore we have to find alternative ways to reduce the caregiver burden and support people living with dementia. One of the successful avenues of support for dementia care is the use of technology or other non-pharmaceutical interventions with the aim of supporting, empowering and increasing the general well-being of people living with dementia (Astell *et al.*, 2010; Brankaert *et al.*, 2014; Cahill *et al.*, 2007; Gibson *et al.*, 2015). Increasingly, commercial parties are entering the domain of dementia with technologies that contribute to the apparent challenges. While these technologies are released with good intentions, e.g. to aid in protection or safety, sometimes the resulting interventions infringe on the integrity of those living with dementia, who depend on the care of others.

technology

To fuel the debate about the ethical consequences of (technology) design in dementia care, Case 14.1-14.3 discuss various forms in which technology and design is used to facilitate white lies in order to support people living with dementia, and their caregivers. All three cases are based on actual, existing interventions – two are based on novel technologies such as AI and robotics, the other one includes the physical redesign of the environment.

Case 14.1. Rosie the robot seal.

Ann, an older lady with middle-stage dementia loses her pet dog Rosie that she loved and cared for over the past sixteen years. After weeks of intense grief, one of her caregivers comes up with a potential solution: they give her an interactive robot seal (Figure 14.1). The experiment works really well, Ann immediately develops a deep attachment to the robot – she cares for, and caresses it continuously and her grief over Rosie is significantly lessened. Over the following period, Ann starts referring to the robot seal as 'her dog' and calls it Rosie. Ann seems to be in a better mood since the robot seal entered her life, and it also seems to provide relief for the caregivers. However, a few weeks later the batteries of the robot run low, Ann is deeply distressed and calls to the caregiver, saying that 'the dog is dying'.

Figure 14.1. An illustration of the Paro robotic seal.

Case 14.2. A daily chat with your son.

Barbara, an older lady in the mid to later stages of dementia – living in a long-term care home – starts to show intermittent episodes of restlessness and emotional distress. Barbara seems to want to call her son, Sander, all the time, up to the point where neither Sander nor the caregivers can provide for this increasing need. The caregivers find a potential solution to this need; they hand her what looks like an old-fashioned dial telephone to call Sander as often as she wants. The new phone is enhanced with Artificial Intelligence technology that connects Barbara to a pre-recorded voice of Sander. The system responds through scripted questions and answers, where the AI-based language recognition and voice stress analysis are able to adapt and tune the conversation to the situation of Barbara. Believing that Barbara is now talking to Sander more often, the conversations have a soothing effect on her, and at the end of the conversations she asks when Sander is coming to visit her again. The digital version of the son responds, in the voice of Sander: 'I'll be over soon' – an answer that puts a big smile on her face.

Case 14.3. Waiting for a bus that never comes.

Carla and Caitlyn are both caregivers who work on a day care programme for people with dementia. Due to recent budget cuts they are the only two people left of an original team of four to take care of a group of fifteen people with dementia. In this group, Tom is especially difficult to handle because he always wants to walk off. To help Carla and Caitlyn, the board decided to install a bus stop in the hallway. It is designed with all the familiar bus stop signs, timetables, an authentic booth, and a bench to sit on. No bus will ever arrive at this stop. The bus stop is fake with the express purpose of attracting people with dementia prone to wandering around, or out of the nursing home grounds – a significant source of stress for caregivers and care facilities. Tom used to take the bus all the time, and is attracted to the bus stop. He sits there all the time, seemingly at peace. He regularly asks Carla and Caitlyn: 'When will the bus finally arrive?'.

benevolent lies

As indicated before, these interventions are designed with the intention of providing support or alleviating caregivers. However, they also facilitate some forms of deceptive practices in order to fulfil that aim. Professionals who design such technologies should be aware of the possible ethical consequences of applying their interventions in practice. The deception in these cases is inherent in the (technology) design itself and further enhanced or diminished by the way they are applied in dementia care practice. Recent developments in nursing and dementia care question and discuss the use, boundaries conditions, and ethical implications of using deception, or benevolent lies, in dementia care practice. The current chapter aims to address this urgent issue and support designers in how to deal with deception in dementia care practice.

14.2 Deception

deception

In this chapter we will review deception in dementia care and discuss the role that designers and their interventions have in this. Deception is 'the act of causing someone to accept as true or valid what is false or invalid' (Merriam-Webster). Deception is in general considered morally reprehensible. From a deontological perspective deception is seen as morally wrong – and promotes a universal denouncement of any form of deception. Instrumentalist views –

more apparent in our society today – attempt to weigh the costs of deception against its potential benefits, creating a more context-dependent view on deception. In this view, there are a variety of cases of deception that are not regarded as clear moral transgressions, such as lying to protect someone from harm.

white lie

It is important to note that not all lies (or deceptions) are necessarily enacted with malicious intent or for personal gain. The best-known example of this is the white lie, which is a lie with often good intentions, generally meant to avoid hurting someone's feelings. White lies are frequent, rather trivial occurrences in day-to-day interpersonal relationships, where they typically support social cohesion.

14.3 White lies in care practice

When reviewing the literature on nursing and dementia care, we can see that white lies are a common strategy in this context, with the general aim of serving the best interests of people with dementia (James *et al.*, 2006). Deception seems to be applied in dementia care primarily in order to manage a person's behaviour and emotional distress. In the later stages of dementia, for example, people will believe something that is untrue. They might be convinced that their husband will soon come to visit or that their children have to go to school. In these cases, caregivers often don't contradict these fabricated realities, but instead affirm them to prevent feelings of distress (Lindholm, 2015). Another example could be when a person with dementia relives an emotional or threatening situation from their past, not unlike PTSD. When this happens, it is common to use a white lie – and, for instance, tell the person 'your mom is fine' – to comfort and distract them (Turner *et al.*, 2017). In dementia care it is considered ethical to 'deceive' someone using white lies if this prevents further harm to self or others. It appears that views on deception in dementia care are instrumentalist.

guidelines

In the literature, several guides and communication strategies have been published, based on qualitative research on how to deal with deception in dementia care practice. One example is the work by Ian A. James *et al.* (2006), that formulates twelve guidelines, including that a lie should only be told in the best interests of people with dementia, that consideration should be given to an individual's ability to discern truth from falsehood, that family consent should be obtained, the lies should be documented, and that caregivers should be reflective on and reflexive towards the practice. By following these guidelines, the authors aim to protect both the interests of the caregivers and the people with dementia being cared for. In all cases, telling a so-called therapeutic lie, should be a last resort after other strategies have been pursued.

However, it is difficult to apply these guidelines and strategies in dementia care practice. Care professionals are continuously seeking a balance between their own virtues of being an honest person, and conducting their caregiving tasks given the circumstances. It is suggested that therapeutic lies should be considered only one of the caregivers' tools and that they should be used with empathy and integrity on a case-by-case basis and providing the caregiver really 'knows' the person (Casey *et al.*, 2019).

14.4 Deceptive practices by (technology) design

In the introduction (Cases 14.1-14.3) we presented three innovative interventions that bring new challenges and complexities to the discussion about deception in dementia care, whether intentionally or unintentionally. The three cases mentioned – the robot seal (Case 14.1), the AI phone conversation (Case 14.2), and the fake bus stop (Case 14.3) – are all cases of technologies and design that are used in dementia care practice today in the Netherlands. The cases are presented to reflect on the ethical challenges and trade-offs inherent in these interventions, so we can be more aware of them when designing technology or other interventions for dementia care, and gain a better understanding of deception mediated by interventions in this context.

People with dementia often find it difficult to distinguish between truth and untruth. That inability to understand what is true or untrue is used in the interventions presented in the cases above to provide comfort and relief. The social simulations presented in Case 14.1 the robot seal and 14.2, the AI enhanced phone, do this in direct interaction, but also Case C, the bus stop, only works if people perceive it as a bus stop. Such simulations or reproductions are designed to encourage connection, stimulate reminiscence, or invoke a sense of familiarity. However, limited research has been done about the extent to which these (technology) interventions might cause confusion, induce distrust or do further harm. For example, if Sander is not coming to visit even though the AI-enhanced phone in Case 14.2. promised that he would, this might hurt the person more than it might comfort them. These cases might therefore cause distortions in reality, and distance people even further from it. It is important that designers are aware of these potential effects of their interventions. Furthermore, a patient's needs might prove more complex than what the technology can accommodate, which could lead to inappropriate functioning of the technology. While the technology is advancing at an unprecedented pace, it is often still unable to fully accommodate human nature, context, life history, or family preferences. Our technologies and interventions have moral implications, but are for the most part not moral agents themselves yet. They often, and especially the technologies presented in the abovementioned cases, don't have the capacity to make informed ethical decisions. The responsibility of the designers who create and the carers who deploy technologies in care should be explored further.

In addition, interventions are often designed with ideal-use cases and scenarios in mind only. We have to ensure that caregivers and especially people with dementia are involved in the design process to influence decision-making (Hendriks *et al.*, 2015). Recent studies in human computer interaction and design research focused on dementia present strategies to accommodate this, such as experience-centred design (Morrissey *et al.*, 2017), involvement of families (Hodge *et al.*, 2019), and person-centred design (Houben *et al.*, 2019; Thoolen *et al.*, 2020). Recently, we introduced Warm Technology that promotes person-centred, non-stigmatising and empowering interventions (IJsselsteijn *et al.*, 2020). These strategies allow for the careful exploration of technology and design in dementia care practice before they are released to mass markets. In these processes it is important to consciously weigh the input from both people with dementia and their caregivers. In Case 14.3, the bus stop provides relief for both caregivers and people with dementia, but only adds confusion and a distortion of reality for the latter.

Warm Technology

Professional caregivers in dementia care may play an active role in choosing to deploy certain technologies and interventions. They can actively assess the ethical pros and cons, and evaluate any unforeseen negative effects. Sometimes technology is more hidden or indirectly applied and, in these cases, the guidelines and protocols that are established around therapeutic lying should also be considered when using these potentially harmful technologies. The costs and benefits of the technologies should be weighed, documented and adjusted in time when needed. In this there is a difference between actively instigating a deception (for example, in Case 14.2. with the AI-enhanced phone) or passively accommodating a deception (for example, in Case 14.1 with the robotic seal). In the first example, the designer of the technology asks the caregivers to actively enact a (white) lie. In the second example, the perception of the robotic seal by the person with dementia is leading. The designer should be aware what role they presume for the caregivers that will deploy the technologies in dementia care. In addition to using technology with good intentions, caregivers need to carefully and continuously evaluate its benefits for their patients.

Also, what works for one person with dementia might not work for another. One of the challenges in designing technologies for people with dementia is that they are still often perceived as a homogenous group characterised by their condition, rather than people with individual situations, wants and needs. This specific input, however, is needed to design appropriate interventions

(Brankaert and Kenning, 2020). This further emphasises the need for a careful and reflective approach that is evaluated regularly. In other words, in the design process of Warm Technology, a permanent feedback loop is a necessity (see also Chapter 4).

person-centred co-design methods

As we have seen, deception in dementia care is mostly applied to defuse potential harmful or stressful situations. When using deception, it is of the utmost importance that this is done with empathy and reflexivity to a person's needs and experiences in accordance with their situation. Casey *et al.* (2019) state in their work that a caregiver *has to really know the person* when they apply therapeutic lies in order to ensure a safe and morally acceptable application. For (technology) design we should consider this in a similar way following person-centred co-design methods, to ensure the interventions are designed with a genuine interest to contribute to their quality of life (IJsselsteijn *et al.*, 2020). Moreover, these (technological) interventions need the continuous attention of caregivers and others to ensure they are appropriate and are applied for the benefit of the person with dementia. The complexities of introducing technologies that contain deceptive elements into care practice therefore emphasises the fact that care technologies should always be designed to supplement rather than replace human care. To strengthen the connection between people, in person or enhanced by technology (Figure 14.2).

Figure 14.2. An illustration of a person with dementia video calling her son.

14.5 Conclusions and way forward

To conclude, the rise of dementia comes with numerous challenges for our society. To address these we need to be innovative and find new and plentiful solutions to support people with dementia and offer relief to caregivers. In this endeavour we have to be careful and aware of the potential effects our interventions have in care practice. Therapeutic lies are sometimes applied in dementia care; while often with good intentions, this should be done under strict guidelines and carefully monitored. When designing technology or other interventions, we have to work with similar guidelines and caution. This to prevent the blind adoption of deceptive practice in design that might cause confusion, invoke distrust or do harm to people with dementia. By increasing the awareness of potential deception in design and research, we can improve the ethical application of interventions in practice. This chapter also demonstrates how a more inclusive and reciprocal way of working with participants in design research helps circumvent potential challenges.

This chapter presents moral considerations regarding deception by design in dementia care. Further research is needed to understand the experiences of people with dementia when interacting with these interventions and the potential consequences they have on people. For designers and design researchers ethical considerations need to become an integral part of their way of working. In the design process they should regularly reflect on the potential impact of their design work. Furthermore, clear ethical guidelines should be developed that can be used as a tool during the design process. In addition, morality and ethical considerations need to be part of the DNA of designers and already included in their education – their design proposals are indirectly vehicles of their own morality. By being aware of both the intended and unintended impact of what we design or deploy, we can better contribute to the lives of those living with dementia.

References

Astell, A.J., Ellis, M.P., Alm, N., Dye, R. and Gowans, G., 2010. Stimulating people with dementia to reminisce using personal and generic photographs. International Journal of Computers in Healthcare, 1: 177-198. https://doi.org/10.1504/IJCIH.2010.037461

Brankaert, R. and Kenning, G., 2020. HCI and design in the context of dementia. Springer Nature, Cham, Switzerland.

Brankaert, R., Snaphaan, L. and Den Ouden, E., 2014. Stay in touch: an in context evaluation of a smartphone interface designed for people with dementia. Lecture Notes in Computer Science (including subseries Lecture Notes in Artificial Intelligence and Lecture Notes in Bioinformatics). Springer, Cham, Switzerland.

Cahill, S., Macijauskiene, J., Nygård, A., Faulkner, J. and Hagen, I., 2007. Technology in dementia care. Technology and Disability, 19: 55-60. http://doi.org/10.3233/TAD-2007-192-302

Casey, D., Lynch, U., Murphy, K., Cooney, A., Gannon, M., Houghton, C., Hunter, A., Jordan, F., Smyth, S., Felzman, H. and Meskell, P., 2019. Telling a 'good or white lie': the views of people living with dementia and their carers. Dementia, 19: 2582-2600. https://doi.org/10.1177/1471301219831525

Gibson, G., Dickinson, C., Brittain, K. and Robinson, L., 2015. The everyday use of assistive technology by people with dementia and their family carers: a qualitative study. BMC Geriatrics, 15: 89. https://doi.org/10.1186/s12877-015-0091-3

Hendriks, N., Slegers, K. and Duysburgh, P., 2015. Codesign with people living with cognitive or sensory impairments: a case for method stories and uniqueness. CoDesign, 11: 70-82. https://doi.org/10.1080/15710882.2015.1020316

Hodge, J., Montague, K., Hastings, S. and Morrissey, K., 2019. Exploring media capture of meaningful experiences to support families living with dementia. Proceedings of the 2019 CHI Conference on Human Factors in Computing Systems. Paper no. 432: 1-14. https://doi.org/10.1145/3290605.3300653

Houben, M., Brankaert, R., Bakker, S., Kenning, G., Bongers, I. and Eggen, B., 2019. Foregrounding everyday sounds in dementia. In: Proceedings of the 2019 on Designing Interactive Systems Conference. ACM Press, New York, NY, USA, pp. 71-83. https://doi.org/10.1145/3322276.3322287

IJsselsteijn, W., Tummers-Heemels, A. and Brankaert, R., 2020. Warm technology: a novel perspective on design for and with people living with dementia. In: Brankaert, R. and Kenning, G. (eds.) HCI and design in the context of dementia. Springer International Publishing, Cham, Switzerland, pp. 33-47. https://doi.org/10.1007/978-3-030-32835-1_3

James, I., Wood-Mitchell, A., Waterworth, A., Mackenzie, L. and Cunningham, J., 2006. Lying to people with dementia: developing ethical guidelines for care settings. International Journal Geriatric Psychiatry, 21: 800-801. https://doi.org/10.1002/gps.1551

Lindholm, C., 2015. Parallel realities: the interactional management of confabulation in dementia care encounters. Research on Language and Social Interaction, 48: 176-199. https://doi.org/10.1080/08351813.2015.1025502

Morrissey, K., McCarthy, J. and Pantidi, N., 2017. The value of experience-centred design approaches in dementia research contexts. In: Proceedings of the 2017 CHI Conference on Human Factors in Computing Systems. ACM Press, Denver, CO, USA. https://doi.org/10.1145/3025453.3025527

Thoolen, M., Brankaert, R. and Lu, Y., 2020. AmbientEcho: exploring interactive media experiences in the context of residential dementia care. In: Proceedings of the 2020 ACM Designing Interactive Systems Conference. pp. 1495-1508. https://doi.org/10.1145/3357236.3395432

Turner, A., Eccles, F., Keady, J., Simpson, J. and Elvish, R., 2017. The use of the truth and deception in dementia care amongst general hospital staff. Aging & Mental Health 21: 862-869. https://doi.org/10.1080/13607863.2016.1179261

15. Ethical usage of data in long-term care: how do we proceed?

Sil Aarts

Living Lab in Ageing & Long-Term Care, Department of Health Services Research, Maastricht University, Duboisdomein 30, 6229 GT Maastricht, the Netherlands; s.aarts@maastrichtuniversity.nl

Abstract

Data, often referred to as 'the new gold', has become invaluable in evidence-based healthcare. To improve the quality of care, large amounts of data are constantly collected. By collecting these vast amounts of data, data analytics (i.e. data-driven exploration methods) can be used for exploratory data research, which is aimed at finding patterns and relationships that are (often) not present in small amounts of data. To date, the large amounts of data collected in long-term care for older adults (LTC) are barely processed or analysed, and therefore not used to their full potential. However, with the rise of data accumulation in LTC, data research and the corresponding ethical issues have become a topic of increasing interest, resulting in an ongoing societal debate. This chapter will focus on the ethical questions that arise when collecting and analysing data in this healthcare echelon, which includes some of the most vulnerable people in society. While it is beyond the scope of the current chapter to discuss all ethical issues related to data in LTC, this chapter aims to espouse multidisciplinary dialogues, among a diverse set of stakeholders, regarding data research. This will be done by discussing the challenges that arise when making data a prominent topic in LTC, as well as possible ways to address these challenges, including the creation of living labs in which scientific research is combined with daily care practice.

Key concepts

- In long-term care for older adults, a diverse set of stakeholders are active. All these stakeholders have their own values, norms and ethical standards

- Ethical standards in scientific research (and data research) in LTC might not have the same connotation as ethical standards in daily care provision
- Living labs that include a network of diverse stakeholders such as scientific researchers, care professionals, clients and their informal caregivers can play a crucial role in analysing and interpreting data, thereby adding to ethical data usage
- Internal scientific committees, initiated by living labs, can assist LTC organisations in discussions about the value of data and the balance between individual well-being (e.g. the privacy of clients) and the added value data can provide in terms of personalised healthcare

15.1 Introduction: data collection in long-term care

At the present time we are confronted with several major challenges within long-term care for older adults: significant demographic changes (i.e. an increase in the number of clients), and a shortage of staff, all while trying to remain financial sustainable. Emerging technological advances and continuous digitalisation have the potential to mitigate, at least partly, these challenges (Aarts and Verbeek, 2020a; Aarts *et al.*, 2020; Kruse *et al.*, 2016). All of this is aimed, either directly or indirectly, at maintaining or improving quality, not only in terms of the care and of the life of people receiving long-term care, but also in terms of the life of their informal caregivers and the quality of work for people working in long-term care (Verbeek *et al.*, 2019).

While technology is advancing rapidly and volumes of data continue to pile up, the implementation of technology in long-term care for older adults is currently focused on the technology itself rather than on the corresponding data it accumulates: much of this data is currently perceived as a by-product of care delivery, rather than a central asset to improve the quality thereof. However, long-term care organisations, aimed at providing care for older people who are experiencing a significant decline in capacity and who receive formal care and support, are constantly collecting data in their daily workflow. Care processes can be supported by collecting data on, for instance, socio-demographic characteristics, health problems, drug dispensing, wearables and other technologies, such as assistive robots.

An important part of long-term care for older adults are nursing homes, defined as facilities with a domestic-style environment that provide 24-hour support and care for persons who require assistance with activities of daily living (e.g. bathing and eating) and who often have complex health needs and increased vulnerability (Sanford *et al.*, 2015). In nursing homes, a lot of quantitative and qualitative information is digitally recorded: data related to clients (e.g. socio-demographic characteristics, health problems and medication usage) and data on various quality indicators (e.g. pressure ulcers, falls, malnutrition) are collected in systems such as Electronic Health Records to map quality of care and life of residents (Hayrinen *et al.*, 2008). This data can be used for data exploration (Lucero *et al.*, 2019). For example, by using the publicly available Medical Information Mart for Intensive Care MIMIC-III database (which contains 1,046,053 nurses' notes from 36,583 unique patients), previously unknown potential fall risk factors, accounting for clinical and contextual features such as prevalence, time before and after fall incident, were detected (Bjarnadottir and Lucero, 2018). In addition, data is collected regarding the experiences and opinions of people close to the client. This not only includes data collected in care organisations, e.g. conversations held with loved ones or surveys conducted, but also public data on websites such as ZorgkaartNederland in which clients and family members share their experiences about care providers and provide corresponding ratings.

Data in nursing home organisations are also collected at the level of care professionals. For example, data about working hours, days of absence, perceived autonomy and roster information can be of added value when making predictions in the area of work quality. Such factors can be related to job characteristics (e.g. perceived work pressure) or outcomes such as burnout symptoms or job satisfaction. All this data might provide novel insights into sustainable employability and vitality in LTC. For example, analysing employee data with exploratory data analytics can help inform and improve tasks such as recruitment, training and daily care provision (Leicht-Deobald *et al.*, 2019).

Since the implementation of technology and corresponding data accumulation is a rather novel approach for long-term care organisations, it poses several new challenges (Case 15.1). In short, the challenge is to embrace new technology and the corresponding data accumulation, while at the same time being 'digitally and ethically responsible' (Cohen and Mello, 2019). However, before discussing possible solutions to overcome this challenge, let us first discuss some of the challenges in more detail. In this chapter when we refer to ethics, we mean big data ethics. Big data ethics, sometimes also just called 'data ethics', refers to the morality of collecting and analysing, sometimes large amounts of (personal) data (Salerno *et al.*, 2017).

Case 15.1. GPS technology in long-term care.

Wandering and getting lost is common among older adults with cognitive problems such as Alzheimer's disease. Wandering isn't necessarily harmful although it can pose several challenges, especially if a person with cognitive problems gets lost and panics.

This not only affects the people that wander, but also their informal and formal caregivers. To date, the use of GPS technology (i.e. global positioning system) is gaining momentum in organisations providing dementia care (Landau and Werner, 2012). GPS technology is implemented to accumulate real-time data using satellites to give information about an individual's exact location. These systems can put out alerts when people with dementia approach restricted areas, i.e. outside the premises of an LTC organisation. By using these technological devices, people with dementia are given more freedom and autonomy, while simultaneously providing relief to formal and informal caregivers. In addition, more and more data are collected through these kinds of technological systems. Apart from pinpointing your location via GPS, a wearable – an item that you often wear on your wrist – can detect all kinds of care-related data, including activity patterns, heart rhythm, blood pressure and sleep rhythm. These types of data can, for example, provide insight into challenging behaviour, sleep patterns, or predict walking patterns of people with dementia. As a result, more and more complete profiles of individuals can be formulated. Hence, digital tools can empower people with dementia by being able to better adapt to individual wants and needs. However, this case also raises several ethical challenges that affect for example, personal liberty, privacy, and, perhaps most importantly, dignity. Needless to say, it is important to carefully consider the moral implications of using such technological devices.

15.2 Challenges in long-term care

15.2.1 Data in long-term care has not yet been used to its full potential

In 1597, Sir Francis Bacon uttered his well-known phrase in 'Meditationes Sacrae': 'Scientia potentia est' ('Knowledge is power'). This is now truer than ever: data has become the driving force behind all kind of industries in our society: while Facebook uses large bulks of data to personalise ads and even news, Netflix uses data to select movies and series tailored to the desires of their viewers, and Tesla uses data to develop highly specified road maps (Neto, 2020). Data is also intensively collected and used in the healthcare domain (Kruse *et al.*, 2016; Ricco *et al.*, 2020; Wasfy and Maccox, 2016). For instance, details about weight, growth and blood results of babies are collected in neonatology departments in order to be able to predict and possibly prevent complications (Bayne, 2018).

At present, large amounts of data collected in long-term care for older adults are not being processed or analysed (i.e. data stored in EHRs or data accumulated by means of in-home sensors), and are therefore not being used to their full potential (Aarts and Verbeek, 2020a,b; Aarts *et al.*, 2020). Since data is known to contain information which can lead to new knowledge, and since society is constantly demanding better care provision, it is reasonable to ask whether it is ethical to leave this data unused. To illustrate, research related to large amounts of data and corresponding data exploration methods can provide important information about diseases and risk factors, outcomes of treatment interventions, functional abilities and care processes. However, to date, there has not been a strong tradition of scientific research in long-term care for older adults (Verbeek *et al.*, 2019). As a result, this care echelon does not yet benefit sufficiently from new advances in data and technology and the best available scientific evidence. It goes without saying that the primary aim is to let society benefit from the advances in research: to discover scientific data-driven solutions that lead to improved care provision. Not using the available data might be considered as doing more harm than good.

15.2.2 Data is never neutral

As discussed in Section 5.2.2, the value neutrality thesis is widely rejected: we assume that technology is morally laden. The same is true for data: it is never neutral (D'Ignazio and Klein, 2020). For example, the technology used to collect and analyse the available data will, in general, shape the data. In addition, people involved in the design process of this technology or in the corresponding data analysis all bring their own subjective experiences, ideas and feelings to the table. Consequently, data and the analysis thereof can also discriminate or sometimes even cause harm. The number of books being published about the harmful implementation of data is on the rise (Domingos, 2015; O'Neil, 2016).

data quality

The quality of data analysis very much depends on the quality, completeness and representativeness of the available dataset. Data contains information that has been collected in certain ways by certain organisations for certain reasons: a possible 'bias' begins with the way in which data is collected and accumulated (Leicht-Deobald *et al.*, 2019). For example, the sample might not be representative of the entire population, i.e. data might not include variables that properly capture the phenomenon or situation, thereby resulting in bias. If in LTC, algorithms are trained on data which do not include people with dementia, these algorithms might perform worse at detecting neuropsychiatric problems in this specific group than in the entire population of people living in a nursing home setting. As Andrew Gelman has postulated, '*The most important aspect of a statistical analysis is not what you do with the data, it's what data you use*' (Gelman, 2018). Hence, one of the first key steps in identifying bias is to understand how the data was generated and for what purpose. The focus should be on providing transparency about what data is collected and how, how this data is used, and if this was indeed appropriate.

data science methods

Data science methods 'learn' from the data they are fed: data that is designed by humans. Not only can the data be biased; so can the people who analyse the data. Not intentionally, but simply because all people harbour beliefs and attitudes about other people (e.g. related to factors such as ethnicity, gender, body weight) and society as a whole. People are often not aware of their own biases that might influence their decisions. This is called *implicit* bias (FitzGerald and Hurst, 2017). Another example is *confirmation bias*, which causes people to interpret data and the results thereof in ways that uphold what one believes to be true (Glick *et al.*, 2017). In other words, people are naturally less critical of data that confirms rather than disproves what they believe or seek to find. Hence, just viewing data as 'being true' might render an organisation at risk of incorrect claims and conclusions. After all, data and data analysis methods always require some degree of human influence, which more often than not reflect one's own values and norms. This does not mean that data or the corresponding analyses are not useful; it does however mean one should be aware of the possible biases that we humans all have. Consequently, working with data requires a constantly reflective mind.

15.2.3 Data as a black box

Data collection and usage in LTC could allow for more personalised healthcare, tailored to the individual needs of clients. By making use of data that are often collected on a day-to-day, real-time basis, new patterns and relationships can be made transparent by means of data exploratory methods such as machine learning (Kruse *et al.*, 2016; Ngiam and Khor, 2019; Pirracchio *et al.*, 2019). In LTC, the amount of data collected about clients has increased tremendously in the last few decades: from descriptive, quantitative data like gender and age to healthcare problems, medication usages and data from in-home sensors, to more qualitative data stored in EHRs or accumulated by means of interviews. Consequently, because of the large amount of data that is accumulated per client, these clients are becoming more and more 'visible' and 'transparent', especially since healthcare data are known to be highly sensitive and private.

In sharp contrast, the collection, usage and analysis of data are becoming decreasingly 'transparent': they become increasingly difficult to comprehend, and are often also referred to as 'the black box' (Price, 2017, 2018). This metaphor dates back to the early days of cybernetics and behaviourism, and typically refers to a system for which we can only observe the inputs and outputs: the internal working mechanism is not observable. In other words, by manipulating the input, one is able to observe the changes in the output, but it is not possible to inspect what is actually going on inside 'the box'. Even if the box could be opened and you could take a look inside, its components would be observable (e.g. 200 different health-related variables that all interact) but we would still be unable to interpret their diverse interactions. How do all these components interact and, more importantly, why? And even more importantly, what does this mean for clients in LTC and the care they receive?

15.2.4 Data sharing

For organisations that collect and store personal data, it is mandatory to protect the privacy of this kind of information. However, the number of reasons why data should be shared between various stakeholders are compelling and ever increasing (Salerno *et al.*, 2017). For example, healthcare professionals not only exchange data among themselves but also with their clients, allowing them to increase their autonomy regarding their own health. Since clients are often dealing with various care providers (e.g. general practitioners, physiotherapists, specialists working in hospitals), exchanging healthcare-related data (which also often include data related to the provision of social services, e.g. municipalities) may allow for more and faster provision of information. Enabling effective data exchange between healthcare organisations is of the utmost importance for securing a future-proof healthcare system: technology, legislation and regulations all need to be properly aligned with daily care practice. Today, many long-term care organisations are already working towards data-sharing in a manner that fully protects the security and privacy of personal information. However, third party access embodies a different type of concern. For example, data may also be shared with scientific researchers who want to conduct scientific studies aimed at improving, directly or indirectly, quality of care and quality of life.

Although there are no ready-made formats or procedures to overcome the abovementioned challenges, several possible solutions can be provided. These will be discussed in the next part of this chapter.

15.3 Possible solutions for ethical data usage in long-term care

In the last few years, data from multiple different sources have become available, and LTC organisations have experimented with ways of using large amounts of data and corresponding innovative data methods to increase their knowledge. However, there is still a huge gap between the world of technology and data on the one hand, and that of healthcare (and more specifically: care for older adults) on the other. It is crucial to be able to guarantee that these large volumes of data are being used in an ethically acceptable manner. In this context, it would be helpful to come up with a clear vision on data in such a way that data usage, directly or indirectly, positively influences the quality of life, the quality of care and the quality of work for those involved.

15.3.1 The need for a vision on data

Within long-term care organisation there are often dozens of different technological products and systems, all of which collect data. While some technology is aimed at autonomy (e.g. wearables as discussed above), other technology is aimed at stimulating social contacts (e.g. social robotics) or specific healthcare problems such as fall prevention (e.g. detection systems such as in-home sensors). Because of this diversity, it is important to make choices and to apply focus. In other words: a clear vision on data is necessary to sustainably organise data ethics in a healthcare organisation. By applying this focus we can ensure that data initiatives do not become fragmented and that the use of data actually results in novel insights and possibilities. Unfortunately, we often see that the technology is the main point of focus rather than the care process or the people involved (i.e. clients as well as care professionals). So, to ensure that technology and data is viewed as a vehicle rather than an outcome in itself, the vision of an organisation on quality of care should be leading, not the technology or data itself. Technology and data should merely be seen as a way to improve this quality.

The purpose of a vision statement on data is to ensure that the organisation is aligned to a common goal. Vision statements are key components of communications in LTC. They convey why each organisation exists and how value is added to their ecosystem. The vision statement regarding data should support and be consistent with an organisation's overall vision on care provision. It should provide answers to questions such as 'Why do we care about data?' and 'What added value does data (science) bring(s) to our organisational decision-making?'. In addition, organisations should ask themselves whether and how they want to contribute to data: is data and the analysis thereof something they want to invest time and effort in? The answer to this last question might not always be a firm 'yes'. This question is part of (and is perhaps the first question that needs to be asked) the ethical reflection of an organisation (Wade, 2007): it enables a discussion about the value of data and the balance between individual well-being (e.g. the privacy of clients), and the added value data can provide in terms of an organisation's vision statement. Since the ethical responsibility for the systematic collection and analysis of (client) data is the responsibility of the organisation concerned, these are the questions that need to be asked and answered first and foremost.

15.3.2 The importance of data ethics

vision statement

Care professionals need to make ethical choices on a daily base (Tauber, 2006). Daily care delivery often requires some challenging judgement calls, and holding yourself to high medical ethics standards may help minimise errors and foster trust, accountability, and respect between care professionals and the people they care for. In doing so, care professionals not only work closely with their clients and other care professionals, but also with other stakeholders such as informal caregivers and volunteers. In this care echelon, in which the role of clients as well as the role of their informal caregiver(s) is extremely important, it is important to achieve a balanced approach towards the various interests of these stakeholders (Sion *et al.*, 2020). In short, in LTC, the primary goal is to promote the well-being of the individual client. Ethically speaking, this implies that the potential benefits of care provided to a particular client must outweigh the risks posed to them. For example, care professionals might feel the necessity to provide care against someone's will, including the use of restraint measures or involuntary treatment (e.g. a person with dementia who refuses to take the prescribed medication), in order to keep clients safe (Mengelers *et al.*, in press). In the Netherlands, these types of complex ethical dilemmas have paved the way for the development of Moral Case Deliberation (MCD) projects in care institutions. In these MCD projects, explicit attention for moral issues related to care provision is provided by facilitating reflection and deliberation on moral issues that arise in daily care practice (Van der Dam *et al.*, 2013).

In contrast to daily care practice, the goals of scientific research are to produce (generalisable) scientific knowledge that will not only improve clinical care for (future) clients but also for society as a whole. In recent years, a growing body of literature has emerged that gives vent to the frustrations of researchers (and care professionals conducting research) when dealing with ethical issues related to scientific research, including ethical review boards (Franeta, 2019; Piasecki and Dranseika, 2019). Ethics committees are sometimes seen as a barrier to scientific research, especially because of the administrative burden they incur. Consequently, ethics related to scientific research (and more specifically to the use of data) may be viewed as a necessary checklist that needs to be 'filled in'. However, ethics and regulations related to scientific research are not alternatives to critical thinking and discussions. It is not helpful to reduce research ethics to essentially a legal issue, because then it diminishes professional responsibility, debases professional ethics, and downplays its primary purpose: a continuous critical reflection on the research being conducted (Franeta, 2019). We should therefore ask ourselves how we can ensure that scientific research and data usage are given the same attention and consideration as ethics related to daily care provision: how can we conduct scientific research aimed at improving personalised care (i.e. care tailored to the specific needs and wants of individual clients), in an ethically sound manner? Continuous ethical reflection and deliberation is necessary to make sure that scientific research and data usage become an important topic in LTC organisations: questions about ethics in research need to be viewed as equally important as ethics related to the provision of care. It is therefore crucial to overcome the reduced awareness of how scientific research can positively contribute to daily care practice, and what ethical issues might arise in this process. A promising pathway ensuring that LTC organisations can be a learning community for their care professionals on this topic, are the so-called living labs.

15.3.3 Living labs as a foundation for data usage in daily care practice

Conducting scientific research is not the core focus of LTC organisations: a strong tradition of scientific research in long-term care is still lacking. Not surprisingly therefore, these organisations often struggle with how to reconcile scientific research with their daily care processes. Networks of researchers and citizens can play a crucial role by analysing and interpreting data from a domain-specific perspective. After all, using data and deploying data analytics without providing a way for end-users (e.g. clients, their informal caregivers) to give feedback would be undesirable. Only by having scientists and data specialists, together with care professionals, older individuals themselves (e.g. residents from nursing homes) and their loved ones, discuss and interpret data and the corresponding issues, can the use of data indeed add value to daily care practice.

This can be accomplished in so-called 'living labs'(Verbeek *et al.*, 2019). A living lab is not a physical space, but a network of researchers, care professionals, healthcare directors, policymakers, teachers, and end-users such as the clients and their informal caregivers (e.g. family members). For instance, a Living Lab in Ageing and Long-Term Care was established by Maastricht University in 1998. To date, Maastricht University, nine large long-term care organisations, Gilde Intermediate Vocational Training Institute, VISTA College (secondary vocational education) and Zuyd University of Applied Sciences participate in this Living Lab. Their mission is to improve by means of scientific research: (1) quality of life for older people and their families; (2) quality of care; and (3) quality of work for those employed in long-term care (Verbeek *et al.*, 2019). At the core of the Living Lab are the so-called 'Linking Pins'. These are senior researchers who hold joint meetings at the University and one long-term care organisation: one day a week they work in one of the long-term care organisations participating in the Living Lab. A Linking Pin focuses on stimulating scientific research within the long-term care organisation in question. Their role is to coordinate scientific research and corresponding teaching activities, lead a multidisciplinary working group of LTC professionals, and together with staff and older people assist in identifying problems in practice that need further investigation by means of scientific research (Verbeek *et al.*, 2019).

Older people and their informal caregivers play a central role within the Living Lab: they participate directly in research projects, either as a member of the project group or as advisor/consultant throughout the project (Verbeek *et al.*, 2019). Hence, these living labs can build a bridge between two different but related worlds: that of science and that of daily care practice. This is done by, for example, a method called 'The World Café' (Brown, 2002). The World Café is a creative, but not yet widely used method aimed at creating conversations that centre around relevant questions in daily care provision. This method contains café-style tables (hence, the term) in which people converse about a certain question or issue. People are given the opportunity to move to another café-table (and thus to another group), with new people, in which possible novel ideas, insights and knowledge are presented. By means of this method, a diverse set of people will have 'a seat at the table' in order to collectively discuss and gain knowledge.

Of course, large amounts of data alone are insufficient to make decisions and predictions regarding daily care: the use of data and innovative methods of analysis do not automatically lead to novel insights and knowledge. If we want to be able to translate the vast amounts of data accumulated in LTC organisations into meaningful information and knowledge, the findings from data science applications must be checked and interpreted in terms of reliability and applicability. An important stakeholder in this process is the so-called 'bridge-builder' (which can, for example, be a Linking-Pin or an innovation manager within an LTC organisation) who speaks the 'language of technology and data' as well as 'the language of care' and can therefore play an important role in connecting these two different, but related, worlds (Aarts and Verbeek, 2020b). The bridge-builder, who has affinity with and knowledge of technology and data, can, in close cooperation with other stakeholders, ensure that themes are researched that are indeed deemed important for daily care practice. These bridge-builders should be able to provide an explanation to various stakeholders about how data can be deployed and how data analytics work (Aarts and Verbeek, 2020b). The main idea is not necessarily to clarify every single piece of data and the corresponding analysis. Instead, the main goal is to understand the most important features of how the inner mechanisms of an analysis work (Burton *et al.*, 2020; Neyland, 2019). This may, directly or indirectly, result in new directions for improving daily care provision.

15.3.4 Intertwining daily practice with scientific, ethically-sound, research

Given the rapid evolution of data and data science, LTC organisations typically need guidance on how to conduct scientific research, and, more specifically, on how to use data to pursue new opportunities aimed at gaining knowledge about daily care processes. Living labs, and their linking-pins, can aid organisations in this process by ensuring that they intertwine scientific research with daily care practice.

For example, an internal scientific committee in a long-term care organisation can be proposed. In these committees a diverse set of research-related questions can be discussed and developed in co-creation, such as a collaboration method in which ideas are shared and improved among a diverse set of stakeholders (De Boer *et al.*, 2020). For example, questions related to participation in scientific studies, questions related to how the results need to be implemented within organisations, but also ethical questions that arise when conducting scientific research need to be discussed. Its primary aim is to initiate scientific research in all 'layers' of these organisations (i.e. from nurses providing daily care to their directors). Hereby facilitating scientific research in daily practice from start to finish. In doing so, organisations not only ensure that their research activities are intertwined with daily care practice (Wade, 2007), but also that employees are constantly involved in this process. Furthermore, in order to create the desired social impact, the results of scientific research need to be translated into that same daily practice in such a way that all those involved can easily comprehend and perceive the benefits.

diversity

A diverse set of people (ideally, from diverse backgrounds in terms of race, gender, age etc.) will need to have a seat at the table: programmers, ICT specialists, researchers, people with knowledge regarding ethics and regulations, care professionals, informal representatives and the clients themselves (Salerno *et al.*, 2017; Wade, 2007). This latter group is especially important, since they represent some of the most vulnerable people in our society. By including these 'ends-users', we not only include the premise of making scientific research a more central theme in daily care practice, but we also ensure that research projects are developed using a 'bottom-up' approach. In doing so, commitment from all included stakeholders is increased. By including a diverse set of stakeholders, public support for research might be increased; people may be more likely to support a research project if they can trust the quality and integrity of the research being conducted.

On the basis of dialogues in these kind of committees (and throughout the organisation), careful decisions need to be made about research in general, and, more specifically, about the use of data. These kinds of committees will need to initiate questions such as: 'what constitutes good research in this situation?' and 'how can we solve ethical dilemmas related to research and data usage?' Hence, these committees facilitate the discussion and resolution of ethical issues related to research and data usage throughout the organisation. Committees need to discuss the values involved, compare different viewpoints and make substantiated choices in the interest of clients, their informal caregivers and care staff. For example, when referring back to the case regarding GPS-trackers for people with dementia (Case 15.1), it might be worthwhile, before providing every person with dementia with this kind of technology, to gain some insight into who would actually benefit from this technology. Data can provide useful insights in this area: questions such as 'which client displays wandering behaviour?' and 'what does this wandering entail?' need to be answered before implementing this rather novel technological product.

By using data, care provision can be personalised to the specific needs and wants of individual clients. In addition, in LTC, changes in the health status of clients, including their mental health and cognitive status, are a recurrent theme. The use of data aimed at the personalisation of care might require the informed consent of the client or the relevant informal caregiver. Since, in LTC, the health status of a client might change quickly, this requires constant attention; several ethical issues related to involving people with dementia as participants in research (or to use their health-related data), including recruiting and the processes of informed consent, arise when wanting these clients to participate in scientific research. The diverse set of stakeholders within the scientific committee could discuss and create new norms, related to these issues, by including a constant learning system aimed at intertwining scientific research with daily care practice. Hence, becoming a more research-minded and 'data-driven organisation' not only requires an organisational vision, but usually also novel, internal procedures. Only then, we can ensure that scientific knowledge leads to improvements in quality of care, quality of life and quality of work in long-term care for older adults.

15.4 Conclusions

Evangelising about new technology and corresponding data collection by emphasising only their positive elements is not the way forward: we must also make sure that privacy, security, and ethics go hand in hand with progress related to technology and data advancements. Hence, ethical responsibility concerning the systematic collection and analysis of patient data for any scientific purpose (including big data research) is the responsibility of the long-term care organisations involved. Since the use of data and data science methods to gain new knowledge is rather novel in LTC, organisations find it difficult to discover ways to use the data they accumulate on a daily base in order to optimise their care processes. Long-term care organisations will need to discuss and overcome the challenges discussed in this chapter in order to be able to develop a data exchange ecosystem that provides ethical and meaningful information. Time, commitment, and open communication are required to make sure that data in LTC leads to the development of novel evidence-based knowledge aimed at improving daily care. So-called Living Labs, initiating close collaboration between science and daily care practice, might be the perfect vehicle to ensure that data usage in long-term care for older adults can indeed be perceived as 'the new gold'.

References

Aarts, S. and Verbeek, H., 2020a. Oproep aan de langdurige ouderenzorg: gebruik die data! Available at: https://tinyurl.com/jc7fpacs.

Aarts, S. and Verbeek, H., 2020b. Vijf handvatten om data in de langdurige ouderenzorg te gebruiken. Available at: https://www.zorgvisie.nl/blog/vijf-handvatten-om-data-in-de-langdurige-ouderenzorg-te-gebruiken/.

Aarts, S., Daniels, R., Hamers, J. and Verbeek, H., 2020. Samenwerkende Academische Netwerken Ouderenzorg (SANO). Academische Werkplaats Ouderenzorg Zuid-Limburg: Data in de langdurige ouderenzorg. Tijdschrift voor Ouderengeneeskunde. Available at: https://tinyurl.com/3km3frk6.

Bayne, L.E., 2018. Big data in neonatal health care: big reach, big reward? Critical Care Nursing Clinics of North America, 30: 481-497. https://doi.org/10.1016/j.cnc.2018.07.005

Bjarnadottir, R.I. and Lucero, R.J., 2018. What can we learn about fall risk factors from EHR nursing notes? A text mining study. eGEMS Generating Evidence & Methods to improve patient outcomes, 6: 21. https://doi.org/10.5334/egems.237

Brown, J., 2002. The World Café: Living knowledge through conversations that matter. ProQuest Dissertations Publishing, Ann Arbor, MI, USA.

Burton, J.W., Stein, M-K. and Jensen, T.B., 2020. A systematic review of algorithm aversion in augmented decision making. Journal of Behavioral Decision Making, 33: 220-239. https://doi.org/10.1002/bdm.2155

Cohen, I.G. and Mello, M.M., 2019. Big data, big tech, and protecting patient privacy. JAMA, 322(12): 1141-1142. https://doi.org/10.1001/jama.2019.11365

D'Ignazio, C. and Klein, L.F., 2020. Data feminism. MIT Press Ltd., Cambridge, MA, USA.

De Boer, B., Bozdemir, B., Jansen, J., Hermans, M., Hamers, J.P.H. and Verbeek, H., 2020. The homestead: developing a conceptual framework through co-creation for innovating long-term dementia care environments. International Journal of Environmental Research and Public Health, 18: 57. https://doi.org/10.3390/ijerph18010057

Domingos, P., 2015. The master algortihm. Penguin Books, London, UK.

FitzGerald, C. and Hurst, S., 2017. Implicit bias in healthcare professionals: a systematic review. BMC Medical Ethics, 18: 19. https://doi.org/10.1186/s12910-017-0179-8

Franeta, D., 2019. Taking ethics seriously: toward comprehensive education in ethics and human rights for psychologists. European Psychologist, 24: 125-135.

Gelman, A., 2018. The most important aspect of a statistical analysis is not what you do with the data, it's what data you use. Available at: https://statmodeling.stat.columbia.edu/2018/08/07/important-aspect-statistical-analysis-not-data-data-use-survey-adjustment-edition/.

Glick, J.B., Craiglow, B.G., Choate, K.A., Kato, H., Fleming, R.E., Siegfried, E. and Glick, S.A., 2017. Improved management of Harlequin Ichthyosis with advances in neonatal intensive care. Pediatrics, 139: e20161003. https://doi.org/10.1542/peds.2016-1003

Hayrinen, K., Saranto, K. and Nykanen, P., 2008. Definition, structure, content, use and impacts of electronic health records: a review of the research literature. Interntional Journal of Medical Informatics, 77: 291-304. https://doi.org/10.1016/j.ijmedinf.2007.09.001

Kruse, C.S., Goswamy, R., Raval, Y. and Marawi, S., 2016. Challenges and opportunities of big data in health care: a systematic review. JMIR Medical Informatics, 4: e38. https://doi.org/10.2196/medinform.5359

Landau, R. and Werner, S., 2012. Ethical aspects of using GPS for tracking people with dementia: recommendations for practice. International Psychogeriatrics, 24: 358-366. https://doi.org/10.1017/S1041610211001888

Leicht-Deobald, U., Busch, T., Schank, C., Weibel, A., Schafheitle, S., Wildhaber, I. and Kasper, G., 2019. The challenges of algorithm-based HR decision-making for personal integrity. Journal of Business Ethics, 160: 377-392. https://doi.org/10.1007/s10551-019-04204-w

Lucero, R.J., Lindberg, D.S., Fehlberg, E.A., Bjarnadottir, R.I., Li, Y., Cimiotti, J.P., Crane, M. and Prosperi, M., 2019. A data-driven and practice-based approach to identify risk factors associated with hospital-acquired falls: Applying manual and semi- and fully-automated methods. International Journal of Medical Informatics, 122: 63-69. https://doi.org/10.1016/j.ijmedinf.2018.11.006

Mengelers, A., Moermans, V.R.A., Bleijlevens, M.H.C., Verbeek, H., Capezuti, E., Tan, F., Milisen, K. and Hamers, J.P.H., ###in press. Involuntary treatment in dementia care at home: results from the Netherlands and Belgium. Journal of Clinical Nursing, ###. https://doi.org/10.1111/jocn.15457

Neto, J.A.R., 2020. Famous big data cases and applications. Tesla – big data success case. Available at: https://medium.com/xnewdata/tesla-big-data-success-case-6429af3cd58c.

Neyland, D., 2019. The Everyday Life of an Algorithm. Palgrave Pivot, London, UK.

Ngiam, K.Y. and Khor, I.W., 2019. Big data and machine learning algorithms for health-care delivery. Lancet Oncology, 20: e262-e273. https://doi.org/10.1016/S1470-2045(19)30149-4

O'Neil, C., 2016. Weapons of math destruction. Penguin Books, London, UK

Piasecki, J. and Dranseika, V., 2019. Research versus practice: the dilemmas of research ethics in the era of learning health-care systems. Bioethics, 33: 617-624. https://doi.org/10.1111/bioe.12571

Pirracchio, R., Cohen, M.J., Malenica, I., Cohen, J., Chambaz, A., Cannesson, M., Lee, C., Resche-Rigon, M., Hubbard, A. and Group, A.R., 2019. Big data and targeted machine learning in action to assist medical decision in the ICU. Anaesthesia Critical Care & Pain Medicine, 38: 377-384. https://doi.org/10.1016/j.accpm.2018.09.008

Price, W.N., 2017. Regulating black-box medicine. Michigan Law Review, 116: 421-474. Available at: https://www.ncbi.nlm.nih.gov/pubmed/29240330

Price, W.N., 2018. Big data and black-box medical algorithms. Science Translational Medicine, 10(471). https://doi.org/10.1126/scitranslmed.aao5333

Ricco, J.B., Guetarni, F. and Kolh, P., 2020. Learning from artificial intelligence and big data in health care. European Journal of Vascular and Endovascular Surgery, 59: 868-869. https://doi.org/10.1016/j.ejvs.2020.01.019

Salerno, J., Knoppers, B.M., Lee, L.M., Hlaing, W.M. and Goodman, K.W., 2017. Ethics, big data and computing in epidemiology and public health. Annals of Epidemiology, 27: 297-301. https://doi.org/10.1016/j.annepidem.2017.05.002

Sanford, A.M., Orrell, M., Tolson, D., Abbatecola, A.M., Arai, H., Bauer, J.M., Cruz-Jentoft, A.J., Dong, B., Ga, H., Goel, A., Hajjar, R., Holmerova, I., Katz, P.R., Koopmans, R.T., Rolland, Y., Visvanathan, R., Woo, J., Morley, J.E. and Vellas, B., 2015. An international definition for 'nursing home'. Journal of the American Medical Directors Association, 16: 181-184. https://doi.org/10.1016/j.jamda.2014.12.013

Sion, K.Y.J., Verbeek, H., De Boer, B., Zwakhalen, S.M.G., Odekerken-Schroder, G., Schols, J. and Hamers, J.P.H., 2020. How to assess experienced quality of care in nursing homes from the client's perspective: results of a qualitative study. BMC Geriatrics, 20: 67. https://doi.org/10.1186/s12877-020-1466-7

Tauber, A.I., 2006. In search of medicine's moral glue. The American Journal of Bioethics, 6: 41-44. https://doi.org/10.1080/15265160500394861

Van der Dam, S., Schols, J.M., Kardol, T.J., Molewijk, B.C., Widdershoven, G.A. and Abma, T.A., 2013. The discovery of deliberation. From ambiguity to appreciation through the learning process of doing Moral case deliberation in Dutch elderly care. Social Science & Medicine, 83: 125-132. https://doi.org/10.1016/j.socscimed.2013.01.024

Verbeek, H., Zwakhalen, S.M.G., Schols, J.M.G.A., Kempen, G.I.J.M. and Hamers, J.P.H., 2019. The living lab in ageing and long-term care: a sustainable model for translational research improving quality of life, quality of care and quality of work. Journal of Nutrition, Health & Aging, 24: 43-47. https://doi.org/10.1007/s12603-019-1288-5

Wade, D., 2007. Ethics of collecting and using healthcare data. Britisch Medical Journal, 334: 1330-1331. https://doi.org/10.1136/bmj.39247.679329.80

Wasfy, J.H. and Maddox, T.M., 2016. Worlds apart: the hype and reality of big data to improve health care. Circulation: Cardiovascular Quality and Outcomes, 9: 495-497. https://doi.org/10.1161/circoutcomes.116.003150

Part 4

A FUTURE WITH MORAL DESIGN

16. For the citizens

Mark J. de Graaf

School of Information Technology, Fontys Hogescholen, P.O. Box 347, 5600 AH Eindhoven, the Netherlands; m.degraaf@fontys.nl

Abstract

In this chapter, we will take the position and interests of the citizen into consideration. Developments in AI and machine learning are moving very fast. An anthology of possibilities is given as a sketch of the possible practical implications of these technologies for society, focusing on what affects the citizen most. The society you look at greatly determines what to expect. In the USA, innovations are primarily driven by businesses. In China, everything begins and ends with the State. Europe is more inclined to put citizens first. That is a challenging proposition and one with which governments are struggling. Data hunger often collides with citizen rights, and algorithms regularly get more authority than they should. Legislation that protects citizens can hinder innovation, while Europe is already lagging the USA and China too much. And yet, this regulatory power combined with technological innovation offers Europe opportunities to move forward.

Key concepts

- The technosphere – the whole of human-made installations and machines – has become more defining for our living environments than the biosphere
- Big data – large quantities of data often gathered from citizens, and necessary in order to train the algorithms we want to use for future society
- Federated machine learning is based on the principle of remote execution: the algorithm is run and trained on the device where the data is held
- Explainable AI is a technology that makes it possible to make visible how algorithms come to their conclusions

16.1 For the citizens

technosphere

In 2020, the technosphere, i.e. the whole of human-made constructions and systems, literally out-weighed the biosphere, the mass of all living organisms on earth (Elhacham *et al.*, 2020).

For most of us, the living environment is what Haff introduced as the technosphere rather than the biosphere (Haff, 2014). It is more important to know how to reset your router or order groceries online than to be able to find water in nature.

The biosphere took time to evolve. Uyeda *et al.* (2011) found that 'for a major change to persist and for changes to accumulate, it took about one million years'. Starting from single-cell life, the biosphere took hundreds of millions of years to become what it is today.

The steam engine took a few centuries to evolve to a more or less stable machine. The internet is about 25 years old. Instagram was started in 2010. If we compare evolution in the biosphere to that in the technosphere, the speeds of change are incomparable. We should therefore not be surprised that it is difficult to adapt to this ever-changing world.

negative and positive liberty

In this chapter we focus on machine learning and big data. These are recent technologies that are taking the world by storm, which is reason enough for many of us to have strong feelings and opinions about them. Some claim that the doors of paradise will open for us; others are sure that we are creating a new Frankenstein. Most of us, however, don't know what to think about all of this, let alone what to do. And yet, we really need to decide what to do with these technologies. That is the ambition of this chapter: to give you a perspective on what to think about and do with machine learning and big data. To get to that point, we will first sketch out the technological developments in a few key areas of change. These are by no means exhaustive – one could write books about each of them, only to discover that developments have outpaced the printing of the book – but they do serve as a basis for discussion. In this discussion, Isaiah Berlin's concepts of negative and positive liberty (Berlin, 1958) provide a frame of reference for the exploration of possible perspectives on the topics. In addition to that, tools and methods will be presented that both technology and legislation could provide to direct the development of technologies.

Dealing with the dilemmas of the use of big data and machine learning can be seen as a responsibility of the collective or the individual. This can be approached from ideological points of view. At the end of this chapter we take a pragmatic approach: taking human nature into consideration, what can be expected from individuals and what should be left to the collective?

If one were alone in the world, liberty would probably mean doing anything you liked. This is the interpretation most people will give when you ask them to describe liberty. However, we are not alone. My freedom to drive a car through town at 200 km/hour would seriously threaten the liberty of other people to live their life without being killed by my car. Berlin calls the first type of liberty, i.e. do anything you like, negative liberty – where negative refers to the absence of constraints. The person that is protected by traffic laws experiences positive liberty and has the possibility to live and achieve their goals. Positive liberty is more often associated with collectives, while negative is associated with individuals.

16.2 Anthology of artificial intelligence and machine learning for society

In this section we focus on the areas that most affect citizens: health, security, and smart city.

health

Developments in the health domain are invading our personal life and care systems. A plethora of wearables that collect health-related data on our bodies are already commercially available to consumers.

Apple Watch can continuously measure ECGs and is certified in the United States for medical use. It is thus possible to detect heart rhythm disorders early and prevent strokes. Further developments of the watch have been announced in the area of early detection of depression and mental decline (Kysliak, 2020). Other wearables use Galvanic Skin Resistance to prevent epileptic attacks but also explore for feedback on stress and emotional reactions, both for critical professional use – e.g. pilots, operators – and for citizens dealing with mental health issues. Blood oxygen sensors are becoming mainstream in smart watches and have, for example, been used by the World Health Organization for the home treatment of Covid-19 in South-East Asia (WHO, 2021).

Lab-on-a-chip technology is another emerging technology that will cause a revolution in health care. And though the collapse of Theranos (Purkayastha, 2020) may have set back the expectations of investors in this technology, the scientific perspectives are still good (Walz, 2017). It is already available for a number of applications, such as diabetes, HIV and, more recently, Covid-19 detection. Researchers at Stanford University (Poudineh *et al.*, 2021) have developed a device that could potentially continuously sense levels of virtually any protein or molecule in the blood. Early detection of any health problem that produces distinguishable proteins could become available to consumers within years.

Another development is the application of machine learning for the diagnosis and treatment of diseases (Goecks *et al.*, 2020). Promising steps are being made in the application of big data to individuals in diseases like cancer, taking into account, e.g. genetic and other physiological aspects.

In elderly care, more and more (ro)bots are appearing on the scene. On the Dutch market, Zora is used to stimulate physical activity. Tessa reminds people of appointments, suggests small activities like drinking coffee and helps to structure the day. Paro mimics a pet and is used as a therapeutic robot for people with dementia. Obi is a robot arm that assists eating. And so on.

Future developments might integrate several of these functions into one. Using these robots to provide data-driven personalised care is still challenging but steady progress is being made in this field (Cortellessa *et al.*, 2021).

security

Humans can't really hide their emotions. Paul Ekman (Ekman, 1999) was a pioneer in the field of emotional expressions, which all humans share. He started to develop the field that has led to face-reading cameras and their applications. A similar development took place in voice and text analysis. These technologies are mature enough to be used commercially. Lemonade.com is a new insurance company that heavily relies on emotion recognition. When you file a claim, you have to talk to your webcam, and within seconds algorithms decide whether you have spoken the truth or tried to cheat. For the War on Terror, the USA government is developing software that can recognise terrorists by their facial expression (Davidrajuh and Try, 2015). Apparently, the intention to blow yourself up and the people around you does leave distinguishing marks on your facial expression. And that's not all. You might have seen the movie Minority Report. In this movie, crimes could be prevented before they happened and the culprits punished for deeds they had not yet carried out. That is exactly what researchers in China are aiming for (Li *et al.*, 2021): to predict and prevent crime using facial expressions.

Smart City

'Could I have a decaf oat milk latte macchiato and that vegan goji berry scone, please?' There is a world of complexity behind this question. Logistics systems that bring millions of products just in time to dense urban areas, energy systems that keep everything running day and night, public services that maintain infrastructures and public safety, and so on. Without advanced technologies, this question would never be asked. We depend on advanced technological systems to manage these issues. Big data and machine learning are now the key technologies.

Take energy transition, for example. We have come from – in hindsight – a fairly simple situation: a few big energy plants, a centralised electricity network, and predictable energy use. What is emerging now is a situation where everyone can be consumer and producer of energy at the same time. Smart grids are continuously redistributing power, our cars and home battery packs become part of the grid, we share data on production and use and increasingly let algorithms decide when to charge or discharge our batteries, to start our domestic appliances, and so on.

Mobility is also changing. Cities around the world are experimenting with self-driving taxis and buses. Singapore is leading the pack in small-scale deployment (Tan and Taeihagh, 2021). Traffic systems gather real-time information from all vehicles on the road and use this data to optimise traffic flow. Satnav systems use this data to suggest the fastest routes.

Smart garbage collection systems with load sensors in containers optimise garbage collection. It is also possible to weigh how much garbage individual citizens are disposing of and to charge them for that automatically.

In 2020, sewage companies in the Netherlands started a national Big Brown Data programme (Big Brown Data, 2020). Sewage is analysed at wastewater treatment plants for the occurrence of Covid-19 virus particles, for example, and has been used in monitoring the pandemic. Future ambitions include monitoring the prevalence of cancer, diabetes, obesity, smoking, allergies, resistant bacteria, medicine and drug use. Brainport Smart District, a high-tech neighbourhood under development in the Netherlands, is planning to collect this brown data per household (Brainport Smart District, 2018). The promise is that people will get access to their own personal data.

High-tech crowd management is also happening. For example, analysis of voices and movements of people can be used to predict possible outbreaks of violence (Galič, 2019). Drones are used for crowd and riot analysis (Castellano *et al.*, 2020). Also, smaller violations of rules can be detected. In Shenzen, China, smart camera systems recognise pedestrians ignoring red traffic lights and add that to their civic track record (Groβe-Bley and Kostka, 2020). If you get a handful of these entries, then life gets complicated. For example, you can be denied the right to rent a house. The number of traffic violations has gone down significantly.

16.3 Business, government and civil rights

In whose interests do all these innovations take place? Which citizens' interests are being served? The answer to that question depends on where you live.

United States

In the United States the answer is clear: big data is big business. Personal data drives their businesses, and the biggest company with the most data wins it all. One can argue Tesla's real advantage is in the massive amount of driving data all drivers are sharing with this company (opting out is not an option), which is used to train the autonomous driving algorithms. Sidewalk Labs, a sibling company of Google, planned to build Quayside in Toronto, a smart city area (Robinson and Coutts, 2019). The city would have been run like a high-tech commercial service, using the abundance of data generated by the customers/inhabitants. Inhabitants – the word citizens is not appropriate here – would influence the services provided, just as customers do. Something like a smart government was not part of the plan. In May 2020, the project was cancelled, partly because there was too much public resistance (Barth, 2020). The public-private value conflict was too big at the time (Frost, 2020). American healthcare, for example, is also a commercial enterprise, and here too the battle is for the data. The usual suspects like Apple Health, Google Health and Amazon Care all promise to revolutionise health care. Smart algorithms fed by massive amounts of sensor data will enable personalised medicine. And everyone that participates in one of these services shares their data. All of that, of course, for the sake of mankind. That is, as long as the lead metric, profit, is served.

From a societal perspective, real problems are being addressed. Mobility becomes better, cleaner and safer. New, better and more efficient solutions to healthcare are developed – in the context of exploding healthcare costs in the USA, approaching 20% of GDP and rising, and epidemic levels of chronic diseases (Tikkanen and Abrams, 2020).

But what does it mean to the individual citizens? What can we see, from the perspective of Berlin's concepts of liberty? Are citizens free of constraints, of barriers? On a superficial level, they are. Government involvement, a very delicate issue for many Americans citizens, is much lower compared to, for example, China. Even in dealing with the Covid-19 pandemic, the government had limited authority. It could not even enforce the wearing of facemasks in public spaces. Does that mean citizens are free to do what they like? Not quite. In many cases, businesses have the real authority. Censorship, i.e. what news is and is not allowed, is for example exercised by Facebook and Twitter (Parker Beard, 2021).

Big companies like Google, Facebook and Netflix decided to demand that their employees be vaccinated in the aftermath of the Covid-19 pandemic – and this did not cause any legal issues. The government can only require this as an employer from its employees and companies working for the government. Most American citizens can't afford health insurance unless it is provided by their employer. It is the employer that has the authority here, too. It is illustrative that Amazon Care is explicitly directing much of its marketing to employers rather than citizens. If you buy a new electric car, data will be gathered and life without a car is not an option for many. Et cetera.

China

Government is the answer in China. That can be very effective. Compare the management of Covid-19 in China to that in Europe or the USA, and it is immediately clear that a strong government is very effective in protecting its population from a pandemic. China is building a data-driven state surveillance infrastructure. Data from social media companies like Wechat, Weibo and Tencent has to be shared with the government. Public space is increasingly supervised by smart camera systems that can identify citizens and their behaviour, as in the Shenzen traffic example above. Data from all these systems is used to generate social credits; positive behaviours and positive messages about the government are rewarded, critical behaviours are discouraged. Recent showdowns between government and big tech companies like Alibaba made clear that the real power is at the state level (Pathak, 2021). From an American or European point of view, this might look like a nightmare. But for many Chinese citizens, the advantages outweigh the disadvantages. For example, all this supervision leads to fewer traffic accidents, lower crime rates, and a much lower impact from a pandemic. Even the growth of income inequality – which is seriously threatening cohesion in many Western countries – might have come to a turning point recently. And is it better to depend on Google Health than to rely on China Health? The former is fundamentally driven by profit, while the latter at least has an interest in keeping the population satisfied.

What is the perspective for individual citizens? Berlin's positive liberty is available in abundance if one doesn't cross certain lines, like criticising the government or showing 'undesirable' behaviour. Then you are usually well protected from disease, crime, and poverty. But there is a catch: if you don't comply with state expectations, life gets harsh. In other words, negative liberty is hard to find. Privacy in China is justified as an instrumental good, rather than an intrinsic good. It is much more acceptable to trade privacy for security than in Western societies (Yao-Huai, 2005).

Europe

Europe has a history of privacy invasions. The availability of highly sensitive data on religion and ethnicity from population registers enabled the Holocaust (Singleton, 1999), and the surveillance state that was set up after World War II in, for example, East Germany extended into every aspect of the lives of normal citizens and was used for the systematic oppression of any form of opposition (Sloan and Warner, 2016). These experiences have led to great sensitivity about privacy issues, which is why Europe makes greater efforts to protect the privacy of its citizens. One problem is that legislation and regulation develop slowly, much slower than technology (Tricoles, 2019); for example, Facebook was founded in 2004, and the first serious response to the privacy challenges, GDPR, was not introduced until 2018, in an attempt to regulate some aspects of privacy protection. The approach taken by GDPR is, however, the subject of some debate. It complicates innovations in big data and machine learning (Zarsky, 2016). Europe is lagging behind China and the USA in these areas. To protect its citizens rights long term, it needs to develop its own way. Possibilities hereto will be discussed in the following.

16.3.1 Innovation and citizen rights are not water and fire

It is possible to develop big data and machine-learning solutions that do respect citizen rights on privacy. And this might even be an opportunity for Europe to take a leading role. What is needed is a combination of regulation and innovation. In terms of regulation, Europe is already emerging as a superpower (Pätsch, 2018). Regulation can – and often does – hamper innovation; however, it can also enhance innovation (Pelkmans and Renda, 2014). GDPR, for example, has slowed down machine-learning innovations in need of quantities of sensitive data, but has also accelerated privacy-preserving technologies (see Chapter 7 for a related debate). Some examples of the latter are given below.

federated machine learning

It is possible to train machine-learning algorithms on distributed datasets. Instead of gathering all data in one place, a hacker needs to break into just one device – it is possible to let the algorithm travel around and train itself on locally stored datasets. Imagine millions of people having smart watches gathering ECG data. The algorithm would be implemented on every device, train itself on the data on the first device, share the training results – but not the ECG data – with the next device, continue training, and so on. ECG data does not leave any device. This technology is called federated machine learning, based on the principle of remote execution – the algorithm is run and trained on the device where the data is located (Hu *et al.*, 2019). It is also possible to train algorithms on specially encrypted data. This data is no longer legible to anyone, yet it is still possible to do computations on it that are needed to train algorithms, for example. A combination with federated learning is also possible.

privacy by design

These are technological examples of privacy by design (Cavoukian, 2009): an approach that anticipates privacy events even before they happen and prevents them from occurring. In the first example, it is not possible to steal the entire data collection, as it is not available on a single device. One would need to break into every single device and steal the data from it to achieve the same results as hacking a centralised data collection. Imagine having to hack every single Apple Watch to obtain a complete set of ECG data instead of breaking into one central computer!

Another example is the fingerprint sensor on Apple devices. The information on the fingerprint is stored inside the device, and cannot be retrieved remotely. The identification of a fingerprint is a completely local process. If biometric data is stored on, for example, a server – a common approach – data is much more likely to fall into the wrong hands.

16.3.2 Governments should not be lazy

Governments tend to love data. The more they get the better. The more accessible the better too. But most of the time, governments are notoriously bad at the responsible use of data, and the protection of data. There are plenty of examples of this. In 2015, the sensitive personal data of millions of US government employees was stolen by Chinese hackers. Data included information on background checks, fingerprints, mental health issues and relatives. The convenience of having all the data in one place came at a high price. There are better examples, like distributed electronic health records (Anderson, 2000). Personal medical information is left locally with the various care providers, like general practitioners, hospitals, dentists and pharmacies. If someone is hospitalised, the hospital can request data on a per case basis from the various providers. Unapproved mining of medical data in such a case is much harder.

In social and fiscal policies, governments often prefer easy data too. In 'Automating inequality', Virginia Eubanks (2018) shows how biased datasets about citizens that were meant to make government operations more efficient led to biased algorithms. The effect is that the poor are policed more and punished more. Simply put: if the police are patrolling your neighbourhood more often than other neighbourhoods, the chance of getting caught for a violation of the law is higher, the negative profile of your neighbourhood is confirmed even more, and so on.

9/11 attack

Enforcement services of all kinds are also fond of data. The best example is perhaps the 9/11 attack. After the attacks, people in the US were determined never to let it happen again. The Patriot Act was introduced, stretching the possibilities to gather private and sensitive information from all citizens. But was this response adequate? According to the 9/11 Commission Report (2004), the problem was not a lack of information and two of the main conclusions were as follows:

'While there were many reports on Bin Laden and his growing Al Qaeda organisation, there was no comprehensive review of what the intelligence community knew and what it did not know, and what that meant.' And 'The U.S. government has access to a vast amount of information. But it has a weak system for processing and using what it has. The system of 'need to know' should be replaced by a system of 'need to share'.'

Dutch childcare benefit scandal

A more recent example of things going wrong is the Dutch childcare benefit scandal. The tax services, in their eagerness to tackle fraud, used data-driven profiling techniques to identify potential wrongdoers. A self-learning algorithm was used to classify the benefit claims. Its role was to learn which claims had the highest risk of being false. The risk-classification model served as an initial filter; officials then scrutinised the claims with the highest risk label. As it turns out, certain claims by parents with double citizenship were systematically identified by the algorithm as high-risk, and officials then hastily marked those claims as fraudulent. These parents came under investigation and even the slightest mistake – a missing signature on a form, forgetting to report a minor change in income – was punished by a full clawback of all childcare benefit over the entire period since the minor mistake. Thousands of families ended up in severe poverty, families were broken, the health of those affected was badly affected. Later it became clear that in most cases there wasn't the slightest bit of evidence.

So, governments should be very critical about what data to use, what to use it for, and how to use it. The easy catches too often turn out to be poisoned. The citizen is the first victim, and trust in government is not far behind.

16.3.3 Trust

Building up trust is essential here. But how? One option is to start certifying algorithms for government use. For example, the cities of Amsterdam and Helsinki both launched algorithm registers, in which approved algorithms are listed. There should be actionable transparency to be able to build trust in the algorithms. Trust requires control and control needs to be transparent in order to be trusted.

Another consideration is to develop explicit policies on the areas in which algorithms may and may not be used. In the cases above, far-reaching decisions about human beings were delegated to algorithms, with devastating outcomes. The datasets used to train them were biased and their use to exclude or blacklist people was irresponsible. The human focus should be brought back here.

Explainable AI is a technology that makes visible how algorithms come to their conclusions. Biases such as those in the abovementioned cases would immediately become visible and could be subject to debate.

Transparency is also a condition for sustained trust. Being honest about how and why AI is used, what went wrong, but also what went well, is important here.

16.4 Conclusions

AI and machine learning have tremendous potential to deal with major challenges in many areas such as health, security, and the management of society. However, solutions often come at the expense of privacy and thus citizens' rights. These are more valued in Europe than in most other countries, and yet many things have gone wrong. There are opportunities for the EU to develop its own way in this field. If the focus of China is on state control, and the USA on business, Europe should focus on citizens' rights. This requires the creation of an ecosystem in which regulation and innovation together shape the future. The EU is already a superpower in regulation, but still lagging behind in technology.

If we want technology to serve the citizens, we have no choice but to act. Karl Popper once said that optimism is a duty. He explained that we are all jointly responsible for what is to come and should support what may lead to a better future.

That is precisely what we should do with AI and machine learning.

References

9/11 Commission Report, 2004. Final report of the national commission on terrorist attacks upon the United States. Available at: https://govinfo.library.unt.edu/911/report/911Report_Exec.htm.

Anderson, J.G. 2000. Security of the distributed electronic patient record: a case-based approach to identifying policy issues. International Journal of Medical Informatics, 60: 111-118. https://doi.org/10.1016/s1386-5056(00)00110-6

Barth, B.J. 2020. Death of a smart city. Available at: https://onezero.medium.com/how-a-band-of-activists-and-one-tech-billionaire-beat-alphabets-smart-city-de19afb5d69e

Berlin, I. 1958. Two concepts of liberty: an inaugural lecture delivered before the University of Oxford on 31 October 1958 (Vol. 31). Clarendon, Oxford, UK.

Big Brown Data, 2020. De potentie van rioolwateronderzoek. Available at: https://magazines.publiekdenken.nl/waterschap-10-2021/cijfers-over-big-brown-data/.

Brainport Smart District Nota van Publiekrechtelijke kaders en uitgangspunten deel C, 2018 (in Dutch). Available at: https://tinyurl.com/26zd3znk.

Castellano, G., Castiello, C., Mencar, C. and Vessio, G., 2020. Crowd detection in aerial images using spatial graphs and fully-convolutional neural networks. IEEE Access, 8: 64534-64544. https://doi.org/10.1109/ACCESS.2017.2712261

Cavoukian, A. 2009. Privacy by design: the 7 foundational principles. Information and privacy commissioner of Ontario, Canada, 5, 12.

Cortellessa, G., De Benedictis, R., Fracasso, F., Orlandini, A., Umbrico, A. and Cesta, A., 2021. AI and robotics to help older adults: revisiting projects in search of lessons learned. Paladyn, Journal of Behavioral Robotics, 12: 356-378. https://doi.org/10.1515/pjbr-2021-0025

Davidrajuh, R. and Try, T.S. 2015. Developing a software application for identifying potential terrorists at airports: application of predictive profiling using fuzzy logic. In: 2015 3rd International Conference on Artificial Intelligence, Modelling and Simulation (AIMS), IEEE, pp. 147-152. https://doi.org/10.1109/AIMS.2015.32.

Ekman, P., 1999. Basic emotions. In: Dalgleish, T. and Power, M.J. (eds.) Handbook of cognition and emotion. John Wiley & Sons, Ltd, Hoboken, NJ, USA. https://doi.org/10.1002/0470013494.ch3

Elhacham, E., Ben-Uri, L., Grozovski, J. Bar-On, Y.M. and Milo, R., 2020. Global human-made mass exceeds all living biomass. Nature, 588: 442-444. https://doi.org/10.1038/s41586-020-3010-5

Eubanks, V., 2018. Automating inequality: how high-tech tools profile, police, and punish the poor. St. Martin's Press, New York, NY, USA.

Frost, R.B., 2020. Public-private value conflict in platform cities. Available at: https://tinyurl.com/9mjpkac5.

Galič, M., 2019. Surveillance, privacy and public space in the Stratumseind Living Lab: the smart city debate, beyond data. Ars Aequi, 2019(juli/augustus): 570-579. Available at: https://arsaequi.nl/product/surveillance-privacy-and-public-space-in-the-stratumseind-living-lab/.

Goecks, J., Jalili, V., Heiser, L.M. and Gray, J.W. 2020. How machine learning will transform biomedicine. Cell, 181: 92-101. https://doi.org/10.1016/j.cell.2020.03.022.

Groβe-Bley, J. and Kostka, G., 2021. Big data dreams and reality in Shenzhen: an investigation of smart city implementation in China. Comparative Political Economy: Comparative Capitalism eJournal. https://doi.org/10.2139/ssrn.3814862

Haff, P. 2014. Humans and technology in the Anthropocene: six rules. The Anthropocene Review, 1: 126-136. https://doi.org/10.1177/2053019614530575

Hu, C., Jiang, J., and Wang, Z. 2019. Decentralized federated learning: a segmented gossip approach. arXiv preprint. arXiv:1908.07782.

Li, Z., Zhang, T., Jing, X., and Wang, Y. 2021. Facial expression-based analysis on emotion correlations, hotspots, and potential occurrence of urban crimes. Alexandria Engineering Journal, 60: 1411-1420. https://doi.org/10.1016/j.aej.2020.10.061

Parker Beard, J.C., 2021. Digital divide: big tech, censorship, and the bridge back to civil discourse. Georgia International Conference on Information Literacy, 34. Available at: https://digitalcommons.georgiasouthern.edu/gaintlit/2021/2021/34.

Pathak, S. 2021. China's tech crackdown: why now. Science, Technology and Security Forum, Manipal University, Karnataka, India.

Pätsch, S. 2018. Including but not limited to: how Brussels is emerging as a global regulatory superpower, establishing its data protection standard worldwide. Lund University Libraries, Lund, Switzerland. Available at: https://lup.lub.lu.se/student-papers/search/publication/8940309.

Pelkmans, J. and Renda, A. 2014. How can EU legislation enable and/or disable innovation. European Commission. Available at: https://tinyurl.com/5u7kc4pr.

Purkayastha, D., Dutta, S. and Chakravarty, S., 2020: Dark side case: Elizabeth Holmes and the rise and fall of Theranos Inc. Proceedings, 2020. https://doi.org/10.5465/AMBPP.2020.12239abstract

Poudineh, M., Maikawa, C.L., Ma, E.Y., Pan, J., Mamerow, D., Hang, Y., Hang, Y., Baker, S.W., Beirami, A., Yokshikawa, A., Eisenstein, M., Kim, S., Vučković, J. Appel, E.A. and Soh, H.T., 2021. A fluorescence sandwich immunoassay for the real-time continuous detection of glucose and insulin in live animals. Nature biomedical engineering, 5: 53-63. https://doi.org/10.1038/s41551-020-00661-1

Robinson, P. and Coutts, S. 2019. The case of Quayside, Toronto, Canada. In: Anthopoulos, L. (ed.) Smart city emergence. Elsevier, Amsterdam, the Netherlands, pp. 333-350.

Singleton, S. 1999. Privacy and human rights: comparing the United States to Europe. Presented At CEI's Conference: The Future of Financial Privacy, 12 January 1999. Available at: https://cei.org/studies/privacy-and-human-rights-comparing-the-united-states-to-europe/.

Sloan, R.H., and Warner, R. 2016. The self, the Stasi, and NSA: privacy, knowledge, and complicity in the surveillance state. Minnesota Journal of Law, Science & Technology, 17: 347. Available at: https://scholarship.law.umn.edu/cgi/viewcontent.cgi?article=1005&context=mjlst.

Tan, S.Y., and Taeihagh, A., 2021. Adaptive governance of autonomous vehicles: accelerating the adoption of disruptive technologies in Singapore. Government Information Quarterly, 38: 101546. https://doi.org/10.1016/j.giq.2020.101546

Uyeda, J.C., Hansen, T.F., Arnold, S.J. and Pienaar, J., 2011. The million-year wait for macroevolutionary bursts. Proceedings of the National Academy of Sciences, 108: 15908-15913. https://doi.org/10.1073/pnas.1014503108

Waltz, E. 2017. After Theranos. Nature biotechnology, 35: 11-15. https://doi.org/10.1038/nbt.3761

World Health Organization, 2021. Managing COVID-19 at home. Available at: https://tinyurl.com/asenspmw.

Yao-Huai, L. 2005. Privacy and data privacy issues in contemporary China. Ethics and Information Technology, 7: 7-15. https://doi.org/10.1007/s10676-005-0456-y

Zarsky, T.Z. 2016. Incompatible: the GDPR in the age of big data. Seton Hall Law Review, 47: 995. https://papers.ssrn.com/sol3/papers.cfm?abstract_id=3022646.

Media

Kysliak, B., 2020, 4 August. UCLA launches major mental health study to discover insights about depression. UCLA Newsroom, Health + Behavior. Available at: https://newsroom.ucla.edu/releases/ucla-launches-major-mental-health-study-to-discover-insights-about-depression.

Tikkanen, R. and Abrams, M. 2020. U.S. Health Care from a Global Perspective, 2019: Higher Spending, Worse Outcomes? The Commonwealth fund. Available at: https://www.commonwealthfund.org/publications/issue-briefs/2020/jan/us-health-care-global-perspective-2019.

Tricoles, R., 2019, 23 March. Smart tech sprints forward, but the law lags behind. Arizona State University, knowledge enterprise. Available at: https://research.asu.edu/smart-tech-sprints-forward-law-lags-behind.

17. AI is not ready for use

We need fair and transparent solutions that are not so data-hungry

Gerard Schouten

School of Information and Communication Technology, Fontys Hogescholen, P.O. Box 347, 5600 AH Eindhoven, the Netherlands; g.schouten@fontys.nl

Abstract

The power of AI is impressive. Algorithms fed with large datasets are able to recognise objects in images, extract words from speech, recommend things to buy or watch, find fraudulent transactions, diagnose a skin lesion, etcetera. Moreover, open-source frameworks are available to build AI for free. In short, AI is already transforming the way we work and live. However, in this chapter we argue that today's data-hungry AI is still in its infancy and not ready for use. Obviously, this technology works well for narrowly defined tasks. However: (1) it is still far from being a versatile natural intelligence; (2) it is not yet a mature engineering discipline; and (3) it does not have a sound approach for tackling moral and ethical issues, which are essential to make it trustworthy. Progress, state-of-the-art, concepts or solutions that are appearing on the horizon with respect to the above three perspectives are discussed in this chapter.

Key concepts

- Artificial Intelligence: machines (computers) that mimic cognitive functions of humans, such as understanding speech and text, interpreting images, planning, reasoning, and problem solving
- Machine learning: the use and development of computer systems that are able to learn and adapt without following explicit instructions, by using algorithms and statistical models to analyse and draw inferences from patterns in data
- Deep learning: a type of machine learning based on artificial neural networks in which multiple layers of processing are used to extract progressively higher-level features from data
- Natural intelligence: the opposite of artificial intelligence, it is all the systems of control present in biology

17.1 Introduction

The morning check of my mailbox invariably results in at least a dozen new messages with AI in the subject line. Apart from messages that contain links to blog posts or newsletters, they comprise announcements of events, proposals for research projects, invitations for guest lectures, information about initiatives to set up coalitions at regional or national level, or requests from companies for students that might be interested in a software engineering and AI internship. What strikes me in particular is that they come from a diverse public and almost any application domain that you could imagine. Their motivation varies: people are both excited and worried about AI.

The Oxford Dictionary defines Artificial Intelligence (AI) as the theory and development of computer systems able to perform tasks normally requiring human intelligence, such as visual perception, speech recognition, decision-making, and translation between languages.

After being in near-silent underground mode for decades, AI gained momentum in 2012, after the landmark publication of Alex Krizhevsky *et al.* (2012). In this paper he showed that a large artificial neural network is on a par with humans with respect to recognising objects in images. And today AI attracts the attention of a wide-ranging audience of professionals. This is largely due to the progress in machine learning and especially deep learning. These techniques rely on algorithms that are able to learn from data – i.e. spot patterns and create models based on labelled historical data that enable generalisation for new, unseen data. Many people and companies perceive it as the new electricity – a powerful and mature technology that is almost a given. And this view is not surprising, because algorithms are already transforming our lives. They help us to get from A to B, recommend things to buy or watch, find fraudulent transactions, and predict how we vote, or even with whom we will fall in love. But can it really be applied and commercialised 'just like that' to create robust and smart solutions for society?

My answer to this question is a firm and clear-cut 'No'. I want to separate the hype from reality, and sketch out what still needs to be done in the field of AI. In my opinion, the current status of AI can at best be described as follows: it is at a trial-and-error stage where a rudimentary form of intelligence, namely data-driven inference, is available to solve a specific task. That is to say, we have the basic ingredients to build narrow AI – i.e. compute power, availability of massive amounts of data, and open-source tools with algorithms wrapped in easy-to-apply function calls. In principle, creating models that can learn from data is within everybody's grasp. However, this is not a sufficient condition for creating future-proof AI innovations. Moreover, I will argue that the capabilities of deep-learning models are limited. AI faces both fundamental and practical challenges that have to be solved before truly intelligent software applications that are socially aware, responsible, robust, and sustainable can become a reality. In other words, AI is not yet ready for use.

In this essay, I provide a cautious outlook that will stretch AI beyond its adolescent stage in three different directions. The first perspective considers AI in depth and touches upon the foundations of intelligence; the second perspective emphasises the need for a sound design process for AI innovations; and the third perspective expresses the necessity to create fair, explainable, and transparent AI. In other words, AI that we can trust.

17.2 Causality and common sense

Picture yourself driving a Tesla Model 3. A 15-inch touchscreen next to the wheel depicts a bird-eye's view of your car on the road. As you drive alongside parked cars or pedestrians, they show up on the screen in a semi-realistic style. Road lanes are detected, even when sparse or no road markings are present. Moreover, in auto-pilot mode the Tesla is automatically kept in the lane at a comfortable distance from the car in front. At regular time intervals the car prompts you to touch the wheel, not out of emergency, but just to make sure that you are still alert. At first sight, technology seems to be well on its way towards driverless navigation – the ultimate goal of self-driving cars.

Despite these impressive results, I wonder how a Tesla car will handle the object-on-the-road problem. The car probably will hit the brakes for a crossing deer and will steer around a slow-moving hedgehog. Objects like mammals, people, trucks, other cars, motorcycles, etc. that appear on the road and are moving on a possible collision course are recognised with vision software, and these events are translated in a split second to meaningful actions by the Tesla. But what about scattered pieces of broken glass on the road. Does the Tesla *know* that glass is sharp and *argue* that it could puncture the tires? And does it *grasp* that a patch of wayward hay on a country road is harmless and that you could safely drive right over it? We all know that a scarecrow or snowman cannot walk by itself. But does a Tesla *understand* that such a human-like figure looming beside the road is not a living creature? And does it *anticipate* that, when a rolling ball is detected on the road, playing children might come chasing after it?

artificial neural networks

Self-driving cars typically solve challenging road situations by using artificial neural networks with an optimised architecture for vision (see also Chapter 4). These deep-learning algorithms construct a mapping that correlates inputs to labelled outputs – in this case, pixel data arranged in images (or video frames for that matter) to object categories. By using massive amounts of examples this mapping can abstract from inter-image variations that are not relevant – such as lighting conditions, viewpoint – and capture the essence of objects. Flaws in all kind of traffic situations are collected systematically and tackled by training these deep-learning models with even more example images and hence improve their predictive power. With this approach, today's self-driving cars can handle the relatively restricted situation of driving on highways in good weather conditions pretty well. However, they still don't have the intelligence to navigate in cluttered and uncontrolled environments, such as busy mixed-traffic crossings in cities.

humans

Humans learn differently from today's data-driven AI. To quote Mitchell (2019): 'Children exhibit an innate drive to actively explore the world, figure out the causes and effects of events, and even enlist adults to teach them what they want to know. The formation of core insights is tightly linked to developing motor skills and awareness of our own bodies. Without being told explicitly, toddlers have grasped the concept of object permanence, i.e. if an object is blocked by another object, the first object still exists, even though it is not visible. They are aware that when rigid objects collide, they don't pass through one another, but their direction of motion changes. And they also know that humans or animals – entities with intentions – can move by themselves and initiate objects' motion.' We are typically social learners that like to experiment with creating and testing hypotheses and generalise experiences fluidly into micro-theories. We seem to have a more or less hard-wired intuition that is able to internalise a rough-cut version of the Newtonian laws of physics. Pearl and Mackenzie (2018) emphasise the importance of causality: 'Humans are causal inference machines. Rather than register facts and process information, we seem to have a strong preference to connect events and to create stories glued together by a web of cause-effect relationships.' This is how we build-up a cognitive model that exhibits a rich understanding about the world and how it works, i.e. we use common-sense knowledge, acquired by experience.

Simply put, AI can't decipher the world with the same nuance and flexibility as a human. Saty Raghavachary (2021) even argues that it is fundamentally impossible to mimic natural intelligence with digital implementations. Wicked problems for self-driving cars will not be solved by giving algorithms more training data. Recognising objects with deep learning is only part of the puzzle. Things that might occur in an open-ended world cannot be enumerated in advance. Next generation self-driving cars should have a form of situational awareness, i.e. understand scenery – what does a composition of objects and actors collectively imply – and assess what possibly can and cannot happen next. Marcus and Davis (2020) discuss many similar scenarios and shortcomings of today's AI systems dealing with either language, vision, or both. Marcus (2018) argues that the emphasis of the current AI-hype is far too much on data-hungry deep learning and that an essential component is missing. The fundamental shortcoming is that it sidesteps the whole process of building and using a cognitive model. Today's narrow-task AI cannot reason about the world, act outside prior trained situations, build micro-theories or use imagination to play with what-if scenarios that are not necessarily backed up by (existent) data. Future-proof autonomous AI systems – like truly intelligent virtual assistants and dexterous household or care robots that interact with us – must go beyond the curve fitting statistics of deep learning and incorporate formats for cognitive models that embody concepts like common sense and causality. This allows these systems to operate safely in an open-ended dynamic world. Representing and encoding these concepts in a self-contained fashion (e.g. inside a machine) is not an easy task and requires a major research effort.

17.3 Big data meets AI

A few years ago, Google instrumentalised millions of Internet search terms to predict flu epidemics. They built a web service – Google Flu Trends – as an alternative for the laborious surveillance data collection performed by national health agencies (Ginsberg *et al.*, 2009). TomTom uses GPS locations of smartphones to draw city maps in near real time. For instance, a complete lay-out of Amsterdam including actual roadblocks, deviations, and traffic congestion zones is generated on any given ordinary morning within a timeframe of only 15 minutes. Philips Healthcare explores the idea of combining time-series data from activity trackers with an MRI or CT scan to compose an improved personal treatment plan and to monitor health progress after a medical intervention.

And my personal 'Big-Data-meets-AI' favourite is summarised by the question: 'Can data produced by an IoT beehive be utilised to gauge local biodiversity in urban areas?' In this concept (Figure 17.1) honeybees are used as biomonitoring agents (Schouten *et al.*, 2021). I do not know the answer, but the idea is worth exploring because the ecological footprint of humans on our planet is growing dangerously large. We are in need of innovative urban layouts in which nature and dense populations of people can co-exist in a sustainable way.

Figure 17.1. Biomass of insects is in rapid decline (Hallmann *et al.*, 2017). Can we measure biodiversity in cities with data produced by an IoT beehive? Painting 'Gezicht op Delft' by Johannes Vermeer, c. 1660-1661. Courtesy Mauritshuis, Den Haag.

What the example described in Case 17.1, as well as thousands of other AI innovations, have in common is that they collect large amounts of data, often in streaming mode, and turn it – 'somehow' – into value. Value can usually be turned into a service that can be commercialised. The steps to do so are by no means trivial. Notice that the aforementioned examples have nothing to do with 'mimicking cognitive functions of humans'. The term AI has become blurry. The human-imitative scope of AI is no longer apt. AI seems to have been rebranded. Everything that can be done with machine learning or deep learning, combined with an automated data collection infrastructure – that is often even beyond our perceptual senses – is now framed as AI.

Case 17.1. Data stories of honeybees.

Honeybees (*Apis mellifera*) are highly sensitive to the quality of the living environment. They live in colonies of up to 50,000 individuals, usually inhabiting man-made hives. In order to feed the colony, including the queen bee and her larvae, the bees collect pollen and nectar from flowers in the environment and bring them back to the hive. Roughly speaking, a single colony of honeybees covers a circular area with a radius of up to three kilometres. Our idea (Schouten *et al.*, 2021) is to build a beehive equipped with IoT sensors and to collect waggle dance data (Wario *et al.*, 2017) that captures the whereabouts of the bees, data related to the health of the colony (e.g. number of offspring they produce) and combine it with open data (satellite images, traffic data, public databases of planted greenery, etc.). In other words, we want to mine the hidden data stories of bees and create an intelligent beehive that functions as a proxy for monitoring local biodiversity in urban areas.

Some of these ideas work, others do not. For instance, TomTom's mapping of cities and traffic jams is a winner; Google's Flu Trends is clearly not. The latter is no longer around, as the service was not able to predict aseasonal influenza (Lazer *et al.*, 2014). Some ideas based on data correlations are successful and profitable, others turn out to be spurious and never make the transition from a proof-of-concept demo (which is often misleading) to the stage of a reliable service or products that can be shipped. It is not always clear why 'Big-Data-meets-AI' projects succeed or fail. There are so many technical as well as business questions to take into account: is an algorithmic 'one-size-fits-all' approach[16] driven by data the best solution? Will users accept an AI service? Which part of a decision-making process can be automated (and why)? And how do humans fit into the mix? What should the algorithm do? Automate a decision, predict an outcome, compute an associative link (match or recommendation), deliver a ranking, or extract a weak signal in noisy data? How can we design edge-cloud interactions that protect privacy? How do we componentise the AI solution and make it safe, secure, responsive and scalable (Sculley *et al.*, 2015)? What are the best programming platforms and libraries for implementing a certain application? How do we handle missing data? How can we solve the difficulties of sharing data across administrative and competitive boundaries?

---

[16] Imagine a situation in which you are diagnosed with a malignant life-threatening tumour. You probably won't want the optimised auto-treatment option proposed by an AI algorithm. I'm guessing you will want to sleep on it for a couple of nights; you will want to weigh up the advantages and disadvantages of all possible options, relate it to quality-of-life aspects, and talk to several specialists and family and friends, before deciding anything.

The abundance of questions is a call for a solid design process. In my view, AI is an alchemy state; it is perceived as a 'magic wand' that is able to solve all possible problems in the world. The field is young and emerging. Experimentation is the norm; algorithms are mixed with a 'potion' of wishful thinking. Jordan (2018) describes this elegantly: 'Just as humans created cathedrals and bridges long before there was civil engineering, humans are proceeding today with building inference-and-decision-making systems that involve machines. Just as early buildings and bridges sometimes fell to the ground – in unforeseen ways and with tragic consequences – our early inference-and-decision-making systems are exposing serious conceptual flaws (e.g. Anh *et al.*, 2015).' There is a clear need for a new engineering discipline and an educational programme dedicated to applied AI. The purpose of this programme is to bring a structured analysis, design and validation approach to the field of AI (Heck and Schouten, 2020). I advocate that it should be built around key concepts – such as *data*, *algorithm*, *inference*, *optimisation* – and professional skills – like *design thinking*, *software engineering*, *prototyping*, *storytelling* – subsequently moulded into well-thought-out student challenges and learning outcomes. Moreover, since much of the focus of such an applied AI programme will be on data from and about humans, it will require input from the social sciences and humanities as well, e.g. on cognition, ethics and law.

17.4 Towards trustworthy AI

The story sketched in Case 17.2 is made up, but it is certainly not science fiction. As we have seen in Chapter 2, it could happen. Face recognition is a mature technology; it is pretty accurate from a statistical point of view (Grother *et al.*, 2019). But it is not flawless. Many people use it to safely unlock their smartphones. Stores are installing the technology to deter and catch shoplifters. Governmental agencies connect millions of cameras in public space with face recognition software for surveillance or crowd-management purposes (see also the Roermond Case discussed in Case 2.3).

Case 17.2. Preventive policing and AI.

Imagine the following fictitious situation: over the last two years you put aside a substantial sum of two hundred euros each month to save money for a once-in-a-lifetime trip to a faraway destination. Finally, the day of departure approaches. You arrive on time at Kastrup airport, Copenhagen and hand in your 19.9 kg carefully packed suitcase at the self-check-in. Next up is the hand luggage and boarding pass control. However, just as you queue up before the conveyer belt and scanning devices, someone in a customs uniform taps you on the shoulder and asks you in a friendly but firm tone to follow him. Without any explanation, you are handed over to the police. I will spare you the details of the long, humiliating, and extensive questioning. The bottom line is that the whole process takes well over four hours. You miss your flight and cannot join the adventurous husky tour that started in Nuuk, Greenland ... just because a camera at the airport picked out your face and raised a red flag. It turns out that you were erroneously identified as a criminal.

responsible AI

The issue that I want to discuss here is not whether these systems are engineered professionally, or whether face recognition should be combined with other easy observable biomarkers – such as gait – to improve its precision and accuracy. The question is: how can we ensure a humane and responsible usage of AI technology? How can we balance privacy and protection, or justice and safety? As pointed out by Fry (2018), smart cameras identified 1,700 suspicious persons in New York in 2015; of which 900 were arrested. However, 5 of them were wrongly accused. Is this acceptable? How many algorithm violations are tolerable for catching criminals? These kinds of non-technical questions have to be answered by society.

Algorithms that use personal data and compute decisions that impact humans are inextricably linked to moral and ethical conundrums (O'Neill, 2016). They have to be considered as moral entities simply because they play a prominent role in complex decision-making processes and influence the autonomy of human beings. Ten years ago, these algorithms were 'sitting' in labs at universities and developed with small, limited datasets. Now they are going through a complex transformation from global principles to a situational fit for society. And as it turned out; they cannot be copied 'just like that' into applications that take in personal data on a large scale. It is worrisome that commercial industries are selling apps that exploit personal data, while a scientific approach to cope with privacy and bias in algorithmic decisions has not yet been settled. Of course, we need AI systems that make accurate and precise predictions based on the data. But above all, we need systems that are fair, accountable, transparent and that can explain why a certain decision is suggested (Janssen and Schouten, 2020). The only way to accomplish this is to internalise societal values in the design process of AI innovation (see also Chapter 4).

In this book we offer possible solutions from a moral design perspective. However, preliminary solutions for more trustworthy AI are a mixed bag. Most of them are mere ideas or just in their infancy. Experts in face recognition advocate regulating face recognition applications in the same way as the FDA regulates the pharmaceutical industry (Gershkorn, 2018): 'Just as a drug might come with a warning ... so could a face recognition system be tagged with do's and don'ts that clearly outline what users can expect and what the intended use is'. A straightforward implementation of this idea is model cards that provide a scheme of meta-data to document trained machine learning models (Mitchell *et al.*, 2019). Ethics frameworks that should guide the potential of AI towards social, economic and environmental benefits are being drafted by national authorities (Australian Government, 2019; European Commission, 2019). Research is beginning to understand the impact of demographic bias (e.g. with respect to gender or ethnicity) in datasets and has started to propose solutions to correct for this imbalance (Abbasi-Sureshjani *et al.*, 2020). Recently, IBM Research (2018) launched an open-source software project titled 'AI Fairness 360' intending to create a toolkit that can help detect and remove bias in models trained with data. And H20.ai (2019) explores the power of visualisations to make machine-learning models more transparent and explainable for scientists, developers, and end-users. In summary, the building blocks are in development. But the principles and guidelines for integrating these blocks into a comprehensive framework are not there yet.

17.5 Conclusions

Perhaps the main reason why AI is so popular is that the subject is suited for narratives. Everybody can imagine the potential and business opportunities of AI, because intelligence is something that is very close to us. The idea that machines have brainpower is mindboggling; it is both exciting and scary. In this chapter we argue that for socially-aware, responsible and truly intelligent AI systems we have to overcome major challenges. These challenges are clustered into three perspectives intended for further research.

The first perspective has to do with a fundamental change in the learning strategy of algorithms. Causality, broadly interpreted, has to do with any kind of knowledge about how the world progresses. Common sense refers to the basic and practical knowledge that we need in order to live in a reasonable and safe way. This will require a deeper investigation into the cognitive mechanisms that enable us to understand the world. Rather than combing through large sets of data, future general AI is less data hungry and teaches itself to observe the world and give meaning to it, much like children do. If we do not invest in this fundamental direction AI will become more and more an exercise of forcing square pegs into round holes. The second perspective advocates that AI should be developed into a new engineering discipline for 21st century millennials, with sound design principles and proven best practices. The third perspective addresses the question: 'Can we trust AI?' It stresses the importance of ethics, the effect that bias in the data can have, and the need to unlock inscrutable black-box algorithms. We do have a major responsibility and colossal task before us in bringing together computers and humans in ways that enhance societal and environmental well-being. In the end, the above question is about the future in which we want to live. Well-designed and maintainable models based on less data and more thought are essential for this much-needed development.

References

Abbasi-Sureshjani S., Raumanns R., Michels B.E.J., Schouten G. and Cheplygina V., 2020. Risk of training diagnostic algorithms on data with demographic bias. In: Cardoso J., Van Nguyen, H., Heller, N., Henriques Abreu, P., Isgum, I., Silva, W., Cruz, R., Pereira Amorim, J., Patel, V., Roysam, B., Zhou, K., Jiang, S., Le, N., Luu, K., Sznitman, R., Cheplygina, V., Mateus, D., Trucco, E. and Abbasi, S. (eds.) Interpretable and annotation-efficient learning for medical image computing. IMIMIC 2020. Lecture Notes in Computer Science, vol 12446. Springer, Cham, Switzerland. https://doi.org/10.1007/978-3-030-61166-8_20.

Anh, N., Yosinski, J. and Clune, J., 2015. Deep neural networks are easily fooled: high confidence predictions for unrecognizable images. Proceedings of the IEEE Conference on Computer Vision and Pattern Recognition.

Fry, H., 2018. Hello world: how to be human in the age of the machine. Transworld Digital, London, UK.

Gershkorn, D., 2018. An AI pioneer wants an FDA for facial recognition. OneZero Online. Available at: https://onezero.medium.com/an-a-i-pioneer-wants-an-fda-for-facial-recognition-cdde309cf553.

Ginsberg, J., Mohebbi, M.H., Patel, R.S., Brammer, L., Smolinski, M.S. and Briljant, L., 2009. Detecting influenza epidemics using search engine query data. Nature, 457: 1012-1014. https://doi.org/10.1038/nature07634

Grother, P., Ngan, M. and Hanaoka, K., 2019. Face recognition vendor test (FRVT) Part 3: demographic effects. National Institute of Standards and Technology, 8280. https://doi.org/10.6028/NIST.IR.8280.

H2O.ai, 2019. Explainable AI: increasing transparency accountability and trustworthiness in AI. Available at: https://www.h2o.ai/explainable-ai/.

Hallmann, C.A., Sorg, M., Jongejans, E., Siepel, H., Hofland, N., Schwan, H., Stenmans, W., Müller, A., Sumser, H., Hörren, T., Goulson, D. and De Kroon, H., 2017. More than 75 percent decline over 27 years in total flying insect biomass in protected areas. PLoS ONE, 12. https://doi.org/10.1371/journal.pone.0185809

Heck, P.M. and Schouten, G., 2020. Turning software engineers into AI engineers. arXiv.org. https://arxiv.org/abs/2011.01590.

IBM Research, 2018. AI fairness 360 open source toolkit. Available at: https://aif360.mybluemix.net.

Janssen, O.T.P. and Schouten, G., 2020. Machine learning, trust, and the whole transparency thing. Medium Online. Available at: https://medium.com/@o.t.a.janssen/machine-learning-and-the-whole-transparency-thing-ac85577be382.

Jordan, M., 2018. Artificial intelligence – the revolution hasn't happened yet. Medium Online. Available at: https://medium.com/@mijordan3/artificial-intelligence-the-revolution-hasnt-happened-yet-5e1d5812e1e7.

Krizhevsky, A., Sutskever, I. and Hinton, G.E., 2012. ImageNet classification with deep convolutional neural networks. Communications of the ACM, 60: 84-90. https://doi.org/10.1145/3065386

Lazer, D., Kennedy, R., King, G. and Vespignani, A., 2014. The parable of Google flu: traps in big data analysis. Science, 343: 1203-1205. https://doi.org/10.1126/science.1248506

Marcus, G., 2018. Deep learning: a critical appraisal. arXiv.org. https://arxiv.org/abs/1801.00631.

Marcus, G. and Davis, E., 2020. Rebooting AI – building artificial intelligence we can trust. Pantheon Books, New York, NY, USA.

Mitchell, M., 2019. How do you teach a car that a snowman won't walk across the road? Aeon Online Magazine. Available at: https://aeon.co/ideas/how-do-you-teach-a-car-that-a-snowman-wont-walk-across-the-road.

Mitchell, M., Wu, S., Zaldivar, A., Barnes, P., Vasserman, L., Hutchinson, B., Spitzer, E., Raji, I.D. and Gebru, T., 2019. Model cards for model reporting. arXiv.org. https://arxiv.org/abs/1810.03993.

O'Neil, C., 2016. Weapons of math destruction. Penguin Random House, London, UK.

Pearl, J. and Mackenzie, D., 2018. The book of why: the new science of cause and effect. Penguin Books, New York, NY, USA.

Raghavachary S., 2021. Intelligence – consider this and respond! In: Samsonovich A.V., Gudwin R.R. and Simões A.S. (eds.) Brain-inspired cognitive architectures for artificial intelligence: BICA*AI 2020. Advances in Intelligent Systems and Computing, vol 1310. Springer, Cham, Switzerland. https://doi.org/10.1007/978-3-030-65596-9_48.

Schouten G., Sangiovanni M. and Van den Heuvel W.J., 2021. IoT beehives and open data to gauge urban biodiversity. In: Arai, K. (ed.) Advances in information and communication. FICC 2021. Advances in Intelligent Systems and Computing, vol 1363. Springer, Cham, Switzerland. https://doi.org/10.1007/978-3-030-73100-7_17.

Sculley, D., Holt, G., Golovin, D., Davydov, E., Phillips, T., Ebner, D., Chaudhary, V., Young, M., Crespo, J-F. and Dennison, D., 2015. Hidden technical debt in machine learning systems. In. Proceedings of the 28th International Conference on Neural Information Processing Systems, pp. 2503-2511.

Wario, F., Wild, B., Rojas, R. and Landgraf, T., 2017. Automatic detection and decoding of honeybee waggle dances. PLoS ONE 12(12): e0188626. https://doi.org/10.1371/journal.pone.0188626

Codified standards and case law

Australian Government, 2019. Artificial intelligence: Australia's ethics framework. Available at: https://consult.industry.gov.au/strategic-policy/artificial-intelligence-ethics-framework/.

European Commission, 2019. Ethics guidelines for trustworthy AI. Available at: https://digital-strategy.ec.europa.eu/en/library/ethics-guidelines-trustworthy-ai.

18. Information ethics as a theoretical foundation for ethical assessment and moral design of AI systems

Frédérick Bruneault[1,2*] and Andréane Sabourin Laflamme[1,3]

[1]Département de philosophie, Collège André-Laurendeau, 1111, rue Lapierre, LaSalle, Montréal, Québec, H8N 2J4, Canada; [2]École des médias, UQAM, 405 rue Sainte-Catherine Est, Montréal, Québec, H2L 2C4, Canada; [3]Faculté de science politique et de droit, UQAM, 400 rue Sainte-Catherine Est, Montréal, Québec, H2L 2C5, Canada; frederick.bruneault@claurendeau.qc.ca

Abstract

AI systems have transformed the way we consume, inform ourselves and interact, but they are entering spheres of activity where the particular vulnerability of users makes it all the more urgent to have adequate ethical evaluation tools. For example, in Quebec, a company that provides information systems and data management solutions for student records to colleges has developed a system that allows them to target students at risk of failing different courses. The aim is to design a tool that would optimise the use of data already recorded in electronic student records in order to add value by analysing them using an AI system. Unlike the manual student tracking program, this software would make it possible to cross-reference a very large amount of available data: high school grade point average, results of ministerial evaluations, absenteeism rates, results of mid-term evaluations, first language, etc. The system would make it possible to automatically identify the student's needs and expectations. It would make it possible to identify the particular needs of students and which students are considered at risk of failing or dropping out of college. They could then get quicker access to success support resources that are appropriate to their needs. Given the high dropout rate among the college population and the political pressure on college administrations to promote success, it is not surprising that several colleges are considering implementing this system. Student associations are concerned about the implementation of such software, given that its use requires the processing of students' personal data. Some professors are also concerned about the psycho-social impact of the implementation of this monitoring program on the student population. In addition, administrators are concerned about the accuracy of the predictions made by this algorithm and whether students from certain groups may be unduly identified as being at risk of failure. It appears that classical ethical theories, stated at a high level of abstraction and focusing on the moral character of individuals' actions, are of little use when we attempt to weigh up the ethical benefits and risks associated with the use of such an AI system in education. Furthermore, while the many charters and declarations that have been published in recent years can be a useful resource for identifying the broad principles that AI systems should respect, the fact that these are not based on a clear ethical theoretical framework that would guide the interpretation of these principles in context, compromises the effectiveness of their application. We believe that the development of an appropriate theoretical framework for these principles is necessary in order to be able to provide practical tools for the ethical assessment and moral design of AI systems.

Key concepts

- Information ethics, as defined by Floridi, is a macro-ethics that considers ethical issues at the informational level of abstraction, thus attributing a minimal intrinsic moral value to all informational entities that populate what Floridi calls the infosphere, namely the digital environment where information has become an integral part of our ecosystem. Unlike classical ethical frameworks, information ethics is a patient-oriented ethics, i.e. it examines moral issues by assessing the effects of actions on the informational entities that receive the action and not by focusing on the agent that produces the action.
- Onlife is a neologism proposed by Floridi that characterises the fact that in our hyperconnected information societies, it is no longer relevant to question whether we are online or offline. This concept also refers to the increasing blurring between reality and virtuality, between human, machine and nature, and to the transition from information scarcity to information overload as well as the substitution of interactions, processes and networks to autonomous and independent things and properties.
- The concept of distributed morality refers to the idea that in our mature information societies, moral obligations cannot be understood only in the individual sense, as is the case in classical ethical frameworks. Indeed, given the increasing presence of intelligent and autonomous artifacts in our various spheres of activity, moral actions increasingly

take on a global dimension, where responsibilities are shared among several agents, whether human or artificial. Distributed morality allows us to consider that actions that individually would have a neutral or negligible moral dimension can have, once aggregated, a considerable moral impact.

- The idea of infraethics is based on the observation that in each society, non-moral factors influence the moral behaviour of agents. Every society is indeed characterised by an implicit and non-neutral ethical infrastructure that, if not moral in itself, can nevertheless constitute a facilitating framework that enables and promotes moral actions and interactions, whether these actions are individual or collective.

18.1 Introduction

Applied ethics emerged late in the history of moral philosophy. In the middle of the 20th century, some philosophers began to focus on ethical issues that emerged from particular contexts, for example professional settings such as medicine, business and law, or related to political issues, such as war, human rights or social justice (Van den Hoven, 2010). Many criticise the highly academic nature of the debates that dominate the academic world and deplore the lack of interest that moral philosophy has shown so far in ethical issues associated with particular social, political and economic phenomena. Indeed, the hermetic character of this discipline, which for some is more interested in the theoretical coherence of its propositions than in the particular moral dilemmas that humans face in their daily lives, is strongly denounced. For many thinkers, it is imperative to reframe the object of ethical reflection: 'The main aim of ethics is not to establish a general theory and set of eternal truth but to provide reasoned solutions and clarification to practical problems' (Van den Hoven, 2010: 65). Many deplore the fact that academic debates are in fact of no use in analysing the ethical issues that arise in our various spheres of activity, and consequently that the moral impact of these discussions is practically zero (Posner, 1998). Proponents of the pragmatist approach had voiced similar criticisms of philosophy, which according to Dewey should cease to be concerned only with the problems of philosophers and devote itself first to those of humans, so as to provide them with conceptual tools enabling them to intervene in the world in a moral manner (Kitcher, 2012; Dewey, 1917; Van den Hoven, 2017: 15).

Many philosophers, convinced that conceptual resources from moral philosophy can contribute to the resolution of concrete ethical issues that emerge from particular social situations, first tried to mobilise classical theoretical frameworks in ethics, such as deontology, consequentialism or virtue ethics, in an attempt to determine what moral action should be taken in a given situation (Van den Hoven, 2017: 13). However, the very abstract nature of these *high-level theories* soon proved to be a major obstacle to their endeavour, and the need to develop a conceptual framework that is adapted to the particular social context in which the ethical issues being analysed are situated became essential. Indeed, Van den Hoven, citing Bernard Gert and Carl Danner Clouser (Van den Hoven, 2010: 72), stresses the importance of *mid-level theories* in applied ethics, whose role is to carry out 'ground preparation', which consists of rethinking in context the principles and concepts set out at a higher level of abstraction in normative ethical theories. The capability approach developed by Sen and Nussbaum (Nussbaum, 2011; Sen, 1992), inspired by virtue ethics and applied in concrete discussions on education (Nussbaum, 1997) or international development (Sen, 1999), is an eloquent example of what Van den Hoven describes as mid-level theory. According to him, conducting ethical reflection at a lower level of abstraction makes it possible to establish a kind of 'middle ground' between the general and the particular, i.e. a conceptual space that would take into account the specific characteristics of the context from which the issues emerge and that would include, unlike classical ethical frameworks, a consideration of the social, economic and institutional conditions necessary for the application of high-level principles and concepts in this particular context. Such an undertaking must obviously rely on a detailed understanding of the particularities of that context, which often requires the reconceptualisation of certain abstract principles in the light of those particularities. This step appears to be essential to any endeavour in applied ethics, since the ultimate objective is to offer possible solutions to moral dilemmas that emerge from particular contexts.

The analysis of ethical issues associated with information and communication technologies (ICTs) and artificial intelligence (AI) obviously requires such work. Given the ubiquity of ICTs in our information societies, the increasing presence of AI systems in our everyday life and the many risks and ethical issues associated with this technology that have been identified in recent years, we believe that the translation of abstract ethical principles into practical evaluation tools is imperative. Indeed, the integration of AI systems in most of our different spheres of activity has profoundly transformed the context in which ethical issues arise, and the reconceptualisation necessary to develop a mid-level theory is all the more necessary.

We will therefore: (1) see what are the characteristics, principles and methods of the information ethics (IE) proposed by Luciano Floridi; (2) understand why it is particularly adapted to the context of our information societies and why it can serve as a foundation for the construction of a mid-level theory necessary for the elaboration of an AI ethics (AIE); and (3) introduce some tools of ethical evaluation and moral design that can be built from such a theoretical approach.

18.2 Information ethics: general framework

18.2.1 ICTs and the philosophy of information

Luciano Floridi lays out the philosophical underpinnings of information ethics in a synthetic form in his book *The 4th Revolution* (2014). In Bruneault and Laflamme (2021), we highlighted the essential aspects of this approach by positioning it in respect to different perspectives. The first feature he mentions about the information society in which we live is what he refers to as ICT ubiquity. The exponential development of ICTs forces us to rethink the philosophical presuppositions from which we understand our societies and ourselves (Floridi, 2013: 8). We are now living in what Floridi, like many others, calls the age of the zettabyte (Floridi, 2013: 5). If, according to estimates, humanity had produced in its entire history (until the mass production of personal computers) a total amount of information of 12 exabytes, it generated 5 exabytes just in 2002 and, since the turn of the 2010s, we have passed the zettabyte mark (1000 exabytes) of information produced annually, 1 zettabyte being thus 1,000,000,000,000,000,000,000 [10^{21}] bytes. In 2019, the estimates for 2020 were around 40 zettabytes, and for 2025, of 463 exabytes/day, for a total of 180 zettabytes (Desjardins, 2019). This information overload, which is connected to the growth of ICTs, emphasises the importance and pervasiveness of the notion of information for individuals and societies. The Copernican revolution challenged the idea that the Earth is at the centre of the universe, the Darwinian revolution called into question our unique status in the world of the living, and the Freudian revolution disputed the Cartesian idea of an individual completely in control of themselves and their thoughts. Floridi claims that the fourth revolution we face now (Floridi, 2013: 13), namely the information revolution, demonstrates that information is at the heart of human identity. Moreover, today we are no longer the only entities capable of producing meaning and information, as we have built machines that can create and manipulate information. Alan Turing could certainly be the emblematic figure of the new paradigm shift that characterises the development of digital technologies. The transformations brought about by the digitalisation of information have not only changed the world we live in, which Floridi calls *extrovert* influence, they have also changed our understanding of ourselves and our place in the universe, which he calls *introvert* influence (Floridi, 2013: 14).

four revolutions

This unprecedented situation, in which we are enmeshed, emphasises the informational character of humans. According to Floridi, humans are and always have been informational entities – and, more accurately, semantic entities producing meaning – but this crucial aspect is presently brought to light because of the development of information storage, analysis, and exchange capabilities, which demands that we rethink the categories from which we

informational entities

interpret our existence (Floridi, 2013: 14). Humanity transitioned from prehistory to history around the time that the first forms of ICTs appeared, with the transition from prehistory to history corresponding to the transition from a society that continually repeats the same life cycles to one that is able to preserve, notably through writing, certain forms of collective memory that allow the development of the complex and elaborate cultures we know today. Now, ICTs have taken on a new significance in the evolution of our society. ICTs have traditionally been used to assemble and document events that would normally occur on their own, whether they were social, economic, or political in nature. ICTs nowadays have become the engine of social, economic and political systems, playing a significant role, at least at the beginning of the 21st century in G7 nations' post-industrial societies, in the creation and replication of information societies. This information revolution has ushered us into a new era, which Floridi (2013: 3) refers to as hyperhistory, a time in which most societies have become vitally dependant on ICTs (2012).

onlife

The gap between 'offline' and 'online' existence, which was still quite obvious a few years ago, is becoming increasingly blurred, if not completely obliterated, so that now, owing to ICTs and all-round technology, our lives are continuously connected to the digital world (on issues related to 'ambient intelligence', see Costa, 2016). Floridi proposes the neologism 'onlife' to describe our contemporary digital interaction (Floridi, 2015). This leads him to focus on what is now known as the 'digital natives', i.e. the generations that were born at the time of (or during, or even after) digitalisation and have therefore never experienced the pre-digital world. It is important to acknowledge the pervasiveness of hyperconnectivity that now prevails in the constitution of the identity of all individuals, and especially of those who have to forge their personal identity in the crucial moments of childhood and adolescence in these circumstances (Floridi, 2014: 65). The specific contexts and practices from which we construct ourselves and become aware of ourselves, which ultimately refer to the informational nature of human beings, are greatly affected by these important changes. Whether one adopts an approach to personal identity (such as John Locke's) as associated with consciousness or prefers an approach to identity that interprets it more as a socio- and/or auto-biographical narrative artefact (Floridi, 2014: 68), it is undeniable for Floridi that the Self has to be interpreted as a complex informational system and that we are ultimately our own information (in the sense that this information is not just an external description).

infosphere

Floridi argues that human beings evolve in this informational universe that he calls the infosphere (Floridi, 2013: 8), a universe one can compare to the biosphere for the living world and that has always been there in the earlier stages of human society's development but is now becoming an essential component of it, as 'it is only very recently that human progress and well-being have begun to depend on information' (Floridi, 2013: 3). Human beings thrive in this universe of meaning and significance, and this becomes even clearer when ICTs release information management capacities beyond previous boundaries, ushering humanity into hyperhistory. In these conditions, particularly after we have entered the zettabyte age, the infosphere, rather than the Newtonian physical world, becomes the essential element of the lived world, what phenomenologists call the *Umwelt*, in which human beings live (Floridi, 2013: 16). Human beings will increasingly interact in the infosphere, regardless of where they are in the physical world. It is therefore important to consider the ethical challenges generated by ICTs by positioning these problems within this realm of meaning.

Floridi adds to this notion that the fourth revolution has resulted in deep changes in our interaction with the world. He claims that the distinctions that formerly organised our engagement with the world are thus increasingly blurred. One could say that these changes have been accentuated by the Covid-19 pandemic and by the lockdown measures that have been imposed on many an essentially virtual personal, professional and social life. As a result, it is vital to consider the philosophical and ethical difficulties associated with the deployment of ICTs, taking into account this 'onlife' and the blurrings that this new dimension of human being entails for many areas of our lives. Four of them are identified by Floridi (2015: 7): (1) obliterating the distinction between real and virtual worlds; (2) blurring distinctions between nature, people and artificial creations; (3) transferring from the information scarcity situation to what is nowadays called the 'infoglut;' overabundant information; (4) transferring from the metaphysics of entities to a metaphysics of processes.

18.2.2 Main characteristics of information ethics

non-standard

Information ethics (IE) is 'a non-standard, ontocentric and e-nvironmental macro-ethics' (Floridi, 2013: 97). First, IE is a patient-oriented ethics, thus non-standard (Floridi, 2013: 62). Unlike standard normative ethics' theoretical frameworks, which assign moral value based on (1) an evaluation of the moral agent (as in virtue ethics) or (2) an evaluation of the agent's actions (as in consequentialist and deontological ethics), IE seeks to morally assess actions or situations according to the impact they have on the 'informational flourishing' of patients (Floridi, 2013: 74). IE thus departs from the classical ADC model.

ontocentric

IE is also ontocentric (Floridi, 2013: 65); according to the degree of informational organisation, IE values informational entities whatever they may be. It is not possible, within the scope of this text, to fully develop the argument that leads Floridi to maintain that the most abstract view we can take of the world, which he calls the informational *Level of Abstraction* (LoAi), allows for all entities to be described in informational terms (see Section 18.2.3 for an explanation of this epistemological approach). It can be stressed, however, that this idea allows for the extension of the idea of intrinsic (but non-absolute) value to all such entities (in contrast to Kant, who notoriously restricts this feature to rational beings only, and in contrast to biocentric ethics, which extends this intrinsic value to sentient beings or to all living beings, excluding inanimate objects) (see Floridi, 2013: 112-124). If all existent entities, however basic, are informational entities, it follows that in ethical assessments of acts or circumstances they should all be considered, at least to a minimum. Floridi proposes, at least infinitesimally, to include all forms of informational entities in ethical assessments, and thus any existing entity, any being, in contrast to anthropocentric ethics, which takes human beings solely into account in ethical evaluation, and to biocentric ethics, which does so only for living (or sentient) organisms.

e-nvironmentalism

IE is also an 'e-nvironmentalism' (Floridi, 2013: 18; Floridi, 2014: 217) which is different from the twentieth century environmental ethics, since this 'e-nvironmentalism' does not oppose a human-made artificial world with the untouched and inviolable nature which would be the human environment. This 'e-nvironmentalism' recognises that the environment in which humans behave and human societies grow is increasingly composed of artificial items manufactured by humans. Therefore, not only should the natural (classical) environment be included in the ethical thinking on human behaviour, but also the fact that this environment is to a great extent the product of human activities and decisions of earlier generations. We must therefore evaluate our own actions and decisions, considering that they will mould the environment in which future generations will live. This feature of IE therefore serves as a link between persons living at the same time, as well as between generations, while also taking into account the social dimension of human existence.

18.2.3 Information ethics: principles and methods

Due to space constraints and since this would take us a long way from our purpose, it is not possible to explain Floridi's epistemological approach in detail in this paper, namely his method of abstraction (MoA). Floridi provides several explanations of this method, one more detailed (Floridi, 2011: 46-79), others more synthetic (Floridi, 2013: 29-52, 2019: 41-47), the general idea being that it is only possible to know any entity (physical or not) from a certain point of view, a certain level of abstraction (LoA). In other words, there is no description of an entity (and thus no knowledge) that comes, to use Nagel's (1986) expression, from *a view from nowhere*. As Floridi (2011: 74) points out, this plurality of LoAs does not imply a relativist conception of knowledge, since at a particular LoA, established according to a certain epistemological goal, some descriptions are true while others are false (although at another LoA, some descriptions might have been different). Following this MoA, it becomes possible – and this is what is particularly interesting here – to assess the ethical impact of ICTs in general and AI in particular not only on individuals, but at different LoAs: interpersonal, generational, societal, etc. We will return to these social dimensions of Floridi's approach when we discuss practical tools for AI ethics (Section 18.4).

Another important notion in IE is that of 'multi-agent system' (MAS). In order to ethically assess actions and interactions between different social actors at different LoAs, Floridi insists that it is not appropriate to restrict this evaluation to the individual agent, as standard macro-ethics does, since 'limiting the ethical discourse to *individual* agents hinders the development of a satisfactory investigation of *distributed morality*, a macroscopic and growing phenomenon of global moral actions and collective responsibilities resulting from the 'invisible hand' of systemic interactions among several agents' (Floridi, 2013: 137-138). We will return to the idea of 'distributed morality' below (Section 18.3.2), but for now we would like to stress that in order to provide a satisfactory explanation, we must first think of agents as always being stakeholders involved in one (or even several) MAS whose actions (at a specific LoA) do not result directly from the decisions of individual agents. Considering these actions at the LoA of different MASs (and interactions between different MASs) must be a fundamental dimension of an adequate understanding of ethical issues, especially those involved in the design, development and use of AI systems (we will return to this issue in Section 18.4). This method leads to the inclusion of the social dimensions of MASs, and consequently to the broadening of ethical assessment, traditionally limited to individual actions.

Floridi proposes interpreting IE through 'informational flourishing' and 'informational entropy,' i.e. respectively what promotes informational development and what leads to informational disorganisation in the infosphere, thus relying on the concept of entropy borrowed first from thermodynamics and then from cybernetics (Wiener, 1959). In this regard, he establishes four essential principles for IE (Floridi 2013: 71), which make explicit reference to Isaac Asimov's renowned robotics laws:

0. that informational entropy ought not to be caused;
1. that informational entropy ought to be prevented;
2. that informational entropy ought to be removed; and
3. that informational flourishing of entities and the whole infosphere ought to be promoted.

These principles should thus apply in a non-absolute way to actions and situations for all informational entities, whatever they may be, while possibly granting a special, perhaps absolute, status to informational and semantic entities, creators of meaning and capable of understanding the meaning of information, namely human beings. Thus, the ability of ICTs and AI systems to promote the 'informational flourishing' or to create 'informational entropy' for specific entities as well as the whole infosphere must be ethically evaluated, considering the effects of these technologies on all information entities, including human beings as informational and semantic entities. Given the pervasiveness of ICTs and AI in our societies, these characteristics of IE allow for a reassessment of the theoretical frameworks of traditional normative ethics.

18.3 Information ethics as a foundation for artificial intelligence ethics

18.3.1 Mid-level theories: from abstract principles to concrete ethical issues

In our information societies, ICTs are omnipresent and mediate our relationship to the world and others: 'much of what we do in the twentieth century we do not do *directly* to each other, but almost always *by way of technology*' (Van den Hoven, 2017: 19). This observation, which was already highly relevant when it was made in 2017, is even more apt in the current context, where the delegation of decisions, previously made by humans, to algorithms is becoming increasingly common in many aspects of our everyday life through AI systems. Because of the predictive power of AI systems, humanity now has more power over its environment and therefore has a greater responsibility. Yet, paradoxically, the attribution of moral responsibility to individual moral agents is proving increasingly difficult and ambiguous (Floridi, 2016: 43). Indeed, it seems that in a globalised world where ICT-mediated interactions occupy a growing place and where complex moral relations involving artificial agents are multiplied (Floridi and Sanders, 2004), a reconceptualisation of the very concept of individual moral responsibility is necessary. However, this endeavour is particularly tricky when it is undertaken within classical ethical frameworks, in which it is difficult, if not impossible, to conceive that non-biological or artificial entities can be worthy of moral considerations (Tavani, 2010: 260). This limitation compromises the possibility of using these theoretical frameworks to develop a mid-level theory that is appropriate to the context of the ubiquity of ICTs and AI, from which to consider possible solutions to the moral dilemmas that emerge in this particular context.

The singular contribution of IE is that it constitutes a high-level ethical theory which, while resting autonomously on solid ontological foundations (Floridi, 2013), is nonetheless particularly appropriate for dealing with the particular issues related to AI and thus provides the adequate theoretical foundations from which it becomes possible to develop a mid-level theory that would be appropriate for the particular context of our information societies. Although the principles outlined in *The Ethics of Information* are set out at a high level of abstraction, the transition to thinking at a lower level of abstraction is remarkably fluid, as we shall see below (Section 18.4.2). IE is undoubtedly a 'macro-ethic', like deontology, consequentialism and virtue ethics, but its non-standard character means that it has considerable advantages over these approaches, particularly in the context of the fourth revolution. For Floridi, classical ethical frameworks have certain limitations that make them inappropriate for thinking about the ethical issues that are emerging in our information societies (Floridi, 2013: 58). Without however disavowing these traditions, IE incorporates some of their essential components, while addressing some of their shortcomings. Some of these traditions, which are well known and have been widely discussed in the history of philosophy, are particularly problematic when we try to use them to address the specific ethical issues related to AI. Since a discussion of the similarities and differences between IE and each of the other ethical doctrines would require an in-depth analysis that is impossible to perform here, we will focus instead on the main characteristics of IE that distinguish it from other macro-ethics, and that make the alternative framework it proposes particularly fruitful for the development of a mid-level theory that allows us to think about the ethical issues associated with AI.

patient-oriented ethics

As discussed above, IE is a patient-oriented ethics, in the sense that it requires a form of decentring of the moral subject, who, in order to evaluate the morality of their actions, must be interested not in the conformity of these actions to a moral duty, virtue or well-being, but in the *action-receiver*, the patient. Floridi proposes nothing less than an alternative or non-standard macro-ethics, which departs from the agent-oriented and action-oriented conceptions of moral responsibility that prevail in standard ethical theoretical frameworks. As mentioned earlier, while the aim is to break free from the limitations of classical frameworks, IE recovers some of their components and redesigns them in the form of a renewed proposal. Indeed, although IE is not incompatible with the idea of a universal moral imperative, the four ethical principles outlined above echoing to some extent the universalism of the Kantian categorical imperative, in IE the determination of the morality of an action is to be carried out not from the point of view of the agent, who would have to determine the conformity of their action to duty, but from the perspective of the patient who receives or suffers the action, contrary to the deontological framework. Floridi notes the similarity of his approach to the prevailing one in medical ethics, bioethics and environmental ethics (Floridi, 2013: 58). However, insofar as IE is ontocentric, unlike anthropocentric or biocentric ethics, the patient's perspective is not limited to an assessment of the repercussions of actions on humans, sentient beings, nature or even the environment, but extends more broadly and fundamentally to Being, considered at the informational level of abstraction (LoAi). Such an approach has the advantage of bringing back into the fold of ethics actions that would otherwise be considered morally neutral.

MINMAX principle

Without being completely at odds with consequentialism, IE, whose four fundamental rules could also be expressed as a variation of the MINMAX principle – 'actions are right in proportion that they tend to increase (ontological) information and decrease (metaphysical) entropy' (Floridi, 2013: 58) – nevertheless differs from it in several ways. Unlike consequentialism, for which an action that at first seems immoral could later, considering a new evaluation of the consequences of the action, be considered moral (and *vice versa*), IE discards immoral actions (that generate entropy) from the start and these actions cannot be rehabilitated. In this sense, IE is closer to deontology. In IE, actions that promote entropy are necessarily considered immoral, but it is sufficient that an action contributes to improving the infosphere, following the four rules stated earlier, for it to be considered moral. This aspect of IE avoids the supererogatory problem often raised by critics of consequentialism, who consider that this theoretical framework is simply too demanding for the moral agent and imposes disproportionate moral requirements. Although driven by considerations of the welfare of those who undergo an action, consequentialism remains an action-oriented moral philosophy whose scope is much narrower than IE, which is problematic, given that many of the issues that emerge from our information societies are thus excluded.

proactive ethics

The particularity of IE is also that it is a proactive ethics, in the sense that it considers the moral agent as a producer of action, a creative agent, who can and must act on the world, and not as an agent who reacts to moral dilemmas that arise in a given situation, and who must decide, drawing on a list of principles, how to conduct themself (Floridi, 2010: 14). IE is, according to Floridi, a 'constructionist' ethics in which the obligation of the moral agent is not reduced to ensuring that their actions conform to the four stated moral imperatives but requires an active contribution to the construction of the environment in which they evolve, so that the latter can promote informational flourishing and limit entropy. Insofar as virtue ethics views moral obligation as a process in which the agent should seek to shape themself in order to become the best possible person, it shares with IE this 'constructionist' conception of ethics (Floridi, 2013: 164). Yet, the fundamental distinction between virtue ethics and IE lies in the fact that, in virtue ethics, the goal of the moral agent always remains, first and foremost, to work towards their own individual fulfilment. Considerations for the community, while not excluded from the outset, necessarily remain secondary. Indeed, even if in Aristotle the development of the individual is closely linked to that of the city, the object of moral action is first and foremost the individual and only secondarily the collective. Virtue ethics certainly has a *sociopoietic* dimension, but it remains above all an *egopoietic* ethics. If it was possible, in the very restricted societies that characterised ancient Greece, to foresee a syncretism between *egopoieis* and *sociopoieis*, the increasing complexity of our information societies, in which digital technologies mediate most social relations and where moral agents are more often than not MASs, compromises the possibility of such a natural integration. For Floridi: 'the kind of ethical constructionism needed today goes well beyond the education of the self and the political engineering of the simple and closed cyberpolis. It must also address the urgent and pressing question concerning the kind of global realities that are being built. This means decoupling constructionism from subjectivism and re-orienting it to the object, applying it *also* to society and the environment, the receivers of the agent's actions' (Floridi, 2010: 17). The moral obligation is defined by Floridi as an *ecopoiesis*: the agent's responsibility, which extends to the whole infosphere as well as to its present or future inhabitants, is to actively participate in the improvement of this environment.

contextual evaluation

In the context of IE, the evaluation of the possible effects of moral action on the patient, from a 'constructionist' perspective, requires a contextual evaluation of these effects, and this evaluation has to be done at different LoAs: informational, of course, but also at individual, interpersonal, group, institutional, sectoral and societal. Thus conceived, moral actions can be seen as a process of construction, which requires a contextual analysis of the social and institutional conditions that may promote such a construction. Although IE is not in itself an applied ethics, but rather a macro-ethics which, as discussed earlier, is expressed at a high level of abstraction, insofar as the aim of the moral agent is to improve the environment in which they operate in a way that promotes informational flourishing, it is yet to be seen how the principles it sets out can be applied to particular situations. Thus, from a 'constructionist' perspective, the moral agent must necessarily evaluate the different options that would allow them to intervene in a committed manner in order to solve ethical problems that arise in our information societies. While IE does not in itself provide the tools for solving particular problems in AI ethics, it does provide the conceptual groundwork for guiding the evaluation of possible solutions to the ethical problems associated with the use of AI systems.

18.3.2 Information ethics as a conceptual framework for a mid-level theory in artificial intelligence ethics

privacy

In order to fully understand how IE can be used to ground a mid-level theory that can deal with issues related to AI ethics (AIE), we will first highlight the specific contribution of this theoretical approach to the interpretation of privacy, before addressing this issue more generally around the idea of distributed morality. The reductionist argument and the argument based on the idea of property are the major classical ethical conceptions of privacy (Floridi, 2013: 240). The reductionist argument asserts that privacy is required to prevent the repercussions of personal information misuse. However, this reasoning is unsatisfying because it is impossible to reduce the value of privacy to the potential abuse associated with the undesired use of personal information. One might undoubtedly consider scenarios where the use of personal information is not acceptable even though it has a benign or a socially beneficial effect. The second argument, based on the concept of property, is frequently advanced in response to the above. Privacy protection is required because personal data is a personal property, and any unwanted use of these data would be equivalent to a kind of theft. In the contemporary technological setting this second argument, which is the classical liberal argument, is unsuitable because of the ubiquity of ICTs and AI systems (Floridi, 2013: 242). Indeed, it is now possible to acquire a staggering amount of personal information about individuals in settings that are far from private, with technologies such as face recognition in public locations – public locations where expectations of information privacy are rather low. Can an individual truly claim private ownership of information acquired about their actions, such as movements, behaviour, buying patterns, and so on, once removed from the private realm and placed in a public context?

The answer, as suggested by IE, is that the private ownership argument is not appropriate and we must rethink the notion of personal data itself (Floridi, 2013: 242). If humans are informational and semantic entities, then personal data is not a property that individuals own in the same way that they do tangible goods. Instead, these personal data embody who these persons are. When we recognise personal data as an integral element of the persons themselves and their personal identity, we can thus understand how important privacy is. Privacy must therefore become a fundamental aspect in any ethical evaluation of AI systems. The harvesting of individuals' personal data is therefore more analogous to kidnapping than stealing (Floridi, 2013: 244), kidnapping being a crime regardless of where it occurs (private or public) and the circumstances in which it takes place (e.g. whether it is committed out of sight or in broad daylight before witnesses). This conception of privacy makes it possible to appreciate the particular relevance of IE in assessing the specific ethical issues that emerge in the context of the 4th revolution and *a fortiori* to address the numerous ethical problems associated with the use of AI systems that are fed with massive amounts of potentially personal data.

More generally, this reasoning should be understood in a broader perspective related to what we explained above (Section 18.2.3), namely what Floridi calls 'multi-agent systems' (MASs). The privacy argument developed by Floridi is an illustration of the change of perspective in the assessment of ethical issues related to ICTs and AI in our information societies. As we have seen, Floridi questions the way traditional ethics evaluates ethical actions primarily in terms of individual actions. He argues that the majority of actions, especially in our hyperconnected information societies, involve MASs and interactions between MASs that may generate decisions or actions that are not equivalent to the sum of individual actions or that do not ultimately rely on individual decisions. This has an important impact on the development of an adequate ethical theory since 'an excessive focus on the moral life of a stand-alone, individual, human agent, [...] might fail to develop a satisfactory investigation of distributed morality (DM) in multi-agent systems (MASs)' (Floridi, 2013: 261). It follows that an understanding of what he calls distributed morality (DM) is imperative if we are to properly consider these ethical issues. It should first be emphasised that in referring to DM, Floridi is pointing to a very specific idea. Of course, collective forms of responsibility have always existed, and some forms of DM have in the past characterised some interactions between human beings. However, Floridi (2013: 262) wants to use DM to characterise a more specific phenomenon. He believes that in the context of the fourth revolution, we are dealing with DM phenomena that are out of all proportion to what may have existed in the past. It is these specific forms of collective responsibility that he calls DM. Several of our actions considered individually and evaluated in isolation may have a negligible or neutral moral value, but once they are aggregated in the general action of a MAS, they can acquire a greater moral value than the sum of the values attributed individually to each of the actions. MASs can be a collection of individuals, a company, an association, or a hybrid of individuals, companies, institutions, and artificial agents. Their actions will acquire a moral value independently of the moral value of the individual actions of the agents. This perspective must therefore be kept in mind in order to understand the current ethical issues related to AI systems.

distributed morality

To clarify the idea of DM, Floridi (2013: 263) identifies three sets of actions: (1) the set of morally neutral actions, (2) the set of morally good actions, and (3) the set of evil actions. This allows him to situate his approach in relation to traditional ethics. It is well known that intentionalist ethics (first and foremost Kant's) tend to shift some of the actions considered morally good at first glance to the set of morally neutral actions, since when the motivation for the action is not duty, but inclinations, desires or motivations other than respect for the moral law, these actions are not deemed to be truly moral. The same is true for morally evil actions. If an action is not intentionally evil, it is relegated to the set of morally neutral actions (Floridi, 2013: 263-265). Conversely, it is well known that consequentialist ethics do the opposite, namely that actions that may appear, at first glance, to be morally neutral will always, in such an approach, have a good or evil moral value. Since an action will always, directly or indirectly, generate consequences that ultimately have some moral value, any action, however small, becomes a morally good or evil action. For Floridi, these tendencies of traditional ethics are problematic, with intentionalist approaches denying moral value to some actions in an unjustified way and consequentialist approaches giving moral value to actions in an inordinate way. Floridi (2013: 266) proposes the idea of a 'threshold' separating morally neutral actions from morally good or evil actions. In doing so, morally neutral actions must pass this threshold of importance in order to transit to the set of morally good or evil actions. This is where the idea of DM comes into play. We must remember that IE is a 'patient-oriented' ethics (Section 18.2.2). It is therefore not on the basis of individual motivations or intentions that the actions can be judged, but on the effect that these actions have on the informational flourishing of the agent who receives this action, i.e. the patient. Thus, not only the actions of individual agents but also those of the MASs must be judged on this basis. It then becomes possible to understand why, once they are part of the decisions and actions of a MAS, several actions that individually seem morally negligible (and therefore neutral) will become sufficiently important to cross the threshold, making them morally good or evil actions. MASs allow for the amplification of actions (otherwise insignificant when considered in isolation), in addition to creating new decisions and actions that cross the threshold and thus have a moral value of their own.

It is the environment, in the sense of the context of action (in particular the social context, which we will come back to), in which individual agents evolve that will allow certain morally neutral (but potentially morally good or evil) actions to be aggregated in the MASs, thus propelling them into the set of morally good or evil actions (Floridi, 2013: 269). This environment surrounding individual action should prevent morally neutral (but potentially evil) actions from crossing the threshold by fragmenting them. Floridi calls this the resilient or 'fault-tolerant' aspect of the environment. Conversely, the environment should promote the amplification or aggregation of the actions of individual agents to bring them out of what Floridi calls moral inertia and thus allow them to cross the threshold towards the set of morally good actions. This environment (cultural, political and social) is established by policies that will affect both the thresholds themselves as well as the ability of individual actions and MAS's to cross these thresholds in one direction or the other. Some of the policies should therefore aim to fragment morally neutral but potentially evil actions so that they cannot amplify or aggregate and thus effectively become evil actions, or if they do, these policies should make sure they do so as little as possible, for as short a time as possible, and under the most unstable conditions possible. Conversely, policies setting the environment for moral actions should seek to amplify and aggregate the actions of individual agents and MASs, so that the combined actions can cross the threshold of morally good actions, allowing them to be more stable.

As mentioned above, DM phenomena (and more generally those of collective responsibility) have always existed. Traditionally, policies were designed to fragment potentially evil actions or to promote the aggregation of potentially good actions through the intervention of political institutions, especially through laws, by encouraging the aggregation of potentially good actions and discouraging the aggregation of potentially evil actions. Political institutions, through laws and policies, still seek to do this, but the exponential development of ICTs has created technological tools that have taken the lead in this function. The peculiarity of the situation in our information societies is that the opportunities for amplification and aggregation of the actions of individual agents and MASs have not only multiplied, but that this environment is constantly mediated by ICTs. These technological developments have indeed created conditions in which these socio-technical devices themselves become first-order factors that enable the amplification and aggregation of the actions of individual agents and MASs, thus allowing morally neutral (and potentially good or evil) actions to cross the threshold towards the realisation of their potential. More specifically, Floridi (2013: 270) tells us that the exponential development of ICTs has increased DM situations in an unprecedented way because they have increased in an entirely unprecedented way the interactions between individual agents and the number of situations in which these individual agents engage with MASs. In our information societies, characterised, as we have seen, by the hyperconnectivity of individuals, the phenomena related to MASs have multiplied to such an extent that today few actions are exclusively the responsibility of the individual agent, and the number of MAS in which we participate on a daily basis is enormous. The globalisation of exchanges, the multiplication of the social groups from which we act and the exponential development of the means of interaction and informational exchange, especially through AI systems, make DM situations ubiquitous and call for a reconceptualisation of our understanding of moral responsibility.

The aggregation of individual actions into MAS and the DM associated with these phenomena also emphasise a dimension of moral action often neglected by traditional ethics. These frameworks have mostly paid little or no attention to the non-moral factors that enable the exercise of moral reflection and moral action, which is hardly surprising given the limited aspect of the actions of individual agents prior to the technological development we have experienced in the twentieth century. Floridi (2013: 272) argues that it is no longer possible today to neglect this essential aspect of a satisfactory ethical theory. Drawing on the idea of infrastructure borrowed from economics and political science, he coins the term 'infraethics' to characterise these elements, namely a 'first order framework of implicit expectations, attitudes, and practices that *can* facilitate and promote morally good decisions and actions' (Floridi, 2013: 272). He thus argues that there are conditions that favour the development of moral action, conditions that are not in themselves moral, but which are essential for the implementation of moral obligations. These conditions are, moreover, part of the environment in which the actions of individual agents and MASs take place. Floridi (2013: 272) identifies some of them, namely trust, respect, reliability, privacy, transparency, freedom of expression, openness, fair competition, etc. These infraethical elements do not have moral value in themselves – in the sense that they can lead to the realisation of both morally good and evil actions – but they do have value as the very conditions of possibility for the exercise of morality, providing the necessary conditions of collaboration, coordination and cooperation that make possible the actions of MASs and that provide, consequently, the necessary environment for those actions to be moral. While recent studies in economics (Floridi, 2013: 273) have often focused on elements of economic infrastructure as facilitators enabling economic development itself, e.g. education, health or security, Floridi argues that elements of infraethics should also be considered as facilitators enabling the development of moral decisions and actions. ICTs and AI systems have a very important role to play as facilitators in this sense (Floridi, 2013: 273). These technologies enable (or hinder) the establishment of the conditions of infraethics by providing an environment in which the actions of individual agents and MASs can subsequently acquire moral value. Furthermore, these technologies can also be evaluated in terms of their potential to amplify, fragment and aggregate the actions of individual agents and MASs, thus transforming morally neutral actions into morally good or evil ones. It is therefore crucial to assess the ethical dimensions of ICTs and AI systems development. In order to morally assess particular AI systems as facilitators that have a major impact on the DM of the many MASs in which our daily actions are embedded, we will now turn to some of the most promising practical tools that highlight the conceptual advantages of IE.

infraethics

18.4 Information ethics and applied artificial intelligence ethics: practical tools

18.4.1 A framework for ethical assessment and moral design of AI systems

One of the tools derived from the IE approach has been developed as part of a reflection on the evaluation of technological devices involving AI in healthcare (Morley *et al.*, 2019). This tool, which takes the form of a table that aims to 'offer a cross-disciplinary systematic review mapping the potential ethical implications of the development of AI-Health in order to support the development of better design practices, and transparent and accountable deployment strategies' (Morley *et al.*, 2019: 4), can in our view help to structure the ethical assessment of AI-based devices, for example in bioethics committees. It has been developed from a review of the main categories of issues in the ethics of algorithms and AI, which is made possible by the patient-oriented approach of IE (Mittelstadt *et al.*, 2016). It should be noted that although it was developed in the context of health-specific issues, we believe that this tool could very well be useful in quite different contexts by structuring the assessment of AI systems in various settings. At the same time, this framework includes a consideration of the conditions of possibility for the exercise of morality that Floridi calls infraethics. The import of the tool is to structure both the evaluation and the questions that can be raised in the ethical assessment and moral design of AI systems. To fully understand the usefulness of this tool, it is first of all necessary to understand the categories drawn from the review of issues in AIE in Mittelstadt *et al.* (2016: 4-12). These categories define the three rows of the table (Morley *et al.*, 2019: 18; Figure 18.1). The first of these lines groups epistemic concerns that generate ethical issues in the application of AI systems. Importantly, this first group of concerns is divided into three sets of issues related to the evidence on which the operation of AI systems relies. A distinction is made between (1) *inconclusive evidence*, (2) *inscrutable evidence* and (3) *misguided evidence*.

inconclusive evidence

The first set of epistemic concerns relates to (1) the inconclusive evidence on which AI systems sometimes rely. AI systems are trained with datasets, and they seek to establish inductive correlations between different elements of these databases. It is therefore important to understand the problems linked with associating causality to what is, until proven otherwise, only a correlation (Mittelstadt *et al.*, 2016: 5). Often, AI systems are implemented and thus generate significant impacts on different actors affected by their use only on the basis of correlations, without any proof of causality (which remains presumed). Furthermore, even if causality could be demonstrated, causality is usually related to whole populations, while actions will affect individuals (for

LoA / Concerns	Individual	Interpersonal	Group	Institutional	Sectoral	Societal
Epistemic (inconclusive, inscrutable and misguided evidence)						
Normative (unfair outcomes and transformative effects)						
Overarching (traceability)						

Figure 18.1. A summary of ethical concerns associated with six levels of abstraction (LoA) (Morley *et al.*, 2019: 18).

which the causal relationship may not necessarily be proven). The inconclusive evidence on which AI systems sometimes rely will therefore lead to unjustified actions that raise concerns about the reliability of the operation of these systems. The second set of epistemic concerns deals with (2) inscrutable evidence. AI systems pose significant limitations on the predictability and explicability of their functioning (Mittelstadt *et al.*, 2016: 6). It is therefore extremely difficult to assess the evidence that leads algorithms to favour certain outcomes over others. This of course leads to numerous issues associated with the transparency of these algorithms. These transparency problems are linked to the very accessibility of the data, which is often protected by trade secrets, but also to the ability to generate an adequate understanding of how algorithms work and, more importantly, how they interpret the data they are fed with. Impenetrable evidence therefore creates a problem of opacity, which is often illustrated with the analogy of the black box. This opacity compromises, as we have just seen, the possibility to understand, explain and interpret the algorithms themselves, and also questions the legitimacy of the decisions that are made by these algorithms (Mittelstadt *et al.*, 2016: 7). These challenges therefore translate into a significant problem of trust in AI systems. Epistemic concerns are also based on (3) misguided evidence. The way in which algorithms are designed and operate reproduces, at least indirectly, certain values held by the people who have to some degree contributed to their conception (Mittelstadt *et al.*, 2016: 7). Furthermore, AI systems, and especially those using *machine learning*, rely on very large datasets that can also be oriented towards certain values, given for example the issues of labelling, those associated with data collection or those referring to the variable accessibility of specific data. The flawed evidence on which AI systems can be based leads necessarily to

inscrutable evidence

misguided evidence

inaccurate results and misleading interpretations (Mittelstadt *et al.*, 2016: 8). This leads to bias that can be transmitted, reproduced and amplified by AI systems and thus raises issues of fairness.

unfair outcomes

The second row of the table groups normative concerns related to AIE issues. Two sets of normative concerns are identified here: (4) unfair outcomes and (5) transformative effects. First, like biases, (4) unfair outcomes are ultimately a question of equity. While biases are epistemic issues concerning the production of knowledge, unfair outcomes are the effects of such biases in practices that lead to discrimination. Profiling and personalisation will create situations in which certain groups will be systematically disadvantaged, given some of the descriptions contained in the datasets used (Mittelstadt *et al.*, 2016: 8). Of course, the most common forms of discrimination are based on gender, cultural or religious groups, race, sexual orientation, etc. Such discriminations are prohibited by law in most jurisdictions. However, two things should be noted: first, there may also be criteria of discrimination which, although not explicitly prohibited by law, are nonetheless quite problematic; second, conversely, certain types of profiling and personalisation explicitly based on neutral characteristics (which do not *prima facie* present discrimination issues) will nevertheless generate unfair results since some of the data (which are not problematic in themselves) will have discriminatory effects when cross-referenced with other data, e.g. the use of postal codes and other 'redlining' practices (Mittelstadt *et al.*, 2016: 8). Unfair outcomes generated by AI systems thus raise questions about fairness and the conditions for fair competition between individuals. AI systems can also lead to (5) transformative effects, which are the second set of normative concerns. Personalisation (the practice behind the notorious filter bubbles), nudging and ultimately the manipulation of individuals through the use of AI systems, of course raise issues concerning the exercise of their autonomy (Mittelstadt *et al.*, 2016: 9). By controlling the information to which individuals have access in the overwhelming mass of information available, it becomes possible to influence (or even direct) the choices they make, which of course conflicts with the exercise of autonomous thinking. We can thus see this problem as a challenge to the respect due to the persons who use (or are used) by AI systems. On the other hand, the transformative effects also lead to a challenge to the idea of privacy, as we have seen above (Section 18.3.2). The use and generalisation of AI systems can not only transform the behaviour of individuals and the very way they construct their self-identity, but also the way they understand their privacy (Mittelstadt *et al.*, 2016: 9-10). These transformative effects therefore raise important questions about these fundamental aspects of individuals' informational identity.

transformative effects

traceability

The third row of the table deals with overarching concerns, more specifically with (6) traceability. The possibility (or impossibility) of tracing the decision-making and impacts of the individual decisions of each of the actors involved in the MAS that designed and developed the AI systems raises important questions about the ability to control the effects of these systems (Mittelstadt *et al.*, 2016: 10). It also raises questions of legal and civil liability, and more fundamentally questions of moral and political responsibility. If the MAS that design, develop and use AI systems (as well as individual actors involved in the MAS) can no longer be held accountable for the functioning and outcomes of these systems (Mittelstadt *et al.*, 2016: 11), then the attribution of responsibility is compromised. This overarching dimension completes the presentation of the different AIE categories.

The different columns of the table, in turn, put into practice the method of abstraction (MoA) explained above (Section 18.2.3) and different levels of abstraction (LoAs) that can be analysed to assess the ethical issues and moral design of a particular AI system (Morley *et al.*, 2019: 6). It is therefore possible to make such an assessment by adopting the LoA of individuals, as well as by broadening the perspective to look at other LoAs: interpersonal relationships, group phenomena, institutional and sectoral practices as well as societal issues. This is a major advantage of a patient-oriented macro-ethics approach, as the focus is on the receiver of the moral action (rather than on the agent or the action). It then becomes possible to assess an AI system not only in terms of its effects on individuals, but also on interactions between individuals, groups to which they belong and their environment (at various levels). While the six broad categories proposed in this version of the table are certainly valid and applicable to different contexts of AI systems use, we can also say that we could modify these LoAs, add some according to the particular situation, or refine them to have a greater (or lesser) granularity between the LoAs (taking into account the objectives of the ethical assessment of a specific AI system). The table, once we understand it this way, thus articulates a set of categories that need to be examined in order to carry out a satisfactory and comprehensive ethical assessment of a particular AI system or to proceed with the moral design of such a system. The objective then is to systematically generate questions in order to complete the table and ensure that all concerns (epistemic, normative and overarching), for each LoA relevant to the situation, have been taken into account in the ethical assessment and moral design of AI systems.

18.4.2 Ethics as a service

Morley *et al.* (2021) have developed a process, inspired by IE, for organising the structures that will frame the ethical assessment and moral design of AI systems. They call this process the 'Ethics as a service' (EaaS) approach.

Considering the difficulties and limitations of *hard governance* (legislation and regulation), they wish to discuss *soft governance* approaches that seek to identify procedures that are closer to practice as part of a design intervention (Morley *et al.*, 2021: 2). Building on the results of an earlier article on the need to move from AIE principles to their application in practice (Morley *et al.*, 2020), the authors argue that the tools for translating principles into practice are often either too flexible or too strict. They are too flexible when they leave room for a multitude of competing (and possibly conflicting) interpretations of the principles that may vary greatly from one context to another. This can lead to 'ethics shopping' or 'ethics washing' (Morley *et al.*, 2021: 3), i.e. respectively the selection of 'à la carte' ethical principles that suit the case being made, without actually testing the AI systems against a full assessment, and the use of ethics and ethics committees as a means of avoiding regulations beyond self-regulation. Conversely, tools are too strict when they are far too remote from the specific context of application of AI systems. When one attempts to impose very abstract principles on reality, the translation of principles into technological practices becomes extremely difficult, if not impossible, for those who design, develop or use these systems.

It is therefore necessary to think of an ethical assessment or a design process that is pro-ethical and that does not fall into either of these problematic categories, while adopting an LoA that is close to practice and avoids the implementation of extremely abstract principles (Morley *et al.*, 2021: 4). However, as we have just pointed out, those tools that are closer to the specific context of practice have some limitations (Morley *et al.*, 2021: 5). They can lead to ethical manipulation, being then created specifically to give certain expected results while avoiding a real ethical assessment of AI systems. These tools are sometimes useful for establishing diagnoses, but their prescriptive dimension is often very limited (if not completely non-existent), thus systematically neglecting the most important normative aspects of any ethical assessment. They therefore provide very little guidance on how to implement ethical principles and on the allocation of responsibility among stakeholders, for lack of a sufficiently clear and appropriate theoretical basis. These tools will often measure mainly (or exclusively) compliance with certain pre-established rules, if not simply compliance with existing laws, in the form of a 'checklist' of criteria. These approaches are of little interest since they do not allow for a thorough ethical assessment, and they completely obfuscate the reflexive dimension of ethics. Moreover, the general impacts, and particularly the societal impacts of AI systems, are rarely taken into account by this type of approaches and this type of tools, which can create a false sense of security, thus leading to 'ethics washing'. It is therefore necessary to design tools that both translate principles into practice and avoid these limitations, without falling back on overly strict approaches.

In order to develop an approach that implements such a pro-ethical practical design, two conditions need to be met (Morley *et al.*, 2021: 8-12). First, the right LoA must be adopted, one that avoids both overly flexible and overly strict approaches. Second, approaches as a 'checklist' of criteria, which would only apply at one point in time, i.e. at the beginning of the AI systems design process, should be avoided. The process 'should not be seen as an end-goal that can be objectively achieved, observed, quantified, or compared' (Morley *et al.*, 2021: 9), but rather as a reflexive development process. In this sense, in addition to the distinction already introduced between overly flexible and overly strict approaches, a second distinction needs to be considered, this time between approaches that seek to devolve ethical assessment and those that seek to restrict it to an internal process within organisations (Morley *et al.*, 2021: 10-12). While the external audit approach can establish clear principles put forward by independent stakeholders regarding the organisation's activities, such an external audit is often limited in several ways. A first limitation is associated with the limited access that the external actors are granted to the AI systems themselves. Since these systems are often at the heart of industrial trade secrets, companies generally provide as little information as possible to third parties. There are also problems associated with the fact that, in most cases, it is only after the system has been not only designed, but also developed and deployed, that the external audit is conducted. This greatly reduces the scope for intervention and, most importantly, it precludes carrying out any ethical assessment before the implementation of AI systems. Moreover, most external audits tend to measure compliance with very specific rules (or simply laws), often without taking into account the foresight analysis of AI systems' deployment in particular contexts. While internal ethics review processes can address some of these issues, they are themselves likely to suffer from significant problems relating to the independence of the process, especially conflicts of interest between those conducting the review and other parts of the organisation. It is therefore necessary to think of an approach that is not completely devolved and not completely centralised in internal organisation mechanisms.

The solution lies, for each particular AI system, in the creation of a MAS that will be responsible for the ethical assessment and moral design, a responsibility that then takes the form of a DM between different agents, namely individual agents, organisations and groups of external and internal auditors, all of whom must act in a coordinated manner (Morley *et al.*, 2021: 11-12). The objective of this MAS should therefore be to aggregate potentially good actions – in the sense that they allow for the informational flourishing of individuals and other MASs affected by this AI system – so that these actions pass the threshold necessary for them to take place. This MAS will also need to seek to isolate or fragment potentially evil actions, 'so that they never reach the necessary

Ethics as a Service

threshold to breach the fault-tolerance level of the overall system' (Morley *et al.*, 2021: 12). To illustrate this approach, the term 'Ethics as a Service' (EaaS), inspired by *cloud computing*, is coined. By comparing three forms of cloud computing, namely Software as a Service (SaaS), Infrastructure as a Service (IaaS) and Platform as a Service (PaaS) (Morley *et al.*, 2021: 12), the authors argue that it is possible to better understand by analogy which form of MAS should be put in place to carry out the ethical assessment and moral design of AI systems.

Software as a Service

Infrastructure as a Service

Platform as a Service

advisory board

While in SaaS users access a service that is almost totally managed by a third party and is very unlikely to be adjusted by the user (it is often impossible to do so and the user is completely dependent on the system vendor), the analogy with the ethical assessment and moral design of AI systems refers to approaches that are totally devolved to external auditors who usually use overly strict tools that are built completely independently of the particular context of the AI systems, and based on extremely abstract principles. In contrast, IaaS cloud computing provides access to an infrastructure that can be controlled at will and can be fully adapted to the particular context of use. The analogy with ethical assessment and moral design of AI systems refers to approaches that are entirely centralised and internal to organisation, being therefore too flexible, in the sense that they are not based on anything external (other than a list of principles that are general enough to be interpreted as one sees fit) and can therefore be modulated at will to the particular contexts of the organisation, thus allowing 'ethics shopping' and 'ethics washing'. The ideal model on which EaaS should be modelled is the PaaS model of cloud computing. In these PaaS tools, we have access to a basic infrastructure – including an operating system and storage resources – established by a third party and which are the same for all users. It is the software and application parts that can be adjusted to the particular context of the users. Similarly, the EaaS process, as proposed from the IE approach, should combine fixed and independently established core principles to support the work of ethical assessment and moral design, while allowing for adjustments of the tools to make the process fit the specific context in which the AI systems are conceived, built and implemented. This implies two things (Morley *et al.*, 2021: 13-15): (1) there must be an independent multidisciplinary advisory board that is able to provide the basic principles from which the ethical assessment and moral design work will be carried out and (2) there must also be internal actors who will actively participate in the development of the tools for ethical assessment and moral design. On the one hand, it will therefore be the task of the advisory board to establish, based on independent work, a code of ethics that is explicitly based on general principles, but which will be interpreted according to discussions with the receivers (or patients) of the actions of the individuals and of the MAS implementing the AI systems. This board should also establish a regular process that will organise the validation, verification and evaluation of AI systems, with a view to implementing pro-ethical design (Morley *et al.*, 2021: 14). Finally, this board should itself carry out audits or commission regular audits, not only of the specific tools under study, but also of the general functioning of the organisation, including the societal effects it generates. The EaaS process also foresees that internal actors will have the task of contextualising the principles defined by the advisory board. They should actively identify appropriate tools and use them in the ethical assessment and moral design, as outlined in the table for structuring such work (Section 18.4.1). Finally, these actors should have the task of documenting the ethical assessment and moral design process, in order to follow up the process internally, especially with regard to stakeholder accountability.

This process seeks to implement different approaches which, while each has advantages and limitations, can reinforce each other when combined (Morley *et al.*, 2021: 15). It is therefore easy to see the value of an approach such as IE in order to establish a satisfactory process of ethical assessment and moral design. It is only once one adopts the IE approach as a mid-level theory that one can consider not only the necessary dual dimension of the ethical assessment process and moral design of AI systems but also a coherent harmonisation of these essential elements that may otherwise appear disparate. The unfolding of IE principles within a mid-level theory adapted to the complexity of the infosphere is thus fruitful. This is probably what has justified the enthusiasm of several Computer Ethics (CE) scholars (Tavani, 2010) who have found in this theoretical framework the foundations to fill the 'conceptual vacuum' that has prevailed in their discipline for many years (Floridi, 2013: 87). Indeed, according to Floridi, being unable to find in classical ethical frameworks the conceptual resources adequate to their object of study and lacking appropriate theoretical foundations to provide a coherent response to the ethical issues related to the use of ICTs in our society, this branch of applied ethics had to resort to an *ad hoc* or casuistic assessment of the ethical issues associated with ICTs. We believe that a similar risk exists in AIE, where the inadequacy of traditional ethical frameworks hinders the development of a coherent and unified analysis of the many ethical issues associated with the design and implementation of this technology. In the field of AIE, more than 160 declarations and codes of ethics from governments, academia, industry, NGOs, civil society or a collaboration between several of these actors have been published in recent years, with the aim of identifying the principles that should frame design, development and use of AI systems. While there is some consensus on these principles (Floridi and Cowls, 2019; Jobin *et al.*, 2019), there have also been many criticisms of the principlist approach to AIE (Mittelstadt, 2019; Haggendorff, 2020). One of the problems is that these principles, stated at a high level of abstraction, are often not defined, nor do they refer to a theoretical framework that could provide a conceptual foundation that would support their interpretation in context. Looking at the broad principles outlined in ethical declarations without looking at the theoretical framework that was used, explicitly or not, in their elaboration may contribute to exacerbating the interpretative difficulties faced by the principlist approach to AIE. For example, the principle that AI should be used for the *benefit of humanity* borders on futility because it is so general (Boddington, 2020). While it is not surprising that there is a consensus on this concept, it is of course the definition that is problematic. Indeed, the question of what should be considered beneficial to humanity in this world radically transformed by technology is far from being a simple one. To answer this question, it is necessary to rely on an appropriate theoretical framework that allows us to rethink the concept of moral responsibility in our complex information societies.

It is clear that the normative effects of AIE declarations remain extremely limited for the time being (Haggendorff, 2020; Fjeld, *et al.*, 2020). Indeed, the transition from principles to practices is proving to be particularly difficult, and while several authors try to solve this problem (Stahl *et al.* 2021; Morley *et al.*, 2019), the relative ineffectiveness of these frameworks and the lack of adapted practical tools that could help translate these principles into technical terms that are intelligible to developers and organisations increases the risk of 'ethics washing' (Floridi, 2019). Some even consider that many AIE initiatives are primarily aimed at promoting self-regulation and discouraging stricter legislative approaches (Calo, 2017). While these criticisms are partly legitimate, as it cannot be denied that some organisations have resorted to this kind of practices, the consequence is that these cases justify what Bietti describes as 'Ethics bashing' (Bietti, 2020), i.e. the disqualification of ethics that contributes to maintaining an extremely limited, if not erroneous, conception of what ethics is and of the normative role it can play in AI governance. The transition from principles to practical tools is probably the most pressing issue in AIE. Although IE is not a panacea that can solve this problem on its own, the fact that it provides a macro-ethics that is particularly appropriate for dealing with the ethical issues that are emerging from our information societies makes it fit as a foundation for the elaboration of a mid-level theory in AIE, from which it is possible to develop ethical assessment and moral design tools that are relevant to the particular context of our information societies.

References

Bietti, E., 2020. From ethics washing to ethics bashing. Proceedings to ACM FAT Conference (FAT* 2020). Available at: https://papers.ssrn.com/sol3/papers.cfm?abstract_id=3513182

Boddington, P., 2020. Normative modes: codes and standards. In: Dubber, M.D., Pasquale, F. and Das, S. (eds.) The Oxford handbook of ethics of AI. Oxford University Press, New York, NY, USA, pp. 125-140.

Bruneault, F. and Laflamme, A.S., 2021. AI ethics: how information ethics provide a framework to avoid usual conceptual pitfalls – an overview. AI & Society, 36: 757-766. https://doi.org/10.1007/s00146-020-01077-w

Calo, R., 2017. Artificial intelligence policy: a primer and roadmap. https://doi.org/10.2139/ssrn.3015350

Costa, L., 2016, Virtuality and capabilities in a world of ambient intelligence – new challenges to privacy and data protection. Springer, New York, NY, USA.

Desjardins, J., 2019, 30 April. How much data is generated each day. World Economic Forum. Available at: https://www.weforum.org/agenda/2019/04/how-much-data-is-generated-each-day-cf4bddf29f/

Dewey, D., 1917. The need for a recovery of philosophy. In: Creative intelligence: essays in the pragmatic attitude. Holt, New York, NY, USA, pp. 3-69.

Fjeld, J., Achten, N. Hilligoss, H. Nagy, A and Srikumar, M., 2020. Principled artificial intelligence: mapping consensus in ethical and rights-based approaches to principles for AI. SSRN Electronic Journal. 1: 1-39. https://doi.org/10.2139/ssrn.3518482

Floridi, L., 2010. The philosophy of information, ten years later. Metaphilosophy, 41: 402-419. https://doi.org/10.1111/j.1467-9973.2010.01647.x

Floridi, L., 2011. The philosophy of information. Oxford University Press, Oxford, UK.

Floridi, L., 2013. The ethics of information. Oxford University Press, Oxford, UK.

Floridi, L., 2014. The 4th revolution: how the infosphere is reshaping human reality. Oxford University, Oxford Press, UK.

Floridi, L., 2015. The onlife manifesto, being human in a hyperconnected era. Springer, Cham, Switzerland.

Floridi, L., 2016. On human dignity as a foundation for the right to privacy. Philosophy & Technology, 29: 307-312. https://doi.org/10.1007/s13347-016-0220-8

Floridi, L., 2019. Translating principles into practices of digital ethics: five risks of being unethical. Philosophy & Technology, 32: 185-193. https://doi.org/10.1007/s13347-019-00354-x

Floridi, L. and Cowls, J., 2019. A unified framework of five principles for AI in society. Harvard Data Science Review, 1. https://doi.org/10.1162/99608f92.8cd550d1

Floridi, L. and Sanders, J., 2004. On the morality of artificial agents. Minds and Machines, 14: 349-379. https://doi.org/10.1023/B:MIND.0000035461.63578.9d

Hagendorff, T, 2020. The ethics of AI ethics: an evaluation of guidelines. Minds and Machines, 30: 99-120. https://doi.org/10.1007/s11023-020-09517-8

Jobin, A., Ienca, M. and Vayena, E., 2019. The global landscape of AI ethics guidelines. Nature Machine Intelligence, 1: 389-399. https://doi.org/10.1038/s42256-019-0088-2

Kitcher, P. 2012. Preludes to pragmatism: toward a reconstruction of philosophy. Oxford University Press, Oxford, UK.

Mittelstadt, B. 2019. Principles alone cannot guarantee ethical AI. Nature Machine Intelligence, 1: 501-507. https://doi.org/10.1038/s42256-019-0114-4

Mittelstadt, B., Allo, P., Taddeo, M., Wachter, S. and Floridi, L., 2016. The ethics of algorithms: mapping the debate. Big Data & Society, 1. https://doi.org/10.1177/2053951716679679

Morley, J., Elhalal, A., Garcia, F., Kinsey, L., Mokander, J. and Floridi, L., 2021. Ethics as a service: a pragmatic operationalisation of AI ethics. https://dx.doi.org/10.2139/ssrn.3784238

Morley, J., Floridi, L., Kinsey, L. and Elhalal, A., 2020. From what to how: an initial review of publicly available AI ethics tools, methods and research to translate principles into practices. Science and Engineering Ethics, 26. https://doi.org/10.1007/s11948-019-00165-5

Morley, J., Machado, C., Burr, C., Cowls, J., Taddeo, M. and Floridi, L., 2019. The debate on the ethics of AI in health care: a reconstruction and critical review. https://doi.org/10.2139/ssrn.3486518

Nagel, T., 1986. The view from nowhere. Clarendon Press, Oxford, UK.

Nussbaum, M., 1997, Cultivating humanity: a classical defense of reform in liberal education. Harvard University Press, Cambridge, MA, USA.

Nussbaum, M., 2011. Creating capabilities: the human development approach. The Belknap Press of Harvard University Press, Cambridge. MA, USA.

Posner, R.A., 1998. The Problematics of Moral and Legal Theory. Harvard Law Review, 111: 1637. Available at: https://heinonline.org/HOL/LandingPage?handle=hein.journals/hlr111&div=55&id=&page=.

Sen, A., 1992. Inequality reexamined. Harvard University Press, Cambridge, MA, USA.

Sen, A., 1999. Development as freedom. Knopf, New York, NY, USA.

Stahl, B.C., Antoniou, J., Ryan, M., Macnish, K. and Jiya, T., 2021. Organisational responses to the ethical issues of artificial intelligence. AI & Society. https://doi.org/10.1007/s00146-021-01148-6

Tavani, H.T., 2010. The foundationalist debate in computer ethics. In: Floridi, L. (ed.) The Cambridge handbook of computer ethics. Cambridge University Press, Cambridge, MA, USA.

Van den Hoven, J. 2010. The use of normative theories in computer ethics. In: Floridi, L. (ed.) Information and computer ethics. The Cambridge handbook of computer ethics. Cambridge University Press, Cambridge, MA, USA.

Van den Hoven, J. 2017. The design turn in applied ethics. In: Pogge, T. (ed.) Designing in ethics. Cambridge University Press, Cambridge, MA, USA.

Wiener, N., 1959. Man and the Machine. An interview with Norbert Wiener. Challenge, 1959, June, 36-4.

19. Towards a moral compass for the future

David A.M. Bamps[1,2]*, Veronique De Roeck[2] and Lize Verbeke[2]

[1]Multidisciplinary Centres of Expertise: Resilience People and Smart Organisations, Engaged and Entrepreneurial European University (E³UDRES²), Human Contribution to Artificial Intelligence; [2]UCLL, Groeneweg 151, 3001 Heverlee (Leuven), Belgium; david.bamps@ucll.be

Abstract

The quest for a moral compass confronts some of humanity's deepest questions. Where do values and norms come from? Who or what is necessary for moral guidance? This chapter deals with those questions from a future perspective. It tells the story of a network society where technology is increasingly important for making (moral) decisions based on complex information exchange. It is important to understand how we can govern the world and how we can predict certain consequences of our behaviour in the short or long term. The aim of this chapter is to offer the reader a dimension that will enable the construction of a moral compass in the future. The starting point is the idea that we should make our preferences clear in accordance with our philosophy of life (and citizenship). After making this choice, technology can accompany us as a personal assistant. Then we can look ourselves in the digital mirror every night before going to sleep and come to the conclusion based on facts: we are on the right track, or we are lost.

Key concepts

- Multiple Morality Approach: differentiation between moral systems available to men. In *The Moral Instinct*, the starting point is to distinguish 5 types of morality (Verplaetse and Delmotte, 2017).
- The moral instinct is a conglomerate of rationality (principles) and emotions (with at least four mechanisms of moral learning based on emotions: care-based – harm, reciprocity; disgust-based – purity; social convention – hierarchy; affect-free morality) (Blair, 2009).
- Normative ethical approach (rationality) is concerned with the criteria of what is morally right and wrong. It requires a justification with reasonable arguments that belong to ethical frames of reference

(deontology, virtue ethics, utilitarianism, …), which does not necessarily coincide with our emotional frameworks and actual morals based on history, tradition, religion, upbringing or personal preferences.

- Ethics of transhumanism is based on the idea that with the help of technology, we should overcome our human, social and moral shortcomings with the aim of achieving an improved version of mankind, society and the management of the ecological system.
- Moral compass: diagonal morality (triangulation) starts from the idea that we can make a connection between what citizens prefer (horizontal) and what is desirable at a higher level (vertical). The diagonal connection is created by using real-time data and artificial intelligence: each measurable moral action is followed by an evaluation. The predetermined moral goal is measured to see if it has been achieved and evaluated to see what can be adjusted to achieve the moral goal.

19.1 Multiple morality approach

descriptive approach

applied ethics

What is morality? The question of the definition of morality is the question of identifying the target of moral theorising or a framework that is justifiable in a certain way. 'There does not seem much reason to think that a single definition of morality will be applicable to all moral discussions' (Gert and Gert, 2020). One reason for this is that 'morality' seems to be used in different ways. One way entails the classic reference to two distinct broad senses: a descriptive sense and a normative sense. The descriptive approach refers to certain codes of conduct put forward by a society or a group, or accepted by an individual for their own behaviour; and the normative approach refers to a code of conduct that, given specified conditions, would be put forward by all rational people (Gert and Gert, 2020). It is not just a theoretical matter. Society demands a certain order and structure. There is a big difference between practice and theory. Even if people say (intentionally) they are doing X, they may actually be doing Y. This is where the importance of applied ethics and the issue of policy starts. Policy, in that context, requires a clear reference. In policy practice (see also Chapter 9), the theoretical approaches refer to the same practical task: are we living in accordance with what we intend? This is not an easy question to answer because it places us at a crossroads of considerations and philosophical ravines. Or to quote Churchland: the 'distinction between descriptions and norms – between what is and what ought to be – is obvious and unbridgeable' (Churchland, 2011: 185).

We can learn from this that we must not start from pure statistics, algorithms or data in general. The issue of morality requires a broader framework of meaning. This does not mean that data do not play a role, but it must be clearly stated which data and algorithms are the subject of ethical consensus. This requires a social debate and a philosophical dispute at the highest level.

The descriptive use of 'morality' is the one used by anthropologists, psychologists and law sociologists when they report on the morality of the societies that they study. Jonathan Haidt and Craigh Joseph (2004) make the complexity of this debate clear by indicating how the various disciplines relate to the object of investigation: 'Anthropologists often begin with sociological facts and then try to work down one level of analysis to psychology. Laws, customs, rituals, and norms, obviously vary, and from that variation it is reasonable to conclude that many psychological facts, such as beliefs, values, feelings, and habits, vary too. Evolutionary psychologists, in contrast, work mostly in the space between psychological and biological levels of analysis.' A similar multiple morality approach can be found in the contribution by James Blair (2009: 87): 'an alternative viewpoint is that there are multiple systems which mediate largely separable forms of social rule related processing that are lumped together within a culture's concept of morality.' Both Haidt and Blair have their own merits in breaking down factors that can play a role in a multiple morality approach. Jonathan Haidt and Jesse Graham (2007) have distinguished five psychological foundations in morality: reciprocity, harm, purity, in-group and hierarchy. It is as if Blair is experiencing a déjà-vu in his analysis. They distinguish exactly the same foundation when it comes to morality. The overlap is complete with one exception: affect-free morality (Blair, 2009: 87). Most remarkable was Jorge Moll's (2009) response to James Blair's argumentation and criticisms. In summary, Blair's qualification of a 'unitary moral system' is not appropriately applicable to the work of Moll and colleagues. Moreover, Moll's research results rather support the idea of a Multiple Morality Approach. Their latest publications on the contribution of attachment to morality clearly reflect their view on how distinct neurobiological components and functions support diverse aspects of human morality (Moll, 2009).

multiple morality approach

unitary moral system

As a final remark, it is important to underline that morality is very much a matter of flesh and blood. The neuropsychological insights bring a counterweight to the philosophical debate, which historically has very often been about the importance of rationality. By the end of this section, it is important to realise that moral decisions require a balance between morality and rationality, in which the functioning of our brain plays a crucial role. Our brains are cognitively limited. There is room for improvement. We need calculating power, an overview and a helicopter perspective. To understand the possibilities, we must first take a closer look at how the moral instinct works.

19.2 The moral instinct

If morality has a biological origin, then apparently different neurobiological components play a role. Such components also express the versatility of moral systems and their underlying nature: the so-called flesh of ethics.

A good example is a feeling like disgust. There is a close connection between disgust, purity and morality. It has also been shown experimentally that a good deed can have a purifying effect. For example, test subjects were instructed to keep an unethical act in mind. After the test, they had to wash their hands. Afterwards, it turned out that test subjects who had an unethical act in mind and were then given the opportunity to wash their hands were less inclined (41%) to help a student with problems than test subjects who had not been given the opportunity to wash their hands (71%) (Verplaetse and Delmotte, 2017: 145; Zhong and Liljenquist, 2006). Those who could wash their hands already felt liberated from the moral burden. Those who could not, sought compensation in some other good deed. This is just one example of what could be called the flesh of morality.

That brings us to the question of which moral instincts can be distinguished. Verplaetse and Delmotte (2017) gives a clear and systematic answer in his major study of the natural origins of morality. His approach integrates findings from other disciplines, and he was particularly encouraged by the views of Haidt and Blair as discussed in Section 19.1. It delegates our moral and immoral behaviour to five different moral systems, four of which are based on intuitions or emotion systems (the ethics of attachment, the ethics of violence, the ethics of cleanliness, the ethics of cooperation), whereas only one is rationally based (the ethics of principles). The forms of morals are briefly summarised in his book *Het Morele Instinct*.

moral systems

In short, a description:

1. The ethics of attachment: it regulates our actions with people we feel connected to. Innate abilities such as imitation, attachment and empathy caused by our dependency are at the basis of this form of morality (Verplaetse and Delmotte, 2017: 7).
2. The ethics of violence: it is perhaps the opposite of the previous form of morality. It regulates the actions that we need when we feel threatened or when we feel the need to protect our group members from potentially dangerous opponents. It dispels fear and the enemy (Verplaetse and Delmotte, 2017: 8).
3. The ethics of cleanliness: it links good and evil with feelings of purity and contamination. From an evolutionary perspective, humans have always been vigilant about coming into contact with pathogenic substances or poisonous plants. We have learned to avoid these dangers. Thus, even immoral acts evoke disgust, and we have a mental avoidance of anything we associate with evil. It encourages the introduction of purification rituals (Verplaetse and Delmotte, 2017: 8).

4. The ethics of cooperation: this form of morality is important during the game of business and negotiation. It offers advantages but also disadvantages, especially when it comes to the free riders who benefit from the system. Nobody likes cheaters. This morality provides the intuitions to avoid bad players and, where possible, to optimise cooperation with good players (Verplaetse and Delmotte, 2017: 8-9).
5. The ethics of principles: this morality distinguishes itself from the morals mentioned above. It is the morality that offers a solution for the disadvantages the other morals have to deal with, which may or may not be a direct consequence of the intuitions. The ethics of principles requires us to argue and not just do things because they strike us from a deep nature. It is rationality that strengthens the power of argument. Within this setting, rationality, and the ethics of principles (5), complements morals (1-4) that are rather based on emotions (Verplaetse and Delmotte, 2017: 9).

These moral systems reveal the underlying forces that make people do what they do for very different reasons and in very different ways. But what is common to all moral systems is not only the fact that they limit our personal and individual freedom in favour of a higher goal, namely, that of the community, but also the power of the cerebral toolbox (Bamps, 2019: 21). The cerebral toolbox here refers to the important role of the brain in making moral decisions. This was discussed in Section 19.1.

Within this framework, it is easy to understand how the development of morality is coloured both rationally and emotionally. It reveals how morals are often intuitive, spontaneous and emotion-driven processes. During the later development of the cortex, the influence of rationality increases, shifting from biological facts to cultural and social diversity. It is not only important to understand the origin and development of morals from a historical point of view. Again referring to the use of the term 'cerebral toolbox', the adjective 'cerebral' is important because it allows the possibility of assuming a non-cerebral toolbox. That has become the thesis of the dashboard society (Bamps, 2021). From a philosophy of transhumanism, we see in current technological developments the prefiguration of a data-driven morality that is generated by non-human decision trees. The human brain is assisted by computers and algorithms, and this gives a new ontological branch to the tree of morality types.

An example is the threshold value of a blood sample with which a certain disease is assumed. The threshold is set by humans, but the medical decision tree is an electronic process; an algorithm:

> One of the many ethical issues raised by such algorithms is the risk of using algorithms that produce false positive and false negative results. By definition, a false positive result occurs whenever the algorithm triggers the system to count something (a cell, a symptom of a disease) in a digital image that is not actually there. A false negative occurs if an algorithm in a similar vein fails to identify a structure in the picture that is actually there (Kraemer *et al.*, 2011).

In the future, it is conceivable that certain thresholds will be determined by technology itself with corresponding moral decision trees. Crisis situations sometimes requires rapid decisions that necessarily use non-cerebral toolkits. A reflection on the previous example is locked in the crucial question: how will morality based on a non-cerebral toolbox develop in the future? And furthermore: what is possible, probable and desirable?

It is not easy to answer these questions. There are various forecasting techniques that can be used, some more abstract than others. An interesting but also very archaic and abstract view can be found in the philosophical doctrine of vitalism, especially in Murphy's interpretation of Charlie Dunbar Broad (1925). In Murphy's study of *Human Potentialities*, there is a basis for drawing the future of morality (Murphy, 1961). *The Dashboard Society* contained the building blocks of a study of social order that assumed the emergence of a new morality caused by the increasing complexity of society (Bamps, 2021). Broad's thinking was particularly inspiring in this respect, with its distinction between 'substantive vitalism', which seeks out new substances as realities become more complex, and 'emergent vitalism' which seeks out new methods of organisation and integration rather than new substance. In simple terms, the complexity of society demands a solution and this can be found in two ways (that are not mutually exclusive).

- A new organisation: with the help of artificial intelligence (see Section 19.3).
- A new substance (after a process of emergence, with the help of a new conscience, baptised with the name 'Moral MindSphere', see Chapter 10).

The importance of these two perspectives has to do with a one-sided interest in a new organisation. Almost all study books, articles or lectures on the subject are about a new organisation. Hardly anyone talks about a new substance. However, it is this ontological paradigm shift that is perhaps even more spectacular as a philosophical starting point. The idea that in the future there will be a non-cerebral structure in the form of a moral cloud that, like a super-consciousness, forms the conscience alter-ego of every citizen and society in the form of an electronic dashboard, defies all (moral) imagination.

19.3 The ethics of transhumanism or supra-humanism

If we extend current technological developments into the future, we can assume, based on functional extrapolation[17] and disruptive cultural-biological changes, that it will also have a very important meaning in the implementation of morality. Human shortcomings and limitations are visible in the moral outcome of processes, and in the social ordering of the world in general. It is not without reason that Hans Boutellier, in his study of social order, came to the conclusion that citizens live in a complex society where they experience unease and disorder (Boutellier, 2008). Transhumanism may offer an interesting solution to certain problems related to human limitations:

> 'The human desire to acquire new capacities is as ancient as our species itself. We have always sought to expand the boundaries of our existence, be it socially, geographically, or mentally. There is a tendency in at least some individuals always to search for a way around every obstacle and limitation to human life and happiness' (Bostrom, 2005: 1).

In recent years, there has also been a tendency to overcome moral obstacles and boundaries from a very new angle: from materiality[18] itself. However, the prevailing view among a large number of sociologists is that objects have no moral dimension (Collins and Kusch, 1998). This premise can be criticised. Following the French philosopher, sociologist and anthropologist Latour, we can see a moral dimension in materiality (Latour, 1992, see also Section 6.2.2 and Chapter 18). A beautiful but at the same time very simple example is the construction of traffic-calming devices (Latour, 1996). 'Traffic devices are not sleeping policemen simply made of concrete instead of flesh and bone' (Latour, 2002: 250). Congealed morality[19] does have an impact: it morally steers our behaviour in the right direction. It is materiality that intervenes in

---

[17] It refers historically to the work of Murphy who speaks of a 'method of stretching' and the idea that it is possible to predict the future on the basis of current developments and trends (Murphy, 1961: 27). The adjective 'functional' was added to indicate that only those trends should be included in the forecast that are sufficiently stable, robust and significant over a long period of time. The described trend of technological development was inspired by Murphy's lecture. At the same time, we must also be aware of the fact that paradigm shifts have a disruptive character, which means that there will be developments that completely break away from existing trends: Kuhn-Loss (Post, 1971).

[18] Referring to Bruno Latour. 'This is followed by a consideration of materiality and power, including claims to transcend materiality, and a consideration of the relativity of materiality where some things and some people are seen as more material than others, leading finally to an exploration of the plurality of forms of materiality' (Miller, 2005: 1).

[19] 'Latour sees this influencing of behaviour through technology as a form of "congealed morality". People who complain about the decay of morality in society should take a better look around them, says Latour. Our material surroundings are bursting with morality, for those who want to see it. Morality is not exclusively human, but material as well' (Verbeek, 2017).

our behaviour so that we drive less fast or recklessly. From a pragmatic point of view, technology is an extension. We get what we want by following an 'alternative route'. Using my own metaphor, I am referring to the term *detour of translation*, which Latour likes to use: 'Technology is the art of the curve, or what, following Serres, I have called translation. If we go in a straight line, as epistemology does, we do not need it, as we know from the Greeks.' (Latour, 2002: 251)

non-cerebral toolbox

This is the first step towards a non-cerebral toolbox. We get what we want, within the realm of morality, by using the art of a curve. With the help of technology, we can thus provide a necessary and meaningful answer to the question: how should we proceed in a reality that is becoming increasingly complex? It will help us translate the complexity of our society. With this philosophical twist, it is expedient to add a hypothetical sixth form of morality: the ethics of transhumanism or trans-humanism (depending on the evolutionary stage) (Bamps, 2021). Following the letter qualification of ethical forms by Verplaetse and Delmotte (Section 19.1):

> The ethics of transhumanism would mean that people can use technology to improve themselves and society. The logic of these ethics does not contain the same emotional and rational components with which we are now familiar. The main reason is that technology is programmed and may in the future develop into self-learning systems with the ability to program (and improve) themselves. The computational power of technology is much greater than that of humans, and is not affected by psychological phenomena (whether or not caused by bias, cognitive failure, sensory impairment, etc.). From a paradigmatic and moral perspective, the hypothesis is that the non-cerebral toolbox could be a turning point in history of mankind. The ontology is no longer the same. The power of change and possible improvement depends on the amount of qualitative data. But with the exponential expansion of the network, and technological monitoring in society, it is only a matter of time.

The consequences will have a major impact, especially on existing reference frameworks: virtue ethics, deontology or consequentialism. The starting point of these frameworks is always a high degree of normative conviction, not a factual or corroborated approach. When morality is driven by technology, facts play a bigger role. It would be an opportunity to bring science into the improved philosophy of democracy (Latour, 2004). And perhaps also honesty and justice?

> In a conversation with the director of a large housing cooperative, the latter gave the example of a discussion with a tenant. Suppose that you, as a housing cooperative, have all the data from a smart home (the method of ventilation, the use of the central heating, cooker hood, etc.) and then you get another discussion in court with the logbook of the technical monitoring as evidence. Where normally it would be his word against hers (persuasion), suddenly it will be a reasoned argument with data on the table in court (facts) (Bamps, 2021: 212).
>
> It can stimulate the development of an empirical ethics in the social sciences where the emphasis is more on a particular form of evidence. According to Borry, five presuppositions lie at the basis of 'empirical ethics':
>
> - Empirical ethics postulates that research into moral attitudes, intuitions behaviour and reasoning generates information that is meaningful for ethics and that this should be the starting point of ethics.
> - Empirical ethics states that the empirical research methodology of the social sciences is a way – and perhaps the best way – to map this reality.
> - Empirical ethics argues that the classical distinction between normative and descriptive ethics should be more flexible: empirical ethics denies the structural incompatibility between empirical and normative approaches and believes in their fundamental complementarity.
> - Empirical ethics is a heuristic term that argues for an integration of the empirical research methodology and results of empirical research into ethical reflection: empirical ethics is in a strict sense not a method for practising ethics but a basic attitude to use the empirical research methodology and results of empirical research in ethical reflection and decision-making processes.
> - Empirical ethics cannot be regarded as an anti-normative approach: if the context alone were to determine what is good or bad right or wrong, then it is no longer ethics.
>
> (These five assumptions were translated and quoted from (Bamps, 2021: 163; Borry *et al.*, 2004).

The result may be a shift in the public and political debate. After all, it is not clear what role technology is allowed to play, even if it is a star player on the social playing field. Moreover, one must take into account the philosophical bifurcation of *what is* and *what ought to be* (Searle, 1964). The danger of a naturalistic fallacy is always present in this kind of discussion. My first attempt to show the fly the way out of the fly bottle (alluding to Wittgenstein, 1958: §309) was a thought experiment as a stepping stone to a possible redefinition of a moral compass for the future.

19.4 Thought experiment of a college girl named 'Laura'

Most ethical reflections are about the choice of a theoretical pathway; a framework of references. This task is similar to an escape room. And most ethical roadmaps involve considerations using different normative theories. This is an intellectual labour (*a priori*). Perhaps this is an impossible task. It presupposes an instrumental act that does not exist. In other worlds, a character should not be able to pull himself out of a swamp by his own hair, as the legendary Baron Münchhausen allegedly did (Blamires, 2009).

That is not what my study is about. It is about what happens after you have chosen to follow path X or Y. It is primarily an ethical or normative *a posteriori* approach. Having said that, I asked myself whether it was possible to design a tool that could help a citizen to live according to their own preferred values and standards. It presupposes an instrumental act that does exist: e.g. monitoring and surveillance.

The thought experiment in Case 19.1 shows that technology could greatly assist us in living up to our own normative and ethical preferences. It is a matter of social ecology. Technology is part of the infrastructure and interwoven with the social and physical environment of the student, in this case Laura. The predefined indicators are followed up by sensors and monitoring present in the materiality of the building. In this example, from a moral design perspective, one could see the moral compass as part of a social ecology: 'the multidisciplinary study of the impact that physical and social environments have on human beings' (Moos and Insel, 1974: ix). But this is only a small piece of the puzzle. Developing a moral compass requires the consideration of more perspectives. The thought experiment in the Laura case gave insight into how specific moral feedback can be given when one has decided to live according to a certain ethical conception.

Case 19.1. The Laura case.[1]

'Laura is a young lady who lives in a dorm. From an early age, she has always been very concerned about the well-being of the planet. Laura is passionate about the environment and wants to help protect it. To put this philosophy into practice it is important to pay attention to some indicators. An example is the level of water consumption, which is measured on a weekly basis. Every week, Laura receives a report of the results: she scores an average consumption quantity compared to the other students. The report also states whether her personal target has been achieved. This personal target is a threshold value that embodies a personal normative objective. She receives a similar report for her study performance, psychological development, conflict behaviour with roommates, and her online conversation with students and teaching staff. Indicators for each life domain are provided. The total report of behavioural indicators can be a good benchmark of moral behaviour in general.'

[1] Recorded during a real conversation with a student on this issue, 2021; The name was replaced by a pseudonym.

19.5 Towards a moral compass for the future

What we can learn from the thought experiment (Case 19.1) is that technology can be designed for everyone, regardless of their values or standards. The moral compass of the future is more a *form* than a *content*: a vehicle with which we can complete a normative path of choice. It contributes not only to a multiple morality approach but also to the idea that we should look in the digital mirror every night.

personal dashboard

From a moral design perspective, I believe that in the future every citizen should have a personal dashboard (Bamps, 2021). Only then, from a transhuman philosophy, we can work on our own cognitive and moral limitations. Moreover, it provides the opportunity to transcend the complexity of the network society to a level where technology can offer a directive. On this assumption, networks are not only a source of problems but also an important key to improvement (Boutellier, 2014: 12). It is a matter of self-optimisation. The improvement lies in the diagonal connection between a horizontal and a vertical morality.

This requires a clear understanding of the concepts used:

- Horizontal forms of morality are present at the level of personal relationships – in friendships, families and households – and at the level of civil society – in social and political movements, in religious communities and philosophical groups; but also within organisations, especially in care and welfare, and in education, sport clubs, within the framework of responsible business, etc. (Kunneman, 2012: 27).
- Vertical forms of morality emphasise the hierarchy and associated authority structures, with the focus on freedom, autonomy and the personal responsibility of individuals and organisations. As dominant values, external supervisors and controlling authorities have been developed (Kunneman, 2012: 25).

According to Harry Kunneman (2012), there is a noticeable shift from a horizontal to a vertical morality, and criticism is justified if the balance is lost. In this context, it is important to define a diagonal morality as a data-driven morality that connects fact-based personal moral convictions with higher-order ideals, or ethical frames of reference (Bamps, 2021). Like a mechanical construction, diagonal morality tries to strengthen the whole construction without destroying any other form of morality (Figure 19.1). It is triangulation in the best sense of the word. From the point of view of moral design, it is very important to recognise and understand the power of this diagonal.

diagonal morality

In the given thought experiment, we were able to imagine a society in which citizens constantly monitor and adjust their own behaviour based on facts, as well as receiving feedback from each other and from artificial intelligence

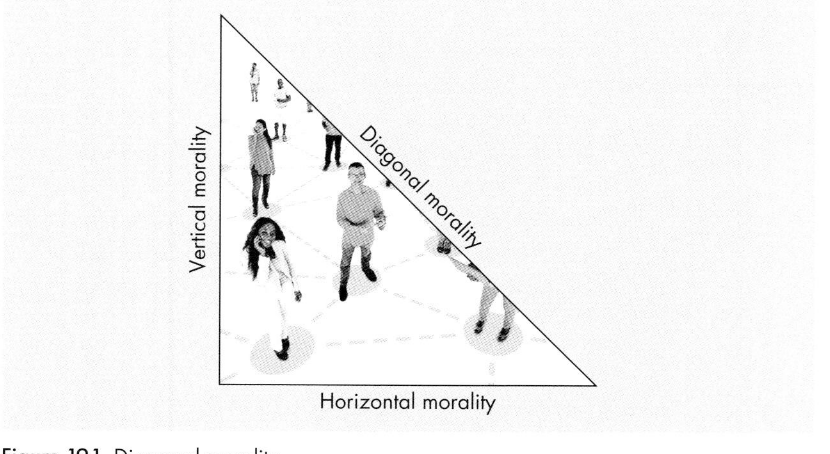

Figure 19.1. Diagonal morality.

residing in the higher layers of the dashboard society. 'Technology is capable of linking horizontal and vertical morality, resolving into a diagonal vector that would point us in the right way, as if it were the hand on our moral compass' (Bamps, 2021).

Double plan of action:

- Bring descriptive and normative ethics together in a diagonal perspective. List all possible actions that play a role in assessing moral behaviour (descriptive ethics). This will also require the definition of a minimum morality per community: what do we expect citizen X to do (normative) and share in order to organise the community according to the moral code. Two important remarks:
  - Different communities may have different moral codes.
  - Sharing information is not necessarily a public matter. It may involve sharing information with one's own monitoring systems (digital mirror in one's own house without third parties), or sharing with personal assistants in the form of artificial intelligence. This technology can also operate autonomously (cybernetics), so that no other people are involved and a different form of privacy experience can arise.
- Improve from a philosophy of transhumanism. This democracy of the future has a number of assumptions that we have discussed in this article. One of the most important is the construction of an empirical ethics with a transition from beliefs to facts. The goal is to improve the world. It is our technological science that can improve democracy; as well as the management of a better living community where citizens say what they do and do what they say. Without this necessary condition, morality is eroded and moral citizens will just wear the ring of Gyges.

19.6 Conclusions

The development of a moral compass as described in this chapter is certainly possible in the future. History shows that good and evil regularly change places. The moral compass adds little to this theoretical discussion. But once the choice has been made, technology can help us achieve what we want to achieve. Moreover, it can do so on the basis of facts and foresight. The latter is particularly important. If we choose a path, and the computer calculations show that the long-term consequences are the opposite of what we intended, we still have time to anticipate.

If, in the future, every citizen has a dashboard, this will lead to practical discussions: not only during the proof of concept phase, with regard to the design in terms of norms and values, but also afterwards when we receive feedback and look in our own digital mirror. Maybe we won't be able to withstand the confrontation: after all, consciousness can cause pain. And if we do gain self-awareness, we will probably want to reorganise our society. This may well lead to a world where people in human communities look for a similar philosophy. 'In human communities intent, belief, resources, preferences, needs, risks and a number of other conditions may be present and common, affecting the identity of the participants and their degree of cohesiveness' (Bulu, 2011: 104). In my opinion, the moral compass of the future as discussed can improve these ingredients of a good society. The power of the normative philosophy in practice within this community is a combination of beliefs and data-driven moralities.

References

Bamps, D., 2019. Puzzelstukken voor een nieuwe moraal over de sociale ordening in de toekomst. LannooCampus, Leuven, Belgium.

Bamps, D., 2021. De Dashboardsamenleving: van improvisatie tot providere. Boom Uitgevers, the Hague, the Netherlands.

Blair, J., 2009. Neuro-cognitive systems involved in moral reasoning. In: Verplaetse, J., De Schrijver, J., Vanneste, S. and Braeckman, J. (eds.) The moral brain: essays on the evolutionary and neuroscientific aspects of morality. Springer, New York, NY, USA, pp. 87-105.

Blamires, D., 2009. Telling tales: the impact of Germany on English children's books 1780-1918. The adventures of Baron Munchausen. Open Book Publishers, Cambridge, UK.

Borry, P., Schotsmans, P. and Dierickx, K., 2004. Empirical ethics: a challenge to bioethics. Medicine, Health Care and Philosophy 7(1): 1-3. Available at: https://philpapers.org/rec/BOREEA-2.

Bostrom, N., 2005. A history of transhuman thought. Journal of Evolution and Technology, 14: 1-25.

Boutellier, H., 2008. Solidariteit en slachtofferschap, de morele betekenis van criminaliteit in een postmoderne cultuur. Amsterdam University Press, Amsterdam, the Netherlands.

Boutellier, H., 2014. Nieuwe tijden, nieuw bestuur. De Wmo in de improvisatiemaatschappij. Verwey-Jonker Instituut, Utrecht, the Netherlands.

Broad, C., 1925. The mind and its place in nature. Harcourt, New Work, NY, USA.

Bulu, M., 2011. City competitiveness and improving urban subsystems: technologies and applications. IGI Global. https://doi.org/10.4018/978-1-61350-174-0

Churchland, P., 2011. Braintrust: what neuroscience tells us about morality. Princeton University Press, Princeton, NJ, USA.

Collins, H. and Kusch, M., 1998. The shape of actions. What hitmans and machines can do. MIT Press, Cambridge, MA, USA.

Gert, B., and Gert, J., 2020. The definition of morality. In Zalta, E.N. (ed.) The Stanford encyclopedia of philosophy (Fall 2020 ed.). Metaphysics Research Lab, Stanford University. Available at: https://plato.stanford.edu/archives/fall2020/entries/morality-definition/.

Haidt, J. and Graham, J., 2007. When morality opposes justice: conservatives have moral intuitions that liberals may not recognize. Social Justice Research, 20: 98-116.

Haidt, J. and Joseph, C., 2004. Intuitive ethics: how innately prepared intuitions generate culturally variable virtues. Daedalus, 133: 55-66.

Kraemer, F., Peterson, M. and Van Overveld, K., 2011. Is there an ethics of algorithms? Ethics and Information Technology, 13: 251-260. https://doi.org/10.1007/s10676-010-9233-7

Kunneman, H., 2012. Het belang van moreel kapitaal in zorg en welzijn – Paul Cremers Lezing 2012. Stichting Paul Cremers, Rijswijk, the Netherlands, pp. 1-60.

Latour, B., 1992. Where are the missing masses? Sociology of a few mundane artefacts. In: Bijker, W. and Law, J. (eds.) Shaping technology – building society. Studies in Sociotechnical Change. MIT Press, Cambridge, MA, USA, pp. 225-259.

Latour, B., 1996. Petites leçons de sociologie des sciences. Seuil, Paris, France.

Latour, B., 2002. Morality and technology: the end of the means. Theory, Culture & Society, 19: 247-60.

Latour, B., 2004. Politics of Nature: how to bring sciences into democracy. Harvard University Press, Cambridge, MA, USA.

Miller, D., 2005. Materiality: an introduction. In: Materiality. Duke University Press, Durham, UK, pp. 1-55.

Moll, J., 2009. Jorge Moll's Response on James Blair's Paper. In: Verplaetse, J., De Schrijver, J., Vaneste, J. and Braeckman, J. (eds.) The moral brain: essays on the evolutionairy and neuroscientific aspects of morality. Springer, Berlin and New York, Germany and USA, pp. 105-107.

Moos, R. and Insel, P., 1974. Issues in social ecology: human milieus. National Press, CA, USA.

Murphy, G., 1961. De geestelijke evolutie van de mens. Het Spectrum, Amsterdam, the Netherlands.

Post, H., 1971. Correspondence, Invariance and Heuristics. Studies in History and Philosophy of Science, 2: 213-255. https://doi.org/10.1016/0039-3681(71)90042-2

Searle, J., 1964. How to derive 'ought' from 'is'. Philosophical Review, 73: 43-58. https://doi.org/10.2307/2183201

Verbeek, P.-P., 2017. Designing the morality of things: the ethics of behaviour-guiding technology. In: Van den Hoven, J., Miller, S. and Pogge, T. (eds.) Designing in ethics. Cambridge University Press, Cambridge, UK. https://doi.org/10.1017/9780511844317.005

Verplaetse, J. and Delmotte, C., 2017. Beginselen van samenleven, handboek ethiek en rechtsfilosofie. Maklu uitgevers, Antwerpen, Belgium.
Wittgenstein, L., 1958. Philosophical investigations. Basil Blackwell Ltd, Oxford, UK.
Zhong, C. and Liljenquist, K., 2006. Washing away our sins: threatened morality and physical cleansing. Science, 313: 1451-1452. https://doi.org/10.1126/science.1130726

20. Looking ahead, and what we can learn from science fiction

Bart F.W. Wernaart

School of Business and Communication, Fontys University, P.O. Box 347, 5600 AH Eindhoven, the Netherlands; b.wernaart@fontys.nl

Abstract

In this book we explored what moral design is (Part One), and how it relates to existing discourse in adjacent disciplines (Part Two); we highlighted various case studies in which moral design plays a central role in innovating technology (Part Three), and discussed what the future could look like when using moral design at the core of technological innovation (Part Four). In this chapter, as a 'dessert' we would like to take a small sidestep and reflect on a literary genre that has always been at the forefront of conceptualising techno-ethical issues: science fiction.

Key concepts

- Moral design is about stretching our moral imagination
- Science fiction conceptualises the moral challenges of a future with technology, and can help stretch our moral imagination.

20.1 Introduction

While it may come across as 'nerdy' to close a book on moral design and technology by musing on science fiction, there are a couple of valid reasons for doing so. First, – but that is a personal consideration – because the sf genre is far from nerdy. Second, and more importantly, science fiction writers have visualised what the future could look like if we do not take moral design seriously, or if we fail to align values in our design processes. They have warned us about how technology can work as a counterproductive force in our societies, but also suggested ways to move forward and create societies in which technology and humans interact in a healthy, constructive way. They did not suggest these things by providing scientific evidence but by using science as a designer's input to create worlds that could be.

20.2 What we can learn from science fiction

Science fiction can be said to have existed since the existence of the Ancient Novel, and fantasies about new technologies are as ancient as literature itself (Roberts, 2016). If there is one literary genre that is involved with the moral side of technology, it is (and always has been) science fiction.

20.2.1 Beyond the utopian and dystopian divide

Also here, a divide between a utopian and a dystopian view on a world with technology can be found. For instance, the techno-optimism of Jules Verne is widely known from his many novels. Mankind could do unimaginable things because of science and technology, such as flying around the earth in 80 days, flying to the moon or travelling to the centre of the earth. The same can perhaps be said of the famous Star Trek franchise, which embraces the idea that technology can free us from many unsustainable habits (such as polluted or profit-driven societies). On the other hand, we see the grim images of a world in which technology leads a life of its own and harms human values. What should we think of the classic 'The Machine Stops' by Edward Morgan Forster (1909), in which mankind is bound to live underground and is fully dependent on 'the machine'? Or consider the HAL computer that decides to kill the crew members of a spacecraft in '2001: A Space Odyssey' An even more futuristic example is the Matrix franchise, in which artificial intelligence has destroyed human civilisation and forced those remaining living men into a virtual reality world.

At the same time, many works in the SF genre offer us a more nuanced approach, sometimes at a second glance. Wells's time machine is exemplary here (Case 20.1): at first glance we would say that the Eloi living on the surface live in a utopian, paradise-like society, and the Morlocks live in a dystopian environment. However, the Eloi have lost any purpose in life since technology can satisfy all their needs, and they just eat fruit or get eaten by Morlocks. The utopian-looking atmosphere is in fact a technology trap. The Time Machine also resonates with other issues in technology that are still the cause of biases in the design of technology: in Well's work, technology is developed and understood by white, male, Western individuals (Hume, 2005). This assumption is still alive today (Auernhammer, 2020; see also Section 2.3.2), and the cause of technological biases in e.g. gender (Leavy, 2018) and race (Garcia, 2016). Also, technology has an effect on the distribution of wealth and power (Frey, 2019).

The Morlocks represent the working class, who have no means to actually enter the paradise-looking world but during nightfall, and spend their lives maintaining machines to make life on the surface possible.

Case 20.1. The Time Machine.

Perhaps one of the first modern science fiction novels was the short story 'The Time Machine' written by the famous Herbert George Wells (1895), years before he published his most influential work of science fiction: The War of the Worlds. The novel, published in 1895, introduces us to a scientist who was able to construct a time machine. After a handful of 'innocent' experiments, the scientist travels to the year A.D. 802,701, and discovers a far-off future society that is divided by the surface of the earth. On the surface live the Eloi, who are fully satisfied as a result of technological progress. They just 'live' to eat their fruit-based diet. Below the surface, (a metaphor for the London subway system at the end of the 19th century) the Morlocks dwell: this ape-like race maintains the underground machines that make life possible on the surface. Since they have lived as cavemen for such a long time, they can only appear on the surface during the night, when they feed themselves on the Eloi.

20.2.2 H.G. Wells and other scientists

What is typical about science fiction is that – the name already suggests it – fiction is pretty close to science. The famous 'War of the worlds' (Wells, 1898) was inspired by the observation of a strange light on the planet mars, which was published in the journal Science in August 1984. Herbert George Wells was not only a SF-author, but also an academic with various interests. He served as a reviewer for Science, published a textbook on biology (1893), and wrote one of the first contemporary encyclopaedia on history (A Short History of the World, 1922). The combination of science and fiction in the work of Wells led to remarkable future insights. Amongst others, he – unknowingly – predicted the invention of the atomic bomb in his novel 'The world set free' (1914) and his critical societal reflection in his last book 'Mind at the end of its tether' (1945). In his works, many moral issues of (then) future technology are discussed. Perhaps most notably the ethical concerns that with bio-engineering in 'the Island of Dr. Moreau' (1896).

three laws of robotics

Wells is not an exception. The famous professor of biochemistry and SF-writer Isaac Asimov coined the three laws of robotics – amongst others in his collection of short stories 'I Robot' (1950):

'1. A robot may not injure a human being or, through inaction, allow a human being to come to harm.
2. A robot must obey the orders given it by human beings except where such orders would conflict with the First Law.
3. A robot must protect its own existence as long as such protection does not conflict with the First or Second Law.'

Laws that were later widely used in the discourse on moral programming of Artificial Intelligence (AI).

While developed in the context of a SF story, the laws have been subject of a wide academic discourse amongst techno-ethicists (e.g. Anderson, 2008; Iosa *et al.*, 2016; Kaminka *et al.*, 2017; McCauley, 2007), and – while mostly rejected – stirred a more advanced debate on moral programming challenges.

More recently, there is a vivid SF-scene in which the future of technology is envisioned. What to think of Steven Hunt[20], a former computer programmer, or Arkady Martine, historian and city planner, who show us worlds in which AI interact with humans in self-learning vehicles and cities, for better or worse. Or how about David Brin, Sir Fred Hoyle, Poul Anderson, Robert L Forward, or Gregory Benford. They are (were) all highly talented academics and used science as a central input for their science fiction-writing career.

What these authors have in common is that through their work, they trigger our moral imagination and visualize moral problems that might be in the (near) future that need to be addressed when we create new technology. And this is exactly what moral design is all about: to create something that is not there yet, to imagine how this creation will affect morality, and in that process to align societal values with the values intended by its creators and the values embodied in the creation itself.

20.3 Stretching our moral imagination

Moral design requires tons of creativity. When we create new technology we do not know what our innovation will become, how it will evolve itself morally and what the societal impact shall be. At the same time, not considering these issues for the mere fact that we cannot foresee all that will happen is not an option. The stakes are simply too high, and making moral blunders can simply be too painful. We need to imagine what technology might become and how it might in other words. We need to make sure that when we build new technologies,

[20] See in particular the Sliding Void series.

we build it in such a way that values are as aligned as possible. For this, we need to fantasize what the future could look like with our innovative ideas. Moral design is about stretching our moral imagination. And that is exactly what sf-authors have been doing all these years, and will continue to do so. Far ahead of all that will be.

References

Anderson, S.L., 2008. Asimov's 'three laws of robotics' and machine metaethics. AI & Society, 22: 477-493. https://doi-org.fontys.idm.oclc.org/10.1007/s00146-007-0094-5

Anonymous, 1984. A strange light on mars. Nature, 50: 319. https://doi.org/10.1038/050319c0

Asimov, I., 1950. I, robot. Gnome Press, New York, NY, USA.

Auernhammer, J., 2020. Human-centered AI: the role of human-centered design research in the development of AI. DRS2020, August 2020, Brisbane, Australia. https://doi.org/10.21606/drs.2020.282

Frey, C.B., 2019. The technology trap, capital, labor, and power in the age of automation. Princeton University Press, Princeton, NJ, USA and Oxfordshire, UK.

Forster, E.M., 1909, The machine stops. The Oxford and Cambridge Review. November edition. Available at: https://manybooks.net/titles/forstereother07machine_stops.html.

Garcia, M., 2016. Racist in the machine: the disturbing implications of algorithmic bias. World Policy Journal, 33: 111-117

Hume, K., 2005. Eat or be eaten: H.G. Wells's time machine. In: Bloom, H. (editor), H.G. Wells. Chelsea House Publishers, Philadelphia, PA, USA, pp.35-51.

Iosa, M., Morone, G., Cherubini, A., and Paolucci, S., 2016 The three laws of neurorobotics: a review on what neurorehabilitation robots should do for patients and clinicians. Journal of Medical and Biological Engineering, 36: 1-11. https://doi.org/10.1007/s40846-016-0115-2.

Kaminka, G.A., Spokoini-Stern, R., Amir, Y., Agmon, N., and Bachelet, I., 2017. Molecular robots obeying Asimov's three laws of robotics. Artificial Life, 23: 343-350. https://doi-org.fontys.idm.oclc.org/10.1162/ARTL_a_00235

Leavy, S., 2018. Gender bias in artificial intelligence: the need for diversity and gender theory in machine learning. Proceedings of the 1st International Workshop on Gender Equality in Software Engineering (GE '18). ACM, New York, pp. 14-16.

McCauley, L., 2007. AI armageddon and the three laws of robotics. Ethics and Information Technology, 9: 153-164. https://doi-org.fontys.idm.oclc.org/10.1007/s10676-007-9138-2

Roberts, A., 2016. The history of science fiction. Palgrave Macmillan, London, UK.

Wells, H.G., 1893. Text-book of biology. University Correspondence College Press, London, UK.

Wells, H.G., 1895. The time machine. William Heinemann, London, UK.

Wells, H.G., 1896. The island of Dr Moreau. Heinemann, London, UK.
Wells, H.G., 1898. The war of the worlds. William Heinemann, London, UK.
Wells, H.G., 1914. The world set free: a story of mankind. Macmillan & Co, London, UK.
Wells, H.G., 1922. A short history of the world. Cassell & Company, London, UK.
Wells, H.G., 1945. Mind at the end of its tether. Heinemann, Portsmouth, UK.

About the editor

Wernaart

Bart Wernaart is professor of Moral Design Strategy for Fontys University of Applied Sciences, the Netherlands. He is also a professional drummer, conductor and composer.

About the authors

Aarts

Sil Aarts is an assistant professor at the Living Lab in Ageing and Long-Term Care at Maastricht University, the Netherlands. Her research is focused on data and innovative methods of analysis (such as text-mining) to gain understanding of quality of care, quality of life and quality of work in long-term care for older adults. She is also a linking-pin at Cicero Zorggroep.

Aliman

Nadisha-Marie Aliman is a visiting scholar at Utrecht University, the Netherlands, and is an independent researcher, transdisciplinary analyst, security artist, autistic systemiser, hyperlexic writer, and poet.

Arets

Danielle Arets is a reader in journalism and responsible innovation at Fontys University of Applied Sciences/School of Journalism. She is also a design researcher at the Design Academy Eindhoven, the Netherlands.

Bamps

David Bamps is professor of (social) ethics, social psychology and juvenile criminology at the University of Applied Sciences, UCLL (Belgium). He is also a principal investigator and responsible for Belgium (region: Limburg) within the domain 'human contribution to artificial intelligence' as an assignment within E^3UDRES^2 (Engaged and Entrepreneurial European University as Driver for European Smart and Sustainable regions), composer, professional musician, director of a private consulting company, THE CONNECTION, and coordinator of a non-profit organisation OLORIN VZW (specialised in the management of exceptional projects for and by young, highly gifted people.

Bastiaanse

Ruben Bastiaanse teaches journalism studies at School of Journalism and is involved as a researcher at the Readership Journalism and Responsible Innovation at Fontys University, Tilburg, the Netherlands.

Brankaert

Rens Brankaert is professor of Warm Technology & Design in the group Health Innovation and Technology at Fontys School of Allied Health Professions, and assistant professor of Healthy and Active Ageing at Eindhoven University of Technology, the Netherlands. He is co-director of the Expertise Center for Dementia & Technology.

Bruneault Frédérick Bruneault is a professor of philosophy at Collège André-Laurendeau, associate professor at the École des Médias of the Université du Québec à Montréal (UQAM), Canada, principal researcher at the Laboratoire d'éthique du numérique et de l'intelligence artificielle (LEN.IA), associate researcher at the Groupe de recherche sur l'information et la surveillance au quotidien (GRISQ), and associate researcher at Observatoire international sur les impacts sociétaux de l'intelligence artificielle et du numérique (OBVIA). He specialises in the ethics of technology.

Cuijpers Colette Cuijpers is Professor of Applied Sciences in Law and Digital Technologies, a specialist in the field of privacy and data protection law, and programme director of the International Master's Programme Law & Technology.

de Cooker Jessy de Cooker is a researcher in the Readership Journalism and Responsible Innovation and tutors journalism techniques at the School of Journalism at Fontys University, Tilburg.

de Graaf Mark de Graaf holds a PhD in physics. Later he was co-founder of the Department of Industrial Design at the Eindhoven University of Technology, the Netherlands, where he worked on two topics: the interaction of people with complex systems, and playful interactions. Currently, he is the head of the Interaction Design research group. His research focuses on adding human value to interactions.

De Roeck Veronique De Roeck is a part-time lecturer in psychology and psychopathology at UCLL, and a part-time psychotherapist.

Giesen Anne Giesen is a student at the School of Journalism and works as a student-assistant in the Readership Journalism and Responsible Innovation at Fontys University, Tilburg, the Netherlands.

IJsselsteijn Wijnand IJsselsteijn is professor of Cognition and Affect in Human-Technology Interaction at Eindhoven University of Technology and scientific director of the interdisciplinary Center for Humans and Technology at TU/e, the Netherlands. He is co-director of the Expertise Center for Dementia & Technology.

Kamp Jo-An Kamp is a lecturer, researcher and coordinator at Fontys University of Applied Sciences, School of Information and Communication Technology. She coaches IT students in the fields of UX, research, (interactive) media, communication, (interaction) design, ethics and innovation. Besides that, she also wants to have an impact on the number of women in digital-focused roles, so she also acts as an ambassador and mentor for RightBrains. She does research on the impact of technology on humans and society, likes to speak at conferences and is co-founder of the Technology Impact Cycle Tool and www.tict.io.

Kester Leon Kester is a senior research scientist in AI(XR) Strategy, AI(XR) Safety, AI(XR) Meta-ethics and AI(XR) Governance for TNO, the Netherlands, and is an internationally active advisor.

Laflamme Andréane Sabourin Laflamme is a professor of philosophy at Collège André-Laurendeau, in Montréal, Canada. She is a doctoral student in law at Université du Québec à Montréal and recipient of the Canada Graduate Scholarship. She is the co-founding member of Len.IA (AI and Digital Ethics Lab). She is a collegiate researcher at OBVIA (International Observatory on the Societal Impacts of AI and Digital Technology). Her current research areas are the normative function of ethics and law in AI governance, advanced training in AI ethics and digital citizenship.

Schouten Gerard Schouten is a professor at the School of ICT, Fontys University of Applied Science and research fellow at Tilburg University in the Netherlands. He holds a Master's degree in physics from the Technical University of Eindhoven and has a PhD in the field of cognitive science. He is a self-made deep learning and machine learning practitioner and is experienced in large-scale software development in the healthcare industry. He initiated a new educational programme on AI at Fontys UAS. His research covers a wide array of topics in the broad field of AI and Big Data, with a special interest in AI solutions that benefit society and foster our natural environment.

Shirbaz Hasib Shirbaz wrote his Master's thesis (Tilburg University, the Netherlands) about audience metrics in journalism and is a researcher-in-residence at the Readership Journalism and Responsible Innovation at Fontys University, Tilburg, the Netherlands.

Tummers Ans Tummers is a PhD candidate in the research group Human-Technology Interaction at Eindhoven University of Technology, the Netherlands, on mediated touch for people with dementia, and is a member of the board in the Alzheimer Forschung Initiative in Germany. She is contact manager in the Expertise Center for Dementia & Technology.

van de Nieuwenhof Jeroen van de Nieuwenhof is a researcher in the Readership Journalism and Responsible Innovation and tutors in online and data journalism at Fontys University, Tilburg, the Netherlands.

van der Vorst Rens van der Vorst is Head of IT Innovation and Technophilosopher at Fontys University of Applied Sciences. He teaches, writes, and conducts research. Rens is a multiple book author and writes columns for a wide range of media. He gives a lot of talks on our relationship with technology. These talks tend to be funny, confronting and interactive, and for everyone. Rens is the owner of www.technofilosofie.com and co-founder of www.tict.io. He is also owner of www.quantifiedstudent.nl, a project to create a RunKeeper App for studying.

van Veen Maarten van Veen is Strategic Policy Advisor Innovation at the municipality of Eindhoven, the Netherlands. He has a PhD in Science and Technology Studies (UvH) and graduated in Information Science (UvA). He previously worked at the Netherlands Defence Academy (NLDA) and the Ministry of Defence as researcher, programme manager and policy advisor in the field of research and innovation.

Verbeke Lize Verbeke is Professor/researcher at the UCLL, Belgium. She teaches psychology and methodology-related courses in the Bachelor course 'Sociale Readaptatiewetenschappen'. In addition, she conducts research on forensic and victimological topics for the UCLL research centre Resilient People.

Visser-Knijff Piek Visser-Knijff works as an independent data ethicist (Filosofie in Actie, since 2011). She advises public and private organisations on how to create a sound and feasible data ethics strategy within the organisation and how to make data ethics work. She has a Master's in Philosophy, a Master's in Dutch language and literature and a Master's in Education (UvA).

Index

N

O

P

R

S

J

K

L

M

W

Z